Metallochemistry of Neurodegeneration

Biological, Chemical and Genetic Aspects

Metallochemistry of Neurodegeneration
Biological, Chemical and Genetic Aspects

Henryk Kozłowski
Faculty of Chemistry, University of Wrocław, Wrocław, Poland

David R. Brown
Department of Biology and Biochemistry, University of Bath, Bath, UK

Gianni Valensin
Department of Chemistry, University of Siena, Siena, Italy

RSCPublishing

ISBN-10: 0-85404-360-8
ISBN-13: 978-0-85404-360-6

A catalogue record for this book is available from the British Library

Published by The Royal Society of Chemistry,
Thomas Graham House, Science Park, Milton Road,
Cambridge CB4 0WF, UK

Registered Charity Number 207890

For further information see our web site at www.rsc.org

Typeset by Macmillan India Ltd, Bangalore, India
Printed by Henry Ling Ltd, Dorchester, Dorset, UK

Preface

The introduction, several years ago, of biological inorganic chemistry provided the amalgamation of concepts at the foundation of Inorganic Chemistry and Biochemistry. The interdisciplinary nature of *Biological Inorganic Chemistry* has the potential of coordinating the efforts of scientists in fields like biochemistry, inorganic and coordination chemistry, molecular and structural biology, enzymology, environmental chemistry, physiology, toxicology, biophysics, pharmacy and medicine. The inorganic biochemist is, therefore, potentially able to combine knowledge about the chemistry of metal ions with consideration of biomolecules with the aim of understanding how complex living systems function.

The progress in experimental investigation of the CNS at molecular level is now opening an unprecedented opportunity for the inorganic biochemist to contribute to the understanding of the roles that metal ions play in synaptic transmission, memory formation, and in the causes and treatments of neurological diseases. These concepts have been recently summarized in introducing the term *metalloneurochemistry* to describe the study of metal ions in the brain and nervous system at the molecular level.[†] Most work in this area is in its early stage, but the progress in the field is continuous and very significant, especially in delineating the role of metal ions in neurodegenerative diseases.

As the world's population lives longer and gets older, brain disorders, generally called neurodegenerative diseases, become one of the most difficult challenges for health care systems and human life protection of elderly. Most neurodegenerative disorders are fatal diseases and some of them like Alzheimer's disease (AD) are major cause of death in the developed societies. AD is the most common cause of dementia affecting several percent of humans of 65 and up to 40 to 50% of those above 90 years old. The annual cost of Alzheimer's in USA is estimated to be around 100 billion dollars. Parkinson's disease (PD) is the second most common brain disease affecting around 1% of those being 65–69 years old and around 3% of those above 80 years old. The increase of longevity of the human race, especially in the industrialized west, is great news, but the value to individuals and society is limited by the brain disorders, which can have great personal toll on the individual, friends and family and are very costly to the health system of any country. Thus, to make life longer and acceptable one needs to resolve several real problems including finding the causes of these age-related diseases and understanding the

---

[†] S.C. Burdette and S.J. Lippard, *Proc. Natl. Acad. Sci. USA*, 2003, **100**, 3605.

mechanisms of particular pathologies. From such findings, it should then be possible to create the effective set of therapeutic agents able to cure or at least slowdown the onset of neurodegenerative disease.

Although AD is certainly the most common cause of death among neurodegenerative disorders the most exciting of these must be prion diseases. Prion diseases are very rare, but captured world attention because of the recent epidemic of bovine spongiform encephalopathy or "mad cow disease." Although prion diseases are rare indeed (around 10,000 times rarer that Alzheimer's) and besides their headline grabbing high profile, they play a fundamental role in learning the neurodegenerative disorders due to very efficient animal models that are not available for other disorder such as Alzheimer's.

Neurodegenerative disorders have different clinical symptoms but they have also mechanistic similarities, *e.g.*, protein misfolding, oxidative stress or a potential role of metal ions. More than 20 diseases could be related to deposition of protein aggregates, often amyloid fibrils, in which is some cases the major components are either short peptide fragments (as in AD) or whole fragments (*e.g.*, prion diseases). Many of these diseases relate to misfolding of a protein expressed normally in the brain. While the role of the protein in disease is known, the cellular activity of the normal isoform often remains poorly understood. We knew of "bad prions" long before we learned that the normal prion protein could be also very useful in a cellular context. The same concerns amyloid precursor protein involved in production of the beta-amyloid peptide in AD or alpha-synuclein implicated in PD pathology. There is a strong research drive going on to establish these protein's biological functions as well as their structure destabilization resulting in protein misfolding, aggregation and fibril formation.

Only recently it was recognized that most, if not all, neurodegenerative disorders are closely related to oxidative stress, essential for development of disease and can cause death of neurons. Oxidative stress results from unequilibrated pro-oxidant-antioxidant homeostasis leading to the over-production of the reactive oxygen and nitrogen species. Both in the oxidative stress as well in protein, regular and pathological functioning metal ions are strongly involved, and therefore understanding neurodegeneration is very closely related to chemistry of essential as well as toxic metal ions involved in brain functioning.

The role of metals in neurodegenerative diseases stems from all these aspects. The main proteins in three neurodegenerative diseases (Alzheimer's, Parkinsons's and prion diseases) have all been shown to bind copper. Metals have been implicated in the misfolding and aggregation of beta-amyloid peptide, alpha-synuclein, and prion protein. Lastly, the metals associated with the pathology of these diseases are linked to the formation of oxidative stress. Therefore understanding the role that metals play in these disorders is becoming more and more central to unraveling their mystery. The genetic, biochemical and chemical aspects of protein behavior, and the impact of oxidative stress and metal ions on protein functioning and misfolding are the major objectives of this book.

Contents

Chapter 8 Alzheimer's Disease: Which Metal Now? 117

Chapter 9 Prion Diseases and Redox Active Metals 141

CHAPTER 1
Introduction

Neurodegenerative disorders are usually fatal diseases. Understanding these dementias and diseases of the nervous system is the leading challenge of sciences related to the quality of human life in the 21st century. In many instances these diseases are associated with changes to particular metals or the pathology resulting in the diseases are associated with proteins that bind metals. Therefore metallochemistry has become an inherent part of the study of neurodegenerative diseases.

More than 20 diseases could be related to the deposition of protein deposits including amyloid fibrils. In some cases (Alzheimer disease, AD) the major components of those deposits are relatively short peptide fragments of proteins. In other cases they are made of the entire or major part of the protein (prion diseases, familial amyloid polyneuropathy). Most of these are fatal diseases and some of them like AD could be a major cause of death in the developed societies due to aging.

1.1   General Model for Protein Misfolding, Aggregation, Amyloid Formation and Neurodegeneration

Recently it has been shown that mutations may cause not only the inactivity or deficiency of a particular protein but also its conformational instability.[1–3] This conformational instability may result in partial protein unfolding and aggregation due to intermolecular bonding usually *via* β-linkage. Serpins, protease inhibitors, were found to provide very useful model to understand the protein aggregation resulting from its mutation and conformational instability.[3]

The basic element of the serpin structure is set of three β-sheets and a mobile reactive peptide sequence used as a pseudosubstrate for a target proteinase.[4] After docking of a psedosubstrate with the enzyme the serpin inserts an extra β-strand into one of its β-sheet. The insertion of this additional β-strand is basic for the antiproteinase activity of serpins. However, this type of linkage may also lead to disease. Point mutation, especially in some key domain of the protein may destabilize β-sheet A to allow the insertion of the reactive peptide loop of another serpin molecule.[5,6] The incorporation of the successive reactive-loop results in the formation of polymeric structures which are retained in the cell leading to cell death and damage of the tissue.[3] The aggregation of neuroserpins results in familial encephalopathy with neuroserpin inclusion bodies (Collins bodies, FENIB) located in deeper layers of cerebral cortex and substantia nigra (*sn=a pigmented band of gray matter in the midbrain, neuronal degeneration in substantia nigra results in Parkinson disease*). In the case of FENIB, there is a direct relationship between the magnitude of the intracellular neuroserpin accumulation and the severity of disease.[7] The most polymerogenic mutation of neuroserpin seems to be G392E.[3,7] This replacement of conserved residue in the shutter region leads to large multiple inclusions in every neuron and caused the death of the members of the affected family by the age of 20 years. Although, the intracellular protein aggregation is by itself sufficient to cause neurodegeneration[7] the role of post-translational degradation of neuroserpin polymers in the development of neurotoxicity has yet to be explained.[3]

The common feature leading to protein aggregation and then to the conformational dementias is the formation of intermolecular interactions, almost always β-linkages, formed by hydrogen bonding between peptide loops and sheets. In larger highly ordered cystatins, the linkages by domain swapping occur.[8–10] In these cases in the formed polymers individual protein molecules substantially retain their ordered structure. However, many proteins may form linkages by realignment of peptide segments to give layered arrays of extended β-sheets called amyloid[11] (Figure 1). In several systemic amyloidoses the end-product the huge tissue-like looking amyloid deposits affect the organ (*e.g.*, heart or liver) functioning having direct impact on particular pathology.[12] The dementias are caused by cumulative loss of neurons and the pathology should be considered at the level of cells. Thus, the earlier forms like fibrils, protofibrils (oligomeric intermediates), oligomers or even protein dimers must be taken into consideration.[3,13]

1.2 Specificity of Molecular Mechanisms in Major Neurodegenerative Diseases

Although neurodegenerative diseases may have common cellular and molecular mechanisms including protein aggregation and inclusion body formation there are some specific features in each pathology, *e.g.*, specific brain regions where degeneration occurs or specificity of protein aggregates.

Figure 1 *Mechanism of fibrils composition and the diverse types of antibodies produced during the exposition. The first step includes the formation of an amyloidogenic intermediate via partly unfolded native state (1) or via partly folded of otherwise naturally unfolded species (2). The second step encompasses the self-association of the amyloidogenic intermediates that eventually leads to the production of amyloid fibrils. The amyloidogenic intermediates poses a high tendency to assembly to one another, and become stabilized by the formation of intermolecular β-sheet. Small oligomers are initiated early and act as the nuclei to induce the further growth of aggregates (the nucleus is for simplicity shown as a dimer). The growth leads to the formation of higher order oligomers referred to as prefibrillar aggregates (PA). These aggregates transform themselves into protofilament (P) directly or indirectly, and eventually into mature fibrils (F). Such fibrils usually consist of two to six protofilaments forming a rope-like structure* (Reproduced with permission from M. Dumoulin *et al.*, *Biochimie*, 2004, **86**, 589.)

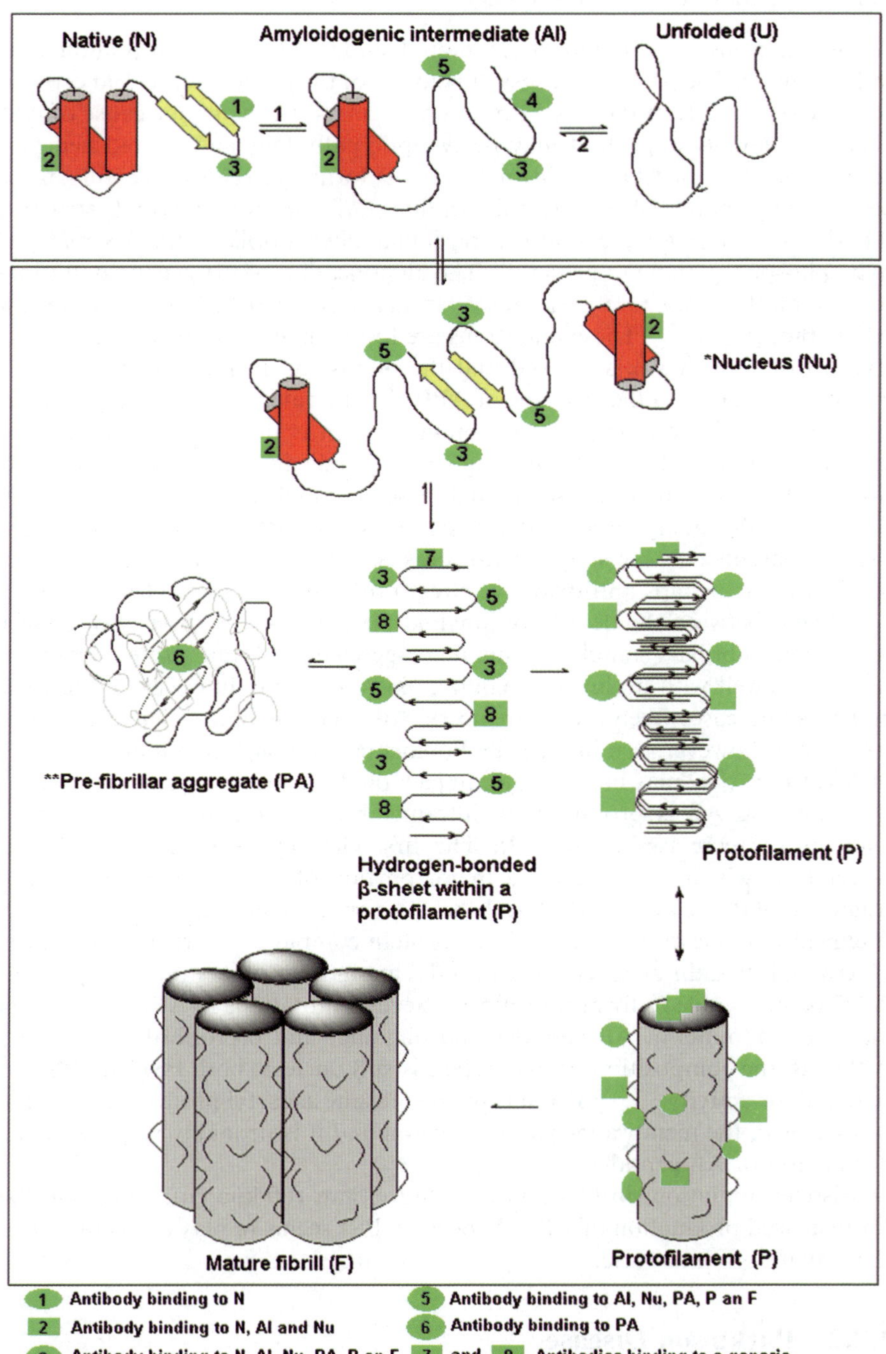

Native (N)
Amyloidogenic intermediate (AI)
Unfolded (U)
*Nucleus (Nu)
**Pre-fibrillar aggregate (PA)
Hydrogen-bonded β-sheet within a protofilament (P)
Protofilament (P)
Mature fibrill (F)
Protofilament (P)

1   Antibody binding to N
2   Antibody binding to N, AI and Nu
3   Antibody binding to N, AI, Nu, PA, P an F
4   Antibody binding to AI
5   Antibody binding to AI, Nu, PA, P an F
6   Antibody binding to PA
7   and   8   Antibodies binding to a generic epitope of P and F

1.2.1   Alzheimer Disease (AD)

AD is a late-onset and the most common disorder with over four million cases in USA alone. The progressive loss of neurons of the hippocampus and cerebral cortex (particularly in the basal forebrain) seems to be the major cause of AD. However, equally important may be synaptic pathology and altered neuronal connections.[14,15] In AD two distinct proteins form aggregates involving extracellular aggregates of β-amyloid peptide (Aβ) having amyloid structure (neuritic or senile plaques) and intracellular neurofibrillary tangles made of hyperphosphorylated tau protein.[16] The identification of the genetic mutations responsible for very rare early-onset familial cases clarified considerably the AD pathogenesis.[17,18] These mutations are found in amyloid precursor protein (APP) from which Aβ is cleaved and the prenisilins 1 and 2 (PS1 and PS2), which are involved in the cleavage of APP.[19–21] Familiar cases are very rare but in both sporadic and familial AD common event is the increased production and accumulation of Aβ. This finding strongly indicates that excessive production of Aβ is the primary cause of the disease (amyloid cascade hypothesis).[21,22]

The most intriguing question in AD pathology is whether plaques and tangles cause disease or they are simply end-products, remains of earlier events that led to disease, or perhaps anti-oxidative stress protective bodies.[23] There is a poor correlation between the density of amyloid plaques and the severity of dementia.[22] Neurofibrillary tangles containing aggregates of tau protein correlate reasonably well with decline of cognitive skills, they are, however, a late event and in some cases likely downstream of Aβ accumulation.[24] There are good indications, however, that in early development of AD oligomers and protofibrils of Aβ40 and Aβ42 may have a major impact on dendritic and synaptic injuries.[17]

Neurotoxic Aβ is produced by intramembrane proteolysis of APP by a complex of secretases (Figure 2). The first cleavage is realized by β- and α-secretase, which are releasing a major portion of extracellular APP as two fragments APP_s-α and APP_s-β and the *C*-terminal, membrane bound protein fragment remains intact. Then a large protein complex, γ-secretase, cleave this *C*-terminal domain at several sites producing among others Aβ40, Aβ42 and Aβ43 peptides able to form protofibrils. Several mutations in APP result in the increased amounts of Aβ peptides and oligomer and protofibril formation.[25] Although the composition of γ-secretase is still unclear, both PS1 and PS2 are required for activity.[17,26] PS1 is a trans-membrane aspartyl protease cleaving its substrates in the membrane-spanning region and it is probably responsible for production of Aβ peptides.

Missense mutations in PS1 and PS2 (more than 100 known) are responsible for increased production of Aβ.[27] Especially PS1 seems to play a pivotal role in the activity of γ-secretase.

1.2.2   Parkinson Disease

Parkinson disease (PD) is the most common disorder of movement. It is characterized by resting tremor, slow movements and rigidity. 95% of PD

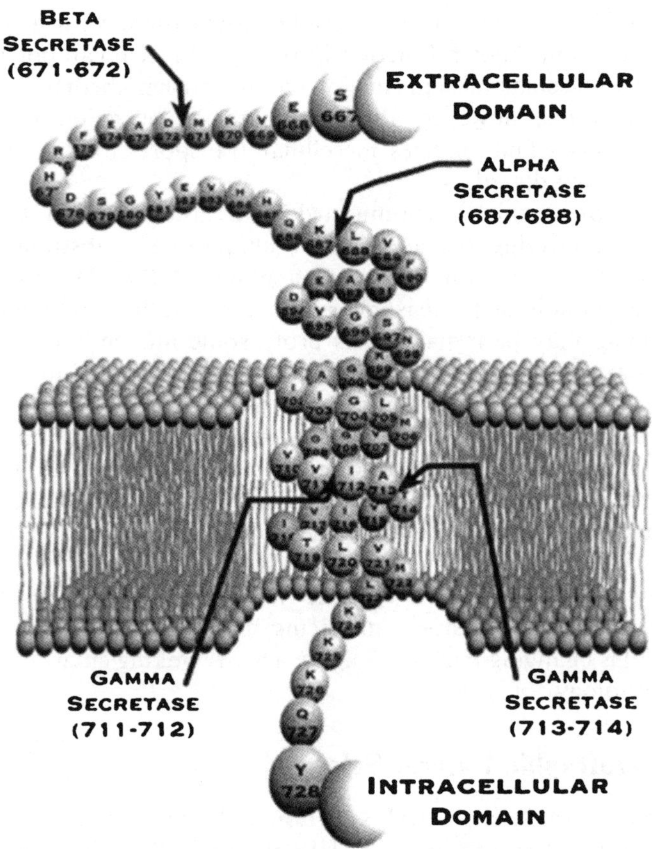

Figure 2 *Schematic view of the membrane spanning and cleavage sites of APP$_{770}$. The Aβ$_{1-40}$ fragment, generated by the cleavage by β- and γ-secretases, comprises residues 672–711 of APP$_{770}$ while the Aβ$_{1-42}$ fragment comprises residues 672–713. The p3 fragment, generated by α- and γ-secretase cleavage includes residues 688–711 and 713*
(Reproduced with permission from P. Turner *et al.*, *Progr. Neurobiol.*, 2003, **70**, 1.)

cases are sporadic and about 1% of population above 65 years may be affected by this disease. The cause of PD symptoms derived from progressive degeneration of dopaminergic neurons in the substantia nigra of the midbrain and monoaminergic neurons in the brain stem.[17–18,28] The important feature of PD are cytoplasmic inclusions (Lewy bodies) containing misfolded fibrillar proteins. The precise composition of these bodies is still not known, although it was shown that they contain parkin protein, ubiquitinated α-synuclein, synphilin and synaptic vesicle proteins.[17] The discovery that some gene mutations may lead to early-onset forms of PD accelerated the research progress distinctly.[29] Point mutations in α-synuclein gene cause autosomal dominant PD *via* toxic gain-of-function mechanisms, resulting in abnormal protein accumulation.[30,31] α-synuclein is normally unstructured with a little of any folding and it has no known function in biology. Protein can polymerize into filaments but molecular mechanisms of the α-synuclein-mediated toxicity are unknown. It seems, however, that the increases of oxidative stress, mitochondrial injuries and changes in cellular transport could results from α-synuclein-involving disorders.

The other protein which when mutated has a basic impact on PD is parkin, an E3 ligase, catalyzing the addition of ubiquitin to substrates aimed for degradation by the ubiquitin–proteasome system (UPS). The parkin protein binds to 26S subunit of proteasome and the mutation within the region of binding domain may be impairing its proteasome interactions.[17] The RING domains of parkin are cysteine-rich fingers involved in recognition of substrates and enzymes that transfer ubiquitin. Missense mutations in RING1 domain alter the protein localization and increase the protein aggregation.[32]

The substrate for parkin is among others α-synuclein. Thus, when E3 ligase activity is lost by mutation α-synuclein can accumulate. Parkin blocks this protein toxicity when overexpressed by promoting its degradation by UPS.

Cysteine residues located in cysteine-rich domains are very sensitive to oxidative and nitrosactive stress and being modified are altering the protein function. NO is changing parkin E3 ligase activity linking environmental stress with PD pathology.[33]

1.2.3   Amyotrophic Lateral Sclerosis

Amyotrophic lateral sclerosis (ALS or Lou Gehrig's disease) involves degeneration of lower motor neurons in the lateral horn of the spinal cord and upper motor neurons of the cerebral cortex resulting in progressive muscle wasting and weakness then paralysis, respiratory failure and death. Around 10% of cases are familial in origin, and less than 5% of those are caused by mutation in Cu/Zn superoxide dismutase (SOD1). As a result of such mutations SOD1 does not loose catalytic function but may gain some toxicity.[17] The impact of SOD1 mutation on pathology is very unclear. In sporadic cases SOD1 does not seem to be deposited in fibrillar structures involved in ALS pathology.[18]

1.2.4   Prions Diseases

Prion diseases (transmissible spongiform encephalopathies, TSEs) may affect both humans (kuru, Creutzfeldt-Jacob disease, CJD, Gerstmann–Sträussler syndrome, GSS, fatal familial insomnia, FFI) and animals (scrapie in sheep, bovine spongiform encephalopathy, BSE in cattle). The main symptoms in humans are dementia, accompanied usually by motor dysfunction (cerebral ataxia, myoclonus). The disorder is characterized by neuronal loss and astrogliosis, spongiform degeneration of the brain and deposits of the plaques containing polymerized prion protein. There are three different possibilities to develop the prion type of disease: infectious, familial and sporadic one. The most unusual among neurodegenerative disorders is the infectious manifestation, illustrated by kuru, a strange neurodegenerative disease found among the Fore tribe in New Guinea and transmitted between humans by ritual cannibalism.[34] BSE and its human version, the so-called variant CJD, are other examples of the infectious prion activity.[35] About 10% of CJD cases and all cases of GSS and FFI are familial inherited in an autosomal fashion and are linked to mutation in the PrP gene.[36] Sporadic prion diseases include most of the CJD cases and have no obvious infectious or genetic etiology. However, all of the diseases are characterized by the generation of an infectious agent. Prusiner, based on earlier speculations and his data, has hypothesized that there is a new class of infectious agents, which he called prions (proteinaceous infectious particles).[37] Biochemical analysis of the infectious agent from scrapie-infected animal brains has revealed that it contains a little if any nucleic acid and is composed largely with 33- to 35-kDa proteinase-resistant fragment called PrP^{Sc}. It was established later on, that PrP^{Sc} is an isoform of PrP^{C}, a protein of unknown biological function, which is widely spread and that is encoded in a cellular gene.[38] PrP^{Sc} and PrP^{C} differ distinctly in their biological behavior with PrP^{Sc} being detergent insoluble and resistant to enzymatic digestion. The two isoforms have no differences in either primary sequence or covalent bond pattern.[35] Although, the molecular structure of PrP^{Sc} is still unknown the major difference between both isoforms seems to be in their three-dimensional conformation with PrP^{Sc} having much higher content of β-sheet than highly helical PrP^{C} (Figure 3). It is generally accepted that PrP^{Sc} molecule propagates itself by changing the PrP^{C} conformation into PrP^{Sc}.[39] Familial prion diseases results from germline mutations in PrP gene, which favor the conformational change of mutant PrP^{C} into PrP^{Sc} without the necessity of the contact with exogenous PrP^{Sc}. Although, sporadic change of PrP conformation is very rare it causes more than 85% of all CJD cases and is

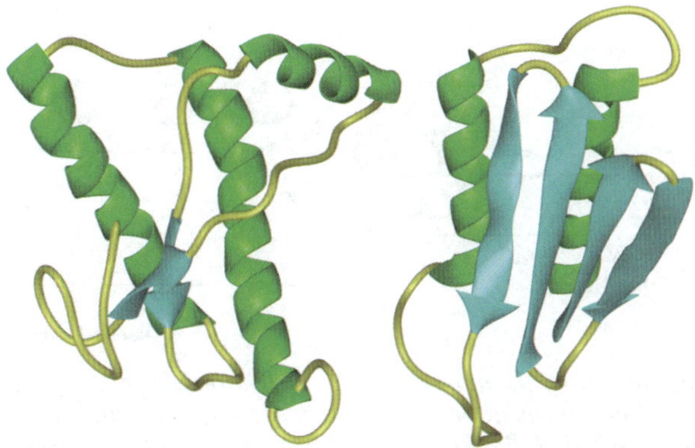

Figure 3 *The comparison of the secondary and tertiary structures of cellular PrP (PrP^{C} – left drawing) and tentative scrapie form of PrP (PrP^{Sc} – right drawing) (Reproduced with permission from ref. 36.)*

assumed to be due to spontaneous conversion of wild-type PrP^C into PrP^{Sc} conformation.

1.3 Models of Amyloid Seeding in Neurodegenerative Diseases

The proteins involved in neurodegeneration forming aggregates and amyloid fibrils can be induced to polymerize *in vitro* forming aggregates indistinguishable from brain-derived fibrils. Thus, much afford was made to model the details of aggregation process and the effect of endogenous molecules involved in the particular pathology. Usually a simple mechanistic model has emerged involving a nucleation-dependent polymerization indicating that aggregation is dependent on protein concentration and time (Figure 4).[40] This process is similar to protein crystallization and is characterized by a slow nucleation phase to form an ordered oligomeric nucleus, then a growth phase, in which the nucleus grows to form larger polymer and finally a steady state phase, in which aggregate and monomer are in equilibrium.[41]

This type of nucleation-dependent mechanism has some basic features

   (i) critical concentration of protein is needed to aggregate;
  (ii) above critical protein concentration there is a lag time before polymerization occurs; and
 (iii) addition of a seed during the lag time results in immediate polymerization.

Aβ amyloid formation (Aβ40 and Aβ42) was studied to some extent according to the above model. Aβ40 has higher critical concentration (around 100 μM at pH 7–8, 5–100 ml ionic strength, 25–37°C) than Aβ42 (about five time

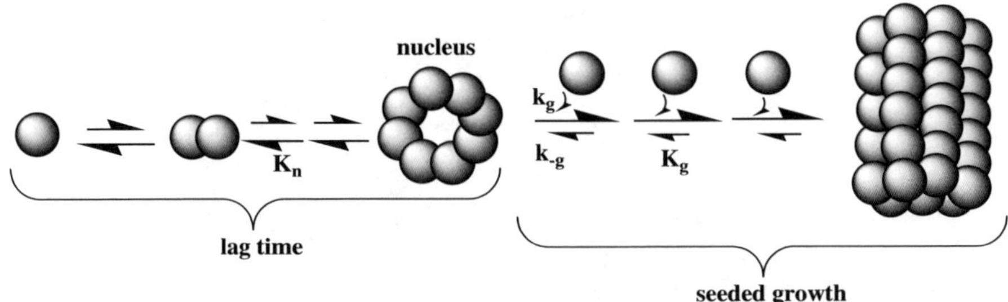

Figure 4 *The mechanism of the most plausible nucleation-dependent process. It represents a series of inconvenient protein–protein association equilibria (K_n) leading to an unstable nucleus, followed by a series of convenient equilibria (K_g), up to fibril formation. The critical concentration phenomenon results from the shift from inconvenient (K_n) to convenient (K_g) equilibria*
(Reproduced with permission from ref. 39.)

lower).[40,41] Aβ42 is much less studied owing to its great insolubility. These *in vitro* studies clearly show that critical Aβ concentration exceed μM concentration, so how is this concentration exceeded in the brain? The normal concentration of Aβ in the brain is in the low nanomolar range.[42] Thus, the aggregation process must occur in the compartment, where a local concentration of Aβ exceeds the μM concentration (Figures 5 and 6), or *in vivo* critical concentration is lowered by an endogenous substance. The studies on Aβ production from APP have demonstrated that proteolysis could occur in a cellular compartment.[40,43] The latter mechanism may be supported by the fact that phospholipid vesicles lower Aβ40 concentration required for a random coil-to β-sheet transition indicative of amyloid formation. In the *in vivo* situation one needs to consider also the impact of other biomolecules on the aggregation process, which usually is a very difficult task. Interfering proteins may increase the critical concentration of aggregating protein, while, *e.g.*, metal ions can lower this concentration. Seeding both homogenous and heterogeneous may be critical for accelerating of amyloid formation in AD.

This book will discuss the role of metals in the nervous system and how metallochemistry gives a greater insight into disease processes such as the formation of amyloid fibrils.

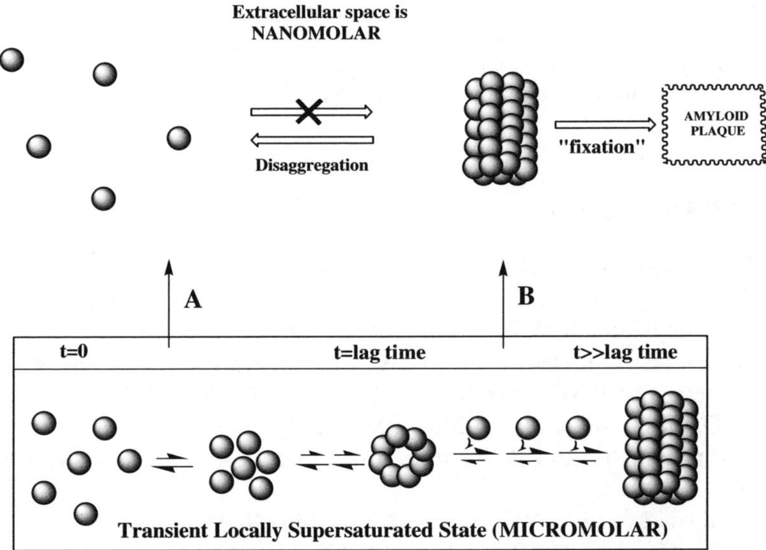

Figure 5 *The formation of amyloid in vivo must involve the local supersaturation in order for amyloid production to occur. This step might be the part of a physiological cellular process which is abnormally elongated in AD. The duration of the supersaturated state might usually be short and the nucleation process does not occur (arrow A). If this time is extended during the disease it may induce the nucleation (arrow B)*
(Reproduced with permission from ref. 39.)

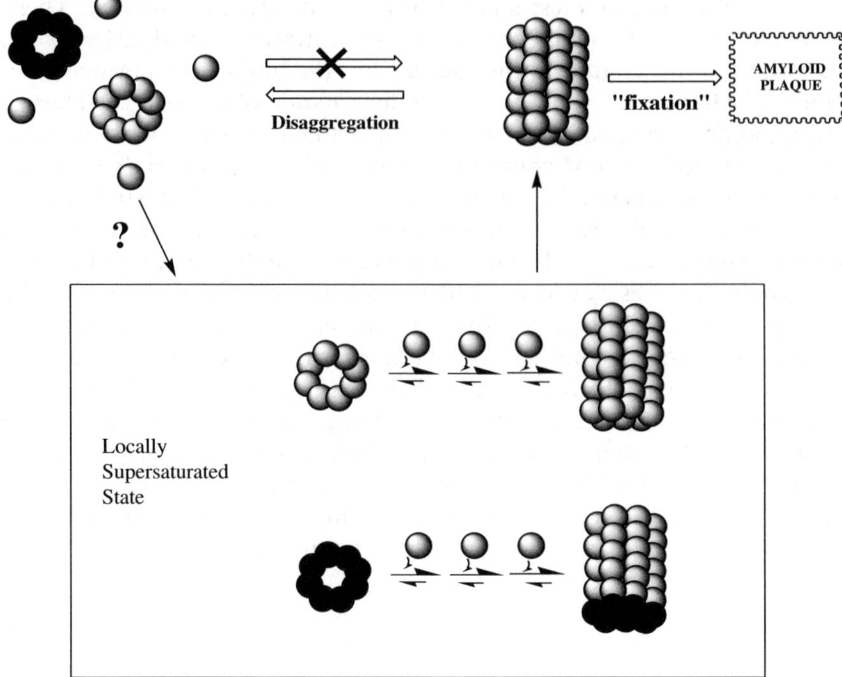

Figure 6 *The seed-depended polymerization of Aβ and/or PrP has to occur in a locally supersaturated environment. Both homogenous (gray seed) and heterogeneous (black seed) seeding might be involved*
(Reproduced with permission from ref. 39.)

References

1. R.R. Kopito and D. Ron, *Nat. Cell Biol.*, 2000, **2**, E207.
2. R.W. Carrell and D.A. Lomas, *N. Engl. J. Med.*, 2002, **346**, 45.
3. D.A. Lomas and R.W. Carrell, *Nat. Genet.*, 2002, **3**, 759.
4. D.A. Lomas, D.L. Evans and R.W. Carrell, *Nature*, 1992, **357**, 605.
5. P.E. Stein and R.W. Carrell, *Nat. Struct. Biol.*, 1995, **2**, 96.
6. R.L. Davies *et al.*, *Lancet*, 2002, **359**, 2242.
7. M. Abrahamson and A. Grubb, *Proc. Natl. Acad. Sci. USA*, 1994, **91**, 1416.
8. R. Jankowski *et al.*, *Nat. Struct. Biol.*, 2001, **8**, 316.
9. R.A. Stanforth, *EMBO J.*, 2001, **20**, 4774.
10. C. Blake and L. Serpell, *Structure*, 1996, **4**, 989.
11. M.B. Pepys, *Phil. Trans. R. Soc. Lond. B. Biol. Sci.*, 2001, **356**, 203.
12. M.S. Goldberg and P.T.J. Lansbury, *Nat. Cell Biol.*, 2000, **2**, E115.
13. D.J. Selkoe, *Science*, 2002, **298**, 789.
14. J. Hardy and D.J. Selkoe, *Science*, 2002, **298**, 962.
15. D.J. Selkoe, *Nature*, 2003, **426**, 900.

16. E. Bossy-Wetzel, R. Schwarzenbacher and S.A. Lipton, *Nat. Med.*, 2004, **10**, S2 (and refs. therein).
17. C.A. Ross and M.A. Poirier, *Nat. Med.*, 2004, **10**, S10 (and refs. therein).
18. W.P. Esler and M.S. Wolfe, *Science*, 2001, **293**, 1449.
19. M. Citron, *Nat. Neurosci.*, 2002, **5**(suppl), 1055.
20. D.J. Selkoe, *Nature*, 1999, **399**, A23.
21. J. Hardy, *Trends Neurosci.*, 1997, **20**, 154.
22. M.E. Obrenovich *et al.*, *Neurobiol. Aging*, 2002, **23**, 1097.
23. A. Delacourte and L. Buee, 2000, **13**, 371.
24. A. Singleton, A. Myers and J. Hardy, *Hum. Mol. Genet.*, 2004, **13**, R123.
25. C. Haasse, *EMBO J.*, 2004, **23**, 483.
26. D. Scheuner *et al.*, *Nat. Med.*, 1996, **2**, 864.
27. L.S. Forno, *J. Neuropathol. Ex. Neurol.*, 1996, **55**, 259.
28. T.M. Dawson and V.L. Dawson, *J. Clin. Invest.*, 2003, **111**, 145.
29. J.L. Erikson *et al.*, *Neuron*, 2003, **40**, 453.
30. M.J. Baptista, M.R. Cookson and D.W. Miller, *Neuroscientist*, 2004, **10**, 63.
31. M.R. Cookson, *Neuromol. Med.*, 2003, **3**, 1.
32. K.K. Chung *et al.*, *Science*, 2004, **304**, 1328.
33. D.C. Gajdusek and V. Zigas, *N. Engl. J. Med.*, 1957, **257**, 974.
34. R. Chiesa and D.A. Harris, *Neurobiol. Disease*, 2001, **8**, 743.
35. K. Young *et al.*, in *Prions: Molecular and Cellular Biology*, D.A. Harris (ed), Horizon Scientific Press, Wymondham, UK, 1999, 139–175.
36. S.B. Prusiner, *N. Engl. J. Med.*, 2001, **344**, 1516.
37. K. Basler *et al.*, *Cell*, 1986, **46**, 417.
38. F.E. Cohen and S.B. Prusiner, *Annu. Rev. Biochem.*, 1998, **67**, 793.
39. J.D. Harper and P.T. Lansbury Jr., *Annu. Rev. Biochem.*, 1997, **66**, 385.
40. J.T. Jarrett and Lansbury Jr., *Cell*, 1993, **73**, 1055.
41. W.A. van Gool *et al.*, *Ann. Neurol.*, 1995, **37**, 277.
42. D. Selkoe, *J. NIH Res.*, 1995, **7**, 57.
43. E. Terzi, G. Holzemann and J. Seeling, *J. Mol. Biol.*, 1995, **252**, 633.

Blood–Brain Barrier and Roots of Entry of Metal Ions into the Brain. Metal Ion Transport and Distribution in the Brain

2.1 General Features of Blood–Brain Barrier

Endothelial cells of brain capillaries create the blood–brain barrier (BBB). The barrier that separates the systemic circulation from the cerebrospinal fluid (CSF) is called the blood–CSF barrier (BCB). The endothelial cells in these barriers differ from those found in capillaries of other organs by having tight junctions between the cells (Figure 1), which prevent transcapillary movement of polar molecules from ions to proteins.

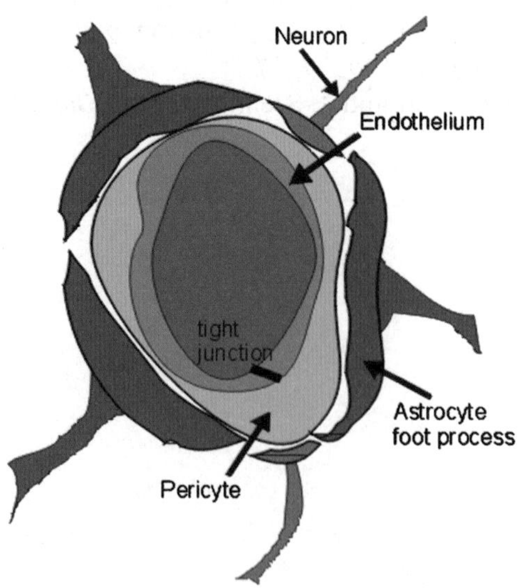

Figure 1 *The four cells comprising the central nervous system microvasculature*

There are no distinct transendothelial pathways, transcellular channels or fenestrations, thus, the endothelial cells in the brain form a continuous and effective cellular barrier between the blood and the interstitial fluid. The capillary endothelial cells are surrounded by collagen containing extracellular matrix. The interactions between astrocytes and endothelial cells seem to be an important factor for the function of BBB.[1]

Not all capillaries of the brain produce BBB. In some areas the tight junctions are discontinuous and endothelial cells show fenestrations. Some of these areas play a role in peptide hormone release.

Substances with high lipid solubility like many therapeutic compounds[2] may move through the BBB by simple passive diffusion. The permeability of some very lipid-soluble compounds including ethanol or diazepam is so effective that they are extracted completely from the blood during a single passage through the brain. The polar molecules, *e.g.* glycine, go through BBB very slowly.

Water enters the brain readily by diffusion, however, the rate of exchange is limited by the permeability of the capillary endothelium and by the rate of cerebral blood flow.

Gases like O_2, CO_2, N_2O or volatile anesthetics diffuse through BBB very rapidly and the equilibrium of their concentrations in the brain with plasma is limited by the cerebral blood flow rate.

Glucose, a basic energy substance of the brain, is a polar molecule and its high permeability through BBB is mediated by a specific carrier. Stereospecific, insulin independent, glucose transporters GLUT-1 is highly expressed in brain capillary endothelial cells and it permits D-glucose but not L-glucose to enter the brain. The inefficient expression of GLUT-1 results in mental retardation, low CSF glucose concentration and seizures in children.

Small water-soluble nutrients and macromolecules, which are necessary for brain functioning, may cross the barrier *via* facilitated diffusion or specific carrier mechanism. These carriers are distributed symmetrically or asymmetrically in the luminal and abluminal membranes of endothelial cells. The asymmetric distribution is observed for Na^+, K^+-ATPase, which is more abundant in the abluminal membrane.

Both macro-metals K^+ and Na^+ enter the brain many times slower than into muscles. Na^+ is transported out of the endothelium into the brain, while K^+ out of the brain into the endothelium. The asymmetric distribution of the ATPase maintains low K^+ concentration in the extracellular fluid of the CNS. K^+ has a powerful impact on the transmission of nerve impulses and neuron firing, thus, the absolute control of its low concentration is critical for the brain functioning.[1] The effective junctions that link brain endothelial cells prevent any paracellular passage in the physiological conditions. This is additionally controlled by transport proteins, which can prevent brain entry or facilitate the brain-to-blood efflux of various compounds. Good examples of such proteins are multidrug resistance (MDR) proteins, p-glyco-proteins and effective ATP-binding cassette transporters, which are responsible for the efflux of a wide range of structurally unrelated compounds from endothelial compartment back to the blood.[3,4] The cerebral capillaries form also an enzymatic barrier containing enzymes able to metabolize many xenobiotics.[5]

This tightness of the junctions in BCB at the choroid plexus is not, however, as perfect as that in BBB, and even very hydrophylic xenobiotics including drugs and intoxicants may get limited access to the brain *via* the choroid plexus. The impact of choroid plexus on the brain bioavailability of xenobiotics, including drugs, results from the morphological and biochemical characteristics of its epithelium forming a monolayer at the interface between the choroidal blood and CSF.[6,7]

Choroid epithelium expresses high levels of various proteins that are able to metabolize drugs, multidrug resistance protein 1 (MRP1), transporters for the organic anions and cations and having the antioxidant activity like glutathione peroxidase.[8,9] These transporters and drug-metabolizing enzymes may create an efficient barrier to a variety of blood-borne compounds, including drugs. They can accelerate the elimination of toxic metabolites out of CSF influencing the overall bioavailability of the brain.[10,11]

Many metals including calcium, magnesium, copper, iron, manganese, zinc or molybdenum are essential elements and are required for normal functioning of the CNS. Being essential cofactors for proteins like metalloproteases, superoxide dismutase, kinases, zinc fingers in transcriptional factors, these metals must be supplied to the brain at an optimal level. The metal deficiency or excess in CNS can result in serious functional aberrations of CNS.[12] Thus, the crossing of the essential metals into CNS is realized by the active or receptor-mediated transport systems able to control their optimal concentrations. Unfortunately, the non-essential metals like lead or mercury are also able to pass CNS barriers readily causing serious neurotoxic effects.

2.2 Iron and Aluminum

The importance of the adequate amounts of iron for brain functioning and iron homeostasis is well established.[13] Receptor-mediated endocytosis of the iron transferrin complex (Figure 2) is the most important physiological mechanism for iron delivery into a particular cell. In the plasma most iron is present as Fe^{3+} and bound in diferric complex to transferrin, (Tf), a glycoprotein of about 80 kD. The Tf receptor, an integral membrane protein of about 180 kD has much higher affinity for the diferric transferrin complex (about two orders) than the metal-free protein. It is generally accepted that through the transferrin receptors on the brain capillary endothelial cells, the iron–transferrin complex enters the brain.[14] However, the brain has several specific features that make it different from the other organs as far as iron transport and metabolism are concerned.[15,16] The brain presents a specific challenge for iron transport because of its separation by vascular barrier from the plasma iron. Cells in the brain without having a direct access to the plasma iron must send signal probably to the endothelial cells that they require iron. It is also likely that iron could be derived from the CSF. Thus, to enter the brain iron must cross the

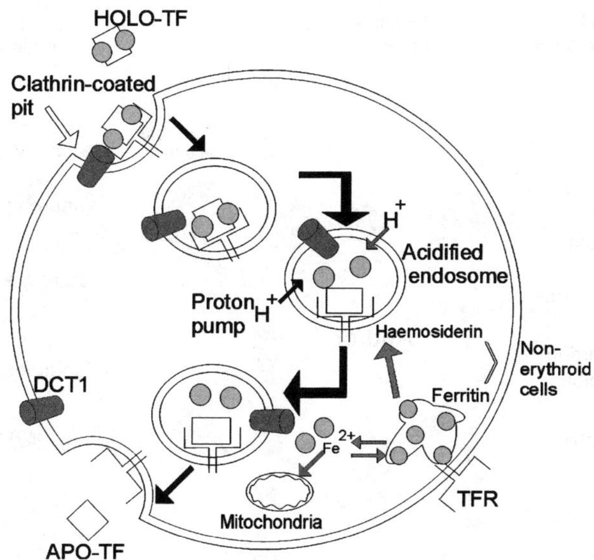

Figure 2 *Receptor-mediated endocytosis of iron transferrin complex. HOLO-TF-holotransferrin, APO-TF-apotransferrin, TFR-transferrin receptor, DCT1-divalent cation transporter*

endothelial cells lining the vasculature or the ependymal cells that line the ventricles. In very few regions BBB is lacking and iron may enter the brain directly.

Although the iron uptake by brain is still not well understood, the transferrin-mediated mechanism seems to be the major root for iron to enter the endothelial layer. Transferrin receptors were found on brain endothelial cells in rat model and human BBB.[17] The next step, the crossing from endothelial cells into the brain is much less clear. Several studies have suggested that iron is transported across the BBB at a greater rate than transferrin.[16,18] The study with hypotransferrinemic mouse, however, has shown clearly that transferrin does cross BBB even if its integrity is not damaged.[19,20] The possible mechanisms of iron uptake into the brain across the BBB is shown in Figure 3. The iron uptake at the endothelial cell involves the internalization of the Tf–transferrin receptor complex according to the general mechanism (Figure 2). Tf–iron complex is transported across the endothelial cell in a transcytotic vesicle. The vesicle is transported to the abluminal membrane and with the assistance of the divalent metal transport protein 1 (DMT1) iron is moved out of endosome into the glial end-foot. The second mechanism (Figure 3 (bottom)) suggests that iron is removed from the endosome within the endothelial cell with DMT1. The removed iron is transported by unknown mechanism to ferritin for storage, for the labile iron pool necessary for the endothelial cell or moved to abluminal membrane for transport into the glial end-foot process. Both mechanisms are likely and they are not mutually exclusive. Although a transport mechanism of iron from plasma to blood is known to some extent,

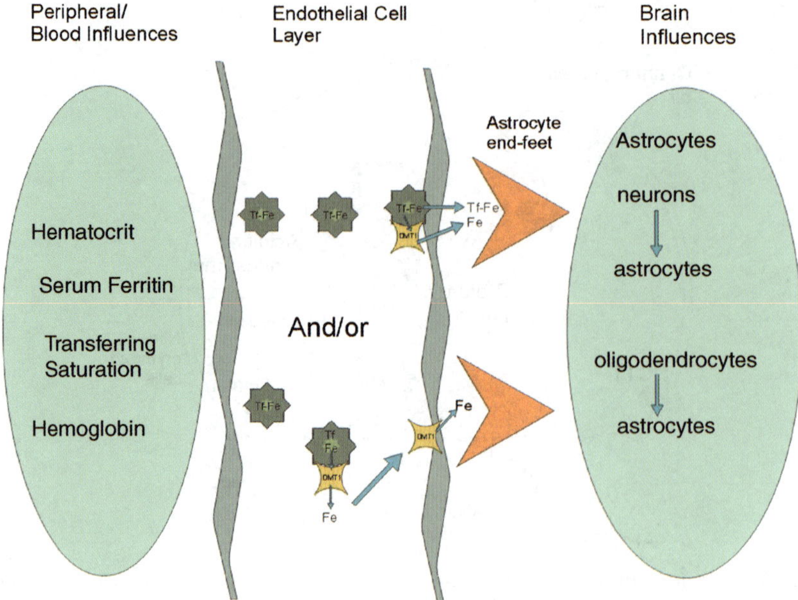

Figure 3   *The possible mechanisms of iron uptake into the brain across the BBB. Tf-transferrin, DMT1-divalent metal transport protein 1*

the iron release into the brain and the regulation of the transport mechanisms are poorly understood.

Transferrin is produced also in the brain. Around 95% of endogenous brain transferrin is located in the oligodendrocytes.[21] Tf produced in the brain could obtain iron transported from the endothelial cells, however, recent study suggests that Tf is not secreted from oligodendrocytes.[16] Tf could be released, however, from choroid plexus into the ventricles. Thus, it may be an important delivery mechanism for iron to the rest of the brain locations.[22] There are also other ligands in the CSF that may be responsible for iron transport when Tf is fully saturated including ferritin,[23] citrate[24] and ascorbate.[25]

Besides Tf, the important transporting protein for iron is DMT1. In CNS it is responsible for movement of iron out of endosomes (Figure 3).[26] DMT1 is located mostly in ependymal cells lining the ventricles, blood vessel endothelial cells and astrocytes associated with these vessels. Thus, this transporter may have a strong influence on the iron transport into and out of the brain.

There are several other proteins that most likely play a role in the iron transport system like melanotransferrin (P97), metal transport protein 1 (MTP1), a membrane protein Hfe and lactoferrin, but their precise function is yet to be learned.[16]

The iron transport system was suggested to be used by aluminum[27] and gallium[28] to enter the brain. Al^{3+} ions are similar to Fe^{3+} ions and both are very effectively bound to transferrin. The speciation calculations indicate that around 91% of plasma Al^{3+} is indeed coordinated to transferrin and only

7–8% is in the form of the citrate complex.[29] In the brain, however, the calculations predict about 90% of Al^{3+} in the form of citrate complex and only 4% as transferrin-bound Al^{3+}. The primary site of aluminum entry into the brain seems to be the BBB.[30] There are at least two different mechanisms of Al transport into the brain. One could involve the receptor-mediated Al–Tf influx as it was described for iron.[31] This mechanism, however, is not well supported. The binding pattern of Al^{3+} to Tf is very similar to that of Fe^{3+} with the first metal ion binding at *C*-lobe and then at *N*-terminal site.[32] However, Tf saturated with Al^{3+} or mixed with the equimolar Al^{3+}/Fe^{3+} load does not interact effectively with the transferrin receptor and even if Tf solubilizes Al^{3+} in the biological fluids it does not imply that the Al transfer from the blood stream into the cell is the receptor-mediated process.[32] The binding of metal ion has a critical impact on the Tf structure and even slight differences between Al^{3+} and Fe^{3+} binding may affect distinctly the interactions of metal loaded Tf with transferrin receptor.

The studies with aluminum citrate, however, strongly indicate that there is likely a second mechanism of the aluminum transport. Very quick appearance of aluminum in the brain after Al^{3+}–citrate injection was suggesting other than the Tf-receptor mediated transport of Al^{3+} into the brain.[30] The Al^{3+}–citrate is transported into the brain much faster than citrate alone and this transport is ATP dependent and sodium independent.[33] The transporter of Al^{3+}–citrate is still unknown, although it could be one of the monocarboxylate transporters, which is still uncharacterized, or one of the member of the organic anion transporters, which is expressed at the BBB.

2.3 Manganese

Manganese is biologically an essential element as is iron but when in excess it is a serious neurotoxicant like aluminum. Like other chemicals, manganese may enter the brain from blood by BBB or choroid plexus. The major route into the brain is BBB, which has much greater opportunity for fast exchange when compared to choroids plexus and CSF compartment.[29] In biological systems manganese is present as Mn^{2+}, Mn^{3+} and Mn^{4+}.[34] The transport mechanisms and specific functions of manganese are not understood well due to its very low concentration in the tissues. Very often the activity of this metal is not manganese-specific, as Mn^{2+} resembles, *e.g.* Mg^{2+} and substitution of the latter metal ion by Mn^{2+} may not change distinctly the activity of biomolecule. There are, however, manganese-specific enzymes like mitochondrial superoxide dismutase[35] or glia-specific glutamine synthetase,[36] and manganese is clearly essential for the development and functioning of the brain.[37,38] This metal, however, when concentration is abnormally high may accumulate in basal ganglia leading to disorders similar to Parkinsons disease.[39]

At early post-natal times, the BBB is quite leaky to metals, including manganese, bound to variety of biomolecules including proteins and smaller ligands. The leaky brain is necessary for the rapid growth of the neonatal brain.

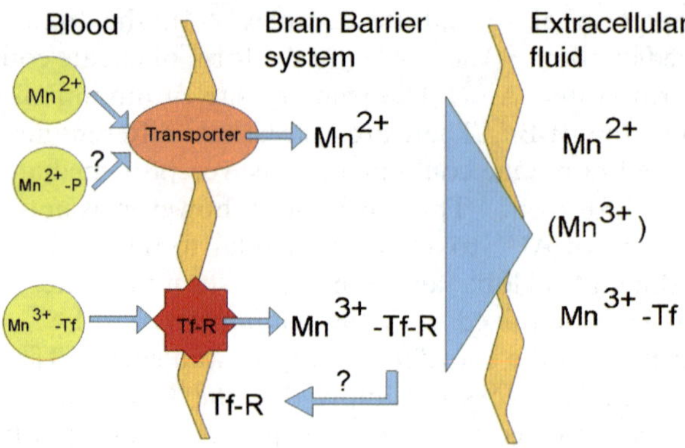

Figure 4 *Manganese transport into the brain across the BBB. Tf- transferring, P- protein, Tf-R- transferring receptor*

At about 4 months after birth, the BBB is formed and essential metals are transported usually by more or less specific transporters.

In plasma, manganese exists as di- or trivalent ion. It is likely that dietary manganese might be divalent and is oxidized by ceruloplasmin.[34] The liver may act as a depot for manganese, which is then transported into the brain. It is likely that a plasma carrier for Mn^{3+} is transferrin.[40,41] Transferrin-bound manganese may be transported into the brain *via* BBB (Figure 4).[42] The mechanisms of the Tf-dependent transport of Mn through BBB could be close to those described above for iron.

Intravenous injection of Mn^{2+} into experimental animal results in the rapid removal of metal from the blood and the manganese is transported into the brain.[43] This may be caused by non-transferrin-mediated manganese uptake mechanism (Figure 4). The species in which Mn^{2+} travels through the blood is unknown. There are suggestions that divalent manganese is just a free ion,[44,45] Mn^{2+} ion may be transported by several systems including active calcium uniporter,[46] calcium channels,[47] Na/Ca exchanger[48] and possibly DMT1.[49] The latter transporter has quite a broad range of substrates including Fe^{2+}, Zn^{2+}, Cu^{2+}, Co^{2+} and Mn^{2+}, and it is expressed in both brain capillary endothelial cells and choroid epithelial cells.

The comparison of Mn^{2+} with manganese citrate has shown that Mn^{2+} citrate may be a major species entering the brain *via* carrier-mediated influx.[50]

The non-transferrin mediated transport of Mn^{3+} is unknown.

When Mn is secreted into the brain extracellular fluid from the capillary endothelial cells and choroidal epithelial cells Mn^{3+} may bind to transferrin, which is secreted from oligodendrocytes.[51] In extracellular fluid manganese may be present also as non-transferrin species. The neurons express Tf receptors on the surface[51] and the Tf-bound Mn^{3+} may enter these cells by receptor-mediated endocytosis (Figure 5). The non-Tf-bound manganese uptake was found in the gial cell culture.[52] Although the specific transporter for non-Tf-bound manganese is not yet known, DMT1 could be one of the candidates, as it is present in high density in several sites of the brain like granule and

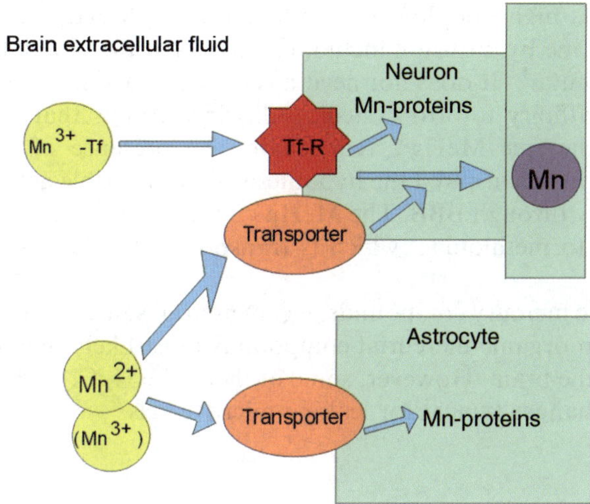

Figure 5 *Manganese transport into neurons and glial cells via the transporter (T) and the receptor-mediated endocytosis*

hippocampal pyramidal cells, cerebellar cells or pyramidal cells of piriform cortex.[49] It is quite likely that this transporter and possibly natural resistance-associated macrophage protein (Nramp) could be responsible for manganese ion or low molecular weight Mn^{2+} complex uptake into neurons. Other details of manganese transport and uptake are given in the chapter devoted to Mn.

2.4   Lead and Mercury

The effect of Pb on the brain is known for a long time owing to very clear clinical manifestations in acute Pb poisoning like brain swelling and ventricular compression, a high incidence of cerebral hemorrhage, thrombosis and arteriosclerosis among the battery workers who are exposed to very high levels of lead.[12] Pb induces microvascular damage ending in the opening of the inter-endothelial tight junctions and increased pinocytotic activity.

Chronic exposure to high levels of Pb is also damaging to cerebral vasculature in animals. Pb has a unique affinity to cerebral endothelial cells, in which it accumulates in much larger concentrations than in the other cells of the brain.[53] Pb is known to accumulate also in the choroid plexus to a greater extent than in endothelial cells. This accumulation can alter distinctly some functions of BCB, *e.g.* depressing of transthyretin, thyroid hormone, causing irreversible mental retardation in children.[54]

The free Pb^{2+} or a low-molecular-weight complex of Pb^{2+} is believed to be transported species. Pb has no affinity to transferrin and may pass through BBB by passive diffusion as Pb^{2+} or $PbOH^{+}$.[55] The Pb efflux from the brain may occur with the help of Ca^{2+}-ATPase localized in luminal and abluminal membranes of CNS capillaries.

Because of its high lipophilicity, methylmercury, $MeHg^+$, can easily cross any cell membrane by diffusion including barriers at the blood–brain interface and at the placenta.[12] It does not need any specific carrier system, although it has very high affinity toward cysteine sulhydryl groups that makes the concentration of the free $MeHg^+$ very small. The presence of MeHg–cysteine complex in blood has inspired the hypothesis about possible active transport of methyl–mercury through BBB. The $MeHg^+$ complex may act as an amino acid analog similar to methionine, which is transported by the carrier for neutral amino acids.[56]

The inorganic mercury compounds, *e.g.* mercurial salts, are usually much less neurotoxic than organic mercurial compounds most likely due to less efficient transport into the brain. However, some of them like $HgCl_2$ induce severe BBB damage with changes typical for leaking microvessels.[57,58]

2.5 The Olfactory Pathway of Metal Entry into the Brain

The BBB, when not damaged, protects brain usually well. However, xenobiotics may have an alternative route to CNS *via* the olfactory pathway.[59] The neurons of the olfactory epithelium have dendrites that are in contact with the nasal lumen and axons, which reach the glomeruli of the olfactory bulbs.

The xenobiotics applied on the olfactory mucosa will enter into the CSF *via* two possible routes.[60] One involves intraaxonal transport along the olfactory neurons to the olfactory bulb (the olfactory nerve pathway) and then *via* diffusion into CSF. The other one is a movement of molecules along the perineural and then subarachnoid space (the olfactory epithelial pathway). The uptake of metals into CSF is less understood, except perhaps manganese.[61] Manganese moves by axonal transport in the long olfactory neurons to the olfactory bulbs. The transfer of other metals from the surface of olfactory mucosa into the CFS could also proceed *via* "the olfactory nerve pathway", but this hypothesis is not yet well documented. An important and unanswered question that remains is the biological relevance of the olfactory transport of metals into the CNS in humans[61] although it seems to be generally accepted that olfactory pathway is involved at least partly in etiology of neurodegenerative disorders.[62,63]

2.6 Astroglia and Metal Accumulation

The ability to concentrate metals by the brain is well established.[64] The metal concentration is necessary for their metabolic use. To manage high levels of metals brains need efficient system to protect cells against metal toxicity. The essential metals like Cu, Fe or Mn, which are redox active, must be especially carefully managed. Also exogenous metals like Pb or Hg, although not directly active in redox reactions, are very damaging neurotoxicants. Essential metals are usually bound to metalloregulatory proteins (*e.g.* metallothioneins) or metal-requiring enzymes. All metals can be also extruded from the cell by diffusion or P-type of ATPases or sent to specific organelles (mitochondrion, lysosome or nucleus).[65] Astroglia are assumed to accumulate selectively several metals including Pb and Hg. Astroglia seem to be also a major localization for brain Cu and Mn.[66] Astroglia have several important features to be considered as the depots for metal ions. Astroglia known as end-feet or footplates surround the vascular surface of the brain forming permeable layers and being the first cells of the brain parenchyma to encounter metals entering through the BBB.[67] Astroglia, in contrast to neurons, have high levels of metallothioneins I and II[68] and can express MT-III during reconstruction of the brain tissue.[69] These proteins allow astroglia to chelate the metal ions and to protect the brain.[70] Some critical metal-dependent enzymes like glutamine synthetase[71] or Cu,Zn-SOD[72] are localized in astroglia as well as other metal proteins including ceruloplasmin[73] and transporter ATP7A (Cu-ATPase).[74] Astroglia are also much richer than neurons in the redox active tripeptide glutathione.[75] These all characteristic properties make astroglia a strong candidate for accumulation and storage of several metals. The existing data indicate that astroglia are principal cells for accumulation for lead and manganese, but in the case of mercury the data are still contradictory.[66] The Hg distribution may be strongly dependent on the mercury species used to treat animals and the animal used for experiment. Monkeys (*Macaca*) treated with $MeHg^+$ or inorganic mercury localize Hg mainly in astroglia and microglia,[76] while in rats that are treated with $MeHg^+$, Hg accumulates predominantly in neurons and at lower levels in glia.[77] In the other animal models the metal distribution could be even less uniform.[66]

Most of the Mn is associated with glutamine synthetase (80%)[78] that is localized predominantly in astroglia.[71] Most experiments with use of animal or cell models clearly show that astroglia is a major localization of Mn.[66] This of course indicates that in the case of Mn overexposure, astroglia are potentially the primary target for Mn-induced damage.[79] The astroglia located damage caused by Mn could be associated with disease similar to idiopathic Parkinson disease.[66] Mn is located in astroglia mitochondria at low level as Mn-SOD.[72] It is also likely that Mn may induce NO synthase to produce toxic levels of NO.[80]

References

1. H. Davson and M.B. Segal (eds), *Physiology of the CSF and Blood–Brain Barriers*, CRC Press, New York, 1996.
2. H. Fisher, R. Gottschlich and A. Seeling, *J. Membrane Biol.*, 1998, **165**, 201.
3. R.B. Kim *et al.*, *J. Clin. Invest.*, 1998, **101**, 289.
4. A.H. Schinkel *et al.*, *J. Clin. Invest.*, 1996, **97**, 2517.
5. N. Strazielle and J.F. Ghersi-Egea, *Molecular Drug Metabolism and Toxicology*, G.M. Williams and O.I. Aruoma (eds), OICA International, London, 2000, 181–200.

6. J. Szmydynger-Chodobska, A. Chodobski and C.E. Johanson, *Am. J. Physiol.*, 1994, **266**, R1488.

7. J.F. Ghersi-Egea, W. Finnegan, J.L. Chen and J.D. Fenstermacher, *Neuroscience*, 1996, **75**, 1271.

8. I. Tarayani, I. Cloez, M. Clement and J.M. Bourre, *J. Neurochem.*, 1989, **53**, 817.

9. J.F. Ghersi and N. Strazielle, *J. Drug Targeting*, 2002, **10**, 353.

10. H. Suzuki, T. Terasami and Y. Sugiyama, *Adv. Drug Deliv. Rev.*, 1997, **25**, 257.

11. N. Strazielle and J.F. Ghersi-Egea, *J. Neurosci.*, 1999, **19**, 6275.

12. W. Zheng, M. Aschner and J.F. Ghersi-Egea, *Toxicol. Appl. Pharmacol.*, 2003, **192**, 1.

13. R.R. Crichton, in *Inorganic Biochemistry of Iron Metabolism. From Molecular Mechanisms to Clinical Consequences*, Wiley, Chichester, 2001.

14. R.D. Klausner, T.A. Rouault and J.B. Harford, *Cell*, 1993, **72**, 19.

15. L. Zecca, M.B.H. Youdim, P. Riederer, J.R. Connor and R.R. Crichton, *Nature Neuroscience*, 2004, **5**, 863.

16. J.R. Burdo and J.R. Connor, *BioMetals*, 2003, **16**, 63.

17. W. Jefferies *et al.*, *Nature*, 1984, **312**, 162.

18. W.A. Banks *et al.*, *Brain Res. Bull.*, 1988, **21**, 881.

19. T.K. Dickinson and J.R. Connor, *J. Com. Neurol.*, 1995, **355**, 67.

20. F. Ueda *et al.*, *J. Neurochem.*, 1993, **60**, 106.

21. J.R. Connor and R.E. Fine, *J. Neurosci. Res.*, 1987, **17**, 51.

22. M.W. Bradbury, *J. Neurochem.*, 1997, **69**, 443.

23. C.J. Earley *et al.*, *Neurology*, 2000, **54**, 1698.

24. A.F. Haerer, *Neurology*, 1971, **21**, 1059.

25. R. Spector, A.Z. Spector and S.R. Snodgrass, *Am. J. Physiol.*, 1977, **232**, R73.

26. J.R. Burdo *et al.*, *J. Neurosci. Res.*, 2001, **66**, 1198.

27. A.J. Roskams and J.R. Connor, *Proc. Natl. Acad. Sci. USA*, 1990, **87**, 9024.

28. V.A. Murphy and S.I. Rapoport, *J. Neurochem.*, 1992, **58**, 898.

29. R.A. Yokel, *Environ. Health Perspect*, 2002, **110**(Suppl. 5), 699.

30. D.D. Allen and R.A. Yokel, *J. Neurochem.*, 1992, **58**, 903.

31. R.A. Yokel, in *Aluminum and Alzheimer Disease*, C. Exley (ed), Elsevier, New York, 2001.

32. M. Hémadi *et al.*, *Biochemistry*, 2003, **42**, 3120.

33. R.A. Yokel *et al.*, *Brain Res.*, 2002, **930**, 101.

34. F.S. Archibald and C. Tyree, *Arch. Biochem. Biophys.*, 1987, **256**, 638.

35. R.A. Weisiger and I. Fridovich, *J. Biol. Chem.*, 1973, **248**, 4793.

36. M.D. Norenberg, *J. Histochem. Cytochem.*, 1979, **27**, 469.

37. J.R. Prohaska, *Physiol. Rev.*, 1987, **67**, 858.

38. A. Takeda, *Brain Res. Rev.*, 2003, **41**, 79.

39. M. Aschner, in *Metal and Oxidative Damage in Neurological Disorders*, J.R. Connor (ed), Plenum Press, New York, 1997, 77.

40. P. Aisen, A. Aasa and A.G. Redfield, *J. Biol. Chem.*, 1969, **244**, 4628.

41. L. Davidsson *et al.*, *J. Nutr.*, 1989, **119**, 1461.
42. M. Aschner and M. Gannon, *Brain Res. Bull.*, 1994, **33**, 345.
43. N. Sotogaku, N. Oku and A. Takeda, *J. Neurosci. Res.*, 2000, **61**, 350.
44. P.M. May, P.W. Linder and D.R. Williams, *J. Chem. Soc. Dalton Trans.*, 1997, **588**.
45. O. Rabin *et al.*, *J. Neurochem.*, 1993, **61**, 509.
46. C.E. Gavin, K.K. Hunter and T.E. Gunter, *Biochem. J.*, 1990, **266**, 329.
47. K. Narita, F. Kawasaki and H. Kita, *Brain Res.*, 1990, **510**, 289.
48. M.D. Frame and M.A. Milanick, *J. Am. Physiol.*, 1991, **261**, C467.
49. H. Gunshin *et al.*, *Nature*, 1997, **388**, 482.
50. J.S. Crossgrove *et al.*, *Neurotoxicol.*, 2003, **24**, 3.
51. J.R. Connor, *Dev. Neurosci.*, 1994, **16**, 233.
52. A. Takeda, A. Devenyi and J.R. Connor, *J. Neurosci. Res.*, 1998, **51**, 454.
53. L. Struzynska *et al.*, *Mol. Chem. Neuropatol.*, 1997, **31**, 207.
54. J.H. Dussault and J. Ruel, *Annu. Rev. Physiol.*, 1987, **49**, 321.
55. R. Deane and M.W.B. Bradbury, *J. Neurochem.*, 1990, **54**, 905.
56. M. Aschner and T.W. Clarkson, *Brain Res.*, 1988, **462**, 31.
57. A.E. Marlin *et al.*, *Neurosurgery*, 1980, **6**, 45.
58. J. Albrecht *et al.*, *Neurotoxicology*, 1994, **15**, 897.
59. H. Tjälve and J. Tallkvist, in *Metal Ions and Deurodegenerative Disorders*, P. Zatta (ed), World Scientific, Singapore, 2003, 67.
60. L. Illum, *Eur. J. Pharm. Sci.*, 2000, **11**, 1.
61. H. Tjälve, C. Mejare and K. Borg-Neczak, *Pharmacol. Toxicol.*, 1995, **77**, 23.
62. D.P. Perl and P.F. Good, *Ann. NY Acad. Med.*, 1991, **640**, 8.
63. P. Zatka, *Metal Ions in Biology and Medicine*, **vol 6**, John Libbey Eurotext, London, 2000, 443.
64. A.I. Bush, *Curry. Opinion Chem. Biol.*, 2000, **4**, 184.
65. C.T. Dameron and M.D. Harrison, *Am. J. Clin. Nutr.*, 1998, **67**, 1091S.
66. E. Tiffany-Castiglioni and Y. Qian, *Neurotoxicology*, 2001, **22**, 577.
67. E. Tiffany-Castiglioni *et al.*, *Neurotoxicology*, 1989, **10**, 383.
68. M. Penkova *et al.*, *J. Comp. Neurol.*, 1999, **412**, 303.
69. L. Acarin *et al.*, *J. Neuropathol. Exp. Neurol.*, 1999, **58**, 389.
70. M. Aschner *et al.*, *Brain Res.*, 1998, **813**, 254.
71. M.D. Norenberg and A. Martinez-Hernandez, *Brain Res.*, 1979, **161**, 303.
72. J. Lindenau *et al.*, *Glia*, 2000, **29**, 25.
73. L. W. Klomp *et al.*, *J. Clin. Invest.*, 1996, **98**, 207.
74. S.G. Kaler and J.P. Schwartz, *Neurosci. Res. Commun.*, 1998, **23**, 61.
75. E. Tiffany-Castiglioni *et al.*, in *Role of Gliain Neurotoxicity*, M. Aschner and H.K. Kimelberg (eds), Boca Baton, CRC Press, 1996, 175.
76. J.S. Charleston *et al.*, *Neurotoxicology*, 1996, **17**, 127.
77. B. Møller-Madsen, *Fundam. Apel. Toxicol.*, 1991, **16**, 172.
78. F.C. Wedle and R.B. Denman, *Curr. Top. Cell Regul.*, 1984, **24**, 153.
79. J. Henriksson and H. Tjälve, *Toxicol. Sci.*, 2000, **55**, 392.
80. M. Spranger *et al.*, *Exp. Neurol.*, 1998, **149**, 277.

CHAPTER 3

Metal Ion-Induced Redox Reactions, Oxidative Stress and Possible Impact on Neurodegeneration

Oxygen is one of the critical molecules for life processes. However, its metabolites, the reactive oxygen species (ROS), are very toxic to cells. Postmortem examination of the brain tissues of the patients with Alzheimer disease (AD) or Parkinson disease (PD) clearly shows the high level of damage caused by ROS-inducing reactions. The experimental data concerning damage of the brain tissues with neurodegenerative disorders do not indicate whether oxidative stress is a major cause of these diseases or it is merely a consequence of pathology.[1] Brain has very high oxygen metabolic rate, relatively poor defense against oxidative reaction and reduced capacity for cellular regeneration. Thus, the brain with its post-mitotic cells is very susceptible to oxidative damage caused by ROS during neurodegeneration process.

3.1 Metal Induced Production of ROS

ROS are usually defined as oxygen derivatives that readily react with cellular components resulting in damage of their function. ROS includes both radical species having very reactive unpaired electrons, like superoxide radical, $O_2^{-\cdot}$, nitric oxide, $NO^{\cdot}$ and most damaging hydroxyl radical $OH^{\cdot}$, and molecules like hydrogen peroxide, H_2O_2 or peroxynitrite, $ONOO^-$. The main source of oxygen species is metabolic reduction of oxygen to water. The general scheme of possible reactions involving reduction of oxygen is given in Figure 1.

The most detrimental effects in the cell are caused by hydroxyl radical, $OH^{\cdot}$, chemically the most aggressive unspecific oxidant. The classical reaction producing $OH^{\cdot}$ is a Fenton reaction that can be summarized as

$$Fe^{2+} + H_2O_2 \rightarrow Fe^{3+} + OH^{\cdot} + OH^- \tag{1}$$

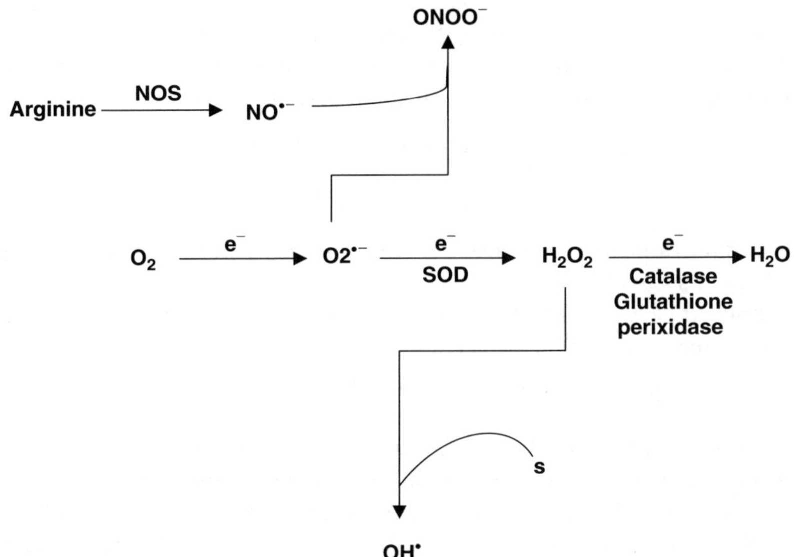

Figure 1　*The scheme of possible products of reactions involving reduction of oxygen*
Adapted from [1]

The reduction of peroxides by any metal species is usually called Fenton-like reaction with rather complex mechanism.[2]

The reaction (1) may be followed by generation of superoxide radical resulting from the oxidation of hydrogen peroxide by hydroxyl radical:

$$HO^{\cdot} + H_2O_2 \rightarrow O_2^{-} + H^{+} + H_2O \qquad (2)$$

Haber–Weiss reaction causes the iron-independent decomposition of H_2O_2 and regeneration of $OH^{\cdot}$:

$$O_2^{-} + H_2O_2 \rightarrow O_2 + OH^{\cdot} + OH^{-} \qquad (3)$$

However, the detailed studies of the reaction mechanisms have shown that reaction (3) does not really occur and superoxide does not reduce peroxide but rather Fe^{3+} ions:[3,4]

$$O_2^{-} + Fe^{3+} \rightarrow O_2 + Fe^{2+} \qquad (4)$$

The reduced forms of other metal ion complexes (like Cu^{+} or V^{4+}) can also be oxidized by hydrogen peroxide producing hydroxyl radical according to general Fenton reaction:

$$M^{n+} + H_2O_2 \rightarrow OH^{-} + OH^{\cdot} + M^{(n + 1)+} \qquad (5)$$

and then

$$M^{(n + 1)+} + O_2^{-} \rightarrow M^{n+} + O_2 \qquad (6)$$

Although reactions (5) and (6) look exactly as reactions (1) and (4), the detailed mechanism of the reactions with different metal ions may not be the same.[2]

In the presence of target molecules other than H_2O_2 (reaction (2)) the produced $OH^{\cdot}$ radical may react with them giving their oxidized forms,[5] *e.g.*

$$RH + HO^{\cdot} \rightarrow R^{\cdot} + H_2O \tag{7}$$

$$R^{\cdot} + O_2 \rightarrow R + O_2^{-} \tag{8}$$

The secondary radicals $R^{\cdot}$ are also able to cause the oxidative type of damage in the biological system.

Fe^{2+} and other metal ion species are also able to activate the hydroperoxides (ROOH) in reaction similar to (1)

$$M^{n+} + ROOH \rightarrow M^{(n + 1)+} + RO^{\cdot} + OH^{-} \tag{9}$$

This reaction is critical in the process of lipid peroxidation, extremely important in the oxidative-stress affecting biological membranes.[6,7]

The metal ions able to catalyze Fenton-like reactions *in vivo* could derive from different sources. The presence of labile iron pool inside the cell seems to be well supported.[8–10] The free ion pool contains most likely the low-molecular-weight (LMW) complexes with amino acids and ATP as ligands.[11] This LMW complexes pool is usually used for synthesis of iron enzymes but its increase or the increase of H_2O_2 or other peroxides may cause the Fenton-type of reactions to occur. The amount of iron may increase due to disorders in the uptake or storage of this metal. The ferritin deficiency causes sensitivity to H_2O_2[12] and ascorbic acid mobilizes iron from ferritin even if it is in sufficient amount.[13] The important source of Fe^{2+} could be iron–sulfur clusters present in many proteins. The impact of oxidants, which may inactivate the proteins containing [4Fe–4S] clusters[14] and then release iron, is well established. The oxidants like O_2^{-} or peroxynitrite are able to react readily with [4Fe–4S] clusters inactivating enzyme and releasing iron.[2,15,16]

Superoxide and hydrogen peroxide species are produced extensively by various enzymes as well as during the oxidation processes of different biomolecules with the help of oxygen in mitochondria and microsomes and autoxidation of flavins or catecholamines.[17–19] Superoxide, which can be an intermediate species during reduction of oxygen to H_2O is dismutated by SOD to H_2O_2 and O_2 or reacts with other molecules inside the cells. The high chemical reactivity of oxygen species, especially that of $OH^{\cdot}$ results in their very low intracellular concentrations.[14,17,20]

The basic role of O_2^{-} in oxidative stress seems to be well supported mainly due to efficient inactivation of Fe–S clusters.[21,22] Mitochondria are producing large amounts of O_2^{-}[23] and they contain a [4Fe–4S] cluster enzyme aconitase, which is O_2^{-} sensitive.[24] As a result the deficiency of MnSOD, which is a mitochondrial enzyme dismutating the superoxide radical, results in very strong reduction of aconitase activity.[25] The inactivation of aconitase results in damage of [4Fe–4S] clusters, LMW iron and production of hydroxyl radical causing serious toxicity resulting in death of cells and then SOD-deficient animal.[25,26]

3.2 Nitric Oxide

Nitric oxide (NO) from the late 1980s have been accepted as one of biological relevant molecules being an endogenous species, which is biosynthesized in many cells.[27–30] NO molecule reacts with superoxide radical very readily at a near diffusion-controlled rate to give peroxynitrite:[31,32]

$$NO + O_2^- \rightarrow ONOO^- \tag{10}$$

While O_2^- is produced by one electron reduction of O_2, NO is synthesized by nitric oxide synthases, which convert L-arginine to NO and L-cytrulline using O_2 and NADPH (Figure 2). The reaction catalyzed by NOS is a stepwise 5 e$^-$ oxidation of one of the guanidinium nitrogens of Arg. Actually there are three major isoforms of NOS: endothelial cell-derived eNOS, neuronal (nNOS) and immunological (iNOS) enzymes. All three isoforms have molecular weights around 125–160 kDa, depending on animal species. Biosynthesis of NO (Figure 2) is relatively well understood although the details of chemical mechanisms are still unclear.

The best understood biological activity of NO is its ability to increase the levels of the second messenger cyclic-GMP (cGMP) resulting in the relaxation of the smooth muscles. Actually NO was used therapeutically (nitroglycerin in the treatment of angina pectoris) already 100 years before it was found to be endogenously produced and essential for biological functioning. The effect of muscle relaxation is due to the interactions of NO with soluble guanylate cyclase a heme protein,[33] that catalyzes the production of cGMP from GTP. It possesses rather low basic activity but in the presence of NO this activity may increase up to 200 fold. The enzyme activation is the result of very unique chemical properties of NO interacting with iron complexes. The coordination of NO to ferrous heme prosthetic group at position of distal ligand (Figure 3) promotes the release of proximal histidine activating the enzyme. Of course,

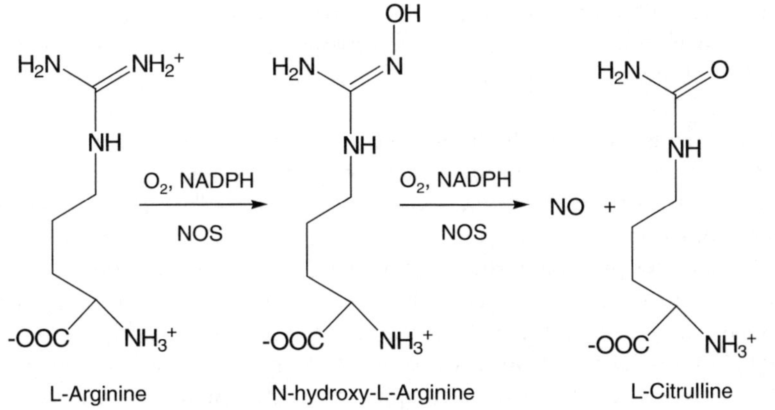

Figure 2 *Biosynthesis of NO*

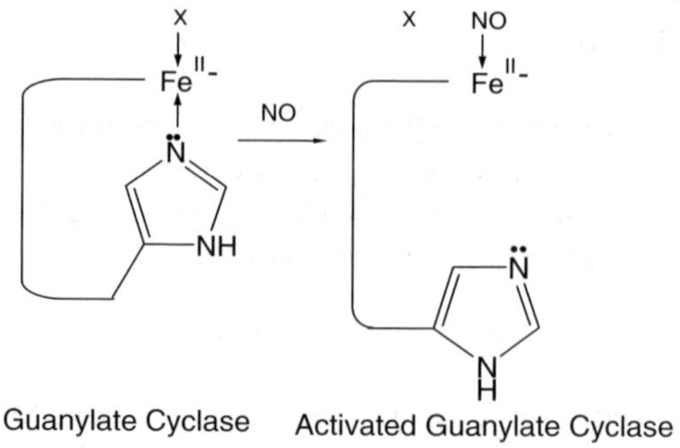

Figure 3 *NO-promoted release of proximal histidine ligand in the activation of guanylate cyclase. Distal ligand is shown as X and this is possibly another histidine* Adapted from [30]

this unique ability to interact with metal species may lead to pathogenic effects. NO is able to interact with the other heme proteins, *e.g.* its interactions with cytochrome P450 changes drug metabolism and hormone biosynthesis[34,35] or inhibit catalase increasing the amount of dangerous hydrogen peroxide.[36] It is also likely that NO inhibits mitochondrial enzymes containing Fe–S clusters by removing iron as the nitrosyl complex.[30,37]

The direct effect of NO can be also seen in radical chemistry. As radical itself, NO may enter the reactions involving other radicals acting as a scavenger, *e.g.*

$$ROO^{\cdot} + NO \rightarrow ROONO \text{ or} \tag{11}$$

$$R^{\cdot} + NO \rightarrow R\text{–}NO \tag{12}$$

In both cases NO terminates the radical reaction acting as the protective species. NO may act as the protective agent also indirectly acting as a signaling molecule inducing the action of protective protein.[38]

However, in the oxidation stress one of the most critical molecule is the product formed between NO and O_2^- according to reaction (10), peroxynitrite, $ONOO^-$. It is anion of a weak peroxynitrous acid with pK_a around 6.5. The chemistry of peroxynitrous acid is clearly summarized in recent review of Koppenol.[39]

Peroxynitrous acid oxidizes effectively many different compounds very effectively. It seems that majority of peroxynitrous acid isomerizes to nitrate and less than 40% carries out oxidation reactions.[40,41] It is necessary for macrophages but far away from macrophages this can be a very dangerous toxin. It activates prostaglandin biosynthesis,[42] and it is good news for biology but kinetic consideration suggest that it competes with hydroxyl radical or perhaps is a major molecule, which damages the cell propagating the oxyradical damage. The oxidation products produced by hydroxyl radical and peroxynitrite could be the same and the clear picture of the impact of both oxidants on initiation and propagation of the tissue damage is still a controversial issue.[39,43]

One of the most important property of peroxinitrite is its diffusibility making this molecule very effective. It can easily diffuse through several cells from its production site to attack particular proteins.[44]

Although the concentration of the particular ROS or kinetic aspects could be critical for their contribution to the oxidative stress effect, all of them may interact with biological molecules causing serious detrimental effects like disrupting the active sites of the enzymes, impact on protein conformation, production of the lipid peroxides resulting in the membrane disorders both of the cell and its organelles, *e.g.* mitochondria. The oxidation of DNA results in mutations, which may have a critical impact on DNA replication.

3.3 Oxidative Stress and Aging

It seems to be generally accepted that age is the biggest risk factor in many neurodegenerative diseases and that oxidative stress increases in brain during the human aging. The oxidative stress results in age-dependent increase of oxidized proteins and oxidative DNA damage.[45] Thus, deterioration in brain functioning is a fundamental feature characteristic for normal aging process and majority of neurodegenerative diseases. The disorders resulting from the oxidative stress damage are best seen in post-mitotic tissues, in which cells damaged irreversibly or lost cannot be replaced by mitosis for new ones. Brain is one such tissue and it plays a basic role in the functioning of whole organism. Besides being post-mitotic tissue, brain contains a high level of easily peroxidizable unsaturated fatty acids, requires very high amounts of oxygen, is able to accumulate redox active metal ions like Fe and Cu and does not have very efficient antioxidant protective defense systems.[45–47] The major source of energy for brain comes from the glucose metabolism. This may explain the high consumption of oxygen by brain. The major fraction of ROS (about 95%) is produced in mitochondria, mostly as H_2O_2.[48–50] During oxygen metabolism ROS are always produced and the oxidation products are always found. The oxidative stress is very dynamic in nature and ROS production, sometimes at low levels sometimes, *e.g.* during a stroke at high levels, always occurs. The cells possess an antioxidant capacity able to protect the oxidation of biomolecules and repair or remove the oxidized species.[51]

3.4 ROS, Protein Oxidation and Aberrant Protein Interactions

Although detailed relations are still uncertain there is solid evidence that increased oxidative changes to proteins are closely related to neurodegenerative disorders and might result in increased protein misfolding and impaired degradation.[1,52–54] Protein oxidation may proceed *via* two possible pathways, the oxidation of the backbone peptide chain or its side-chains. The backbone oxidation occurs by reactions mediated by Cu^+ and Fe^{2+} or the hydroperoxyl radical (HOO·) and often results in formation of the carbonyl groups that are used as a marker of protein oxidation (Figure 4).[52]

The most of amino acid side-chains may undergo the oxidation but the products of such oxidation are well characterized only for several amino acids including tyrosine, methionine, cysteine, histidine, tryptophan, lysine, arginine, phenylalanine, threonine, proline and glutamic acid.[52] The critical oxidation reaction occurring with involvement of amino acid side-chain is the reversible nitration of Tyr residue by peroxynitrite.[55] The product of this reaction, 3-nitrotyrosine (3-NT, Figure 5) may block processes involving 4-OH function of Tyr including protein activation *via* phosphorylation/dephosphorylation reactions. The insertion of nitro group into phenyl ring increases considerably the hydrophylicity of the aromatic ring and as a result alters the protein tertiary structure. Both sulfur containing amino acids Cys and Met undergo easily reversible oxidation in rather mild conditions with formation of disulfides and methionine sulfoxide, respectively. Met oxidation is one of the protective mechanisms against oxidation damage.[56] This is supported among others by the presence of the methionine sulfoxide reductases.[57] The amino acid side-chains can be also modified by the products of the lipid peroxidation, various reactive aldehydes, including malodialdehide or 4-hydroxynonenal.[58] The residues attacked by the aldehydes are those with electron-rich side-chains like His, Cys or Lys. The following reactions induce severe changes of the membrane protein conformation.[59–61]

Protein carbonyls introduced to proteins *via* direct oxidation of the proteins or by reaction with alkenals are markers of protein oxidation and can be evaluated by spectrophotometric[52] or immunochemical methods.[62,63] Additionally the formation of 3-NT by reaction of Tyr with peroxinitrite as well as covalently bound dityrosine products, that are formed due to ROS attack on Tyr can also be determined.[64–66]

The products of protein oxidation processes may accumulate in the tissues and their increased amount is generally considered as a hallmark of aging.[67,68] The results obtained clearly show that there is a strong increase of protein oxidation products in human cerebral cortex with age. The amount of oxidized protein evaluated by analysis of protein carbonyl content in older humans (around 70 years) is 2.6 times higher than that found in younger subjects (around 29 years).[68] Also, the amount of oxidized form of Tyr (dityrosine and *ortho*-tyrosine) was found to be much higher in older animal subjects in heart and skeletal muscle but not in the brain.[69,70] It is likely that oxidative reactions may inactivate the enzymes critical for cell protection against oxidative stress. As mentioned above Met can be easily oxidized to sulfoxide, which then is reduced back to methionine by methionine sulfoxide reductase. The activity of the latter enzyme was shown to decrease in the aging brain especially in the final few months of a rat's life.[71] The decrease of activity of the other enzymes, *e.g.* creatine kinase and glutamine synthase were also suggested. The enzyme inactivation is even more pronounced in the brain affected by neurodegenerative disorder.[72]

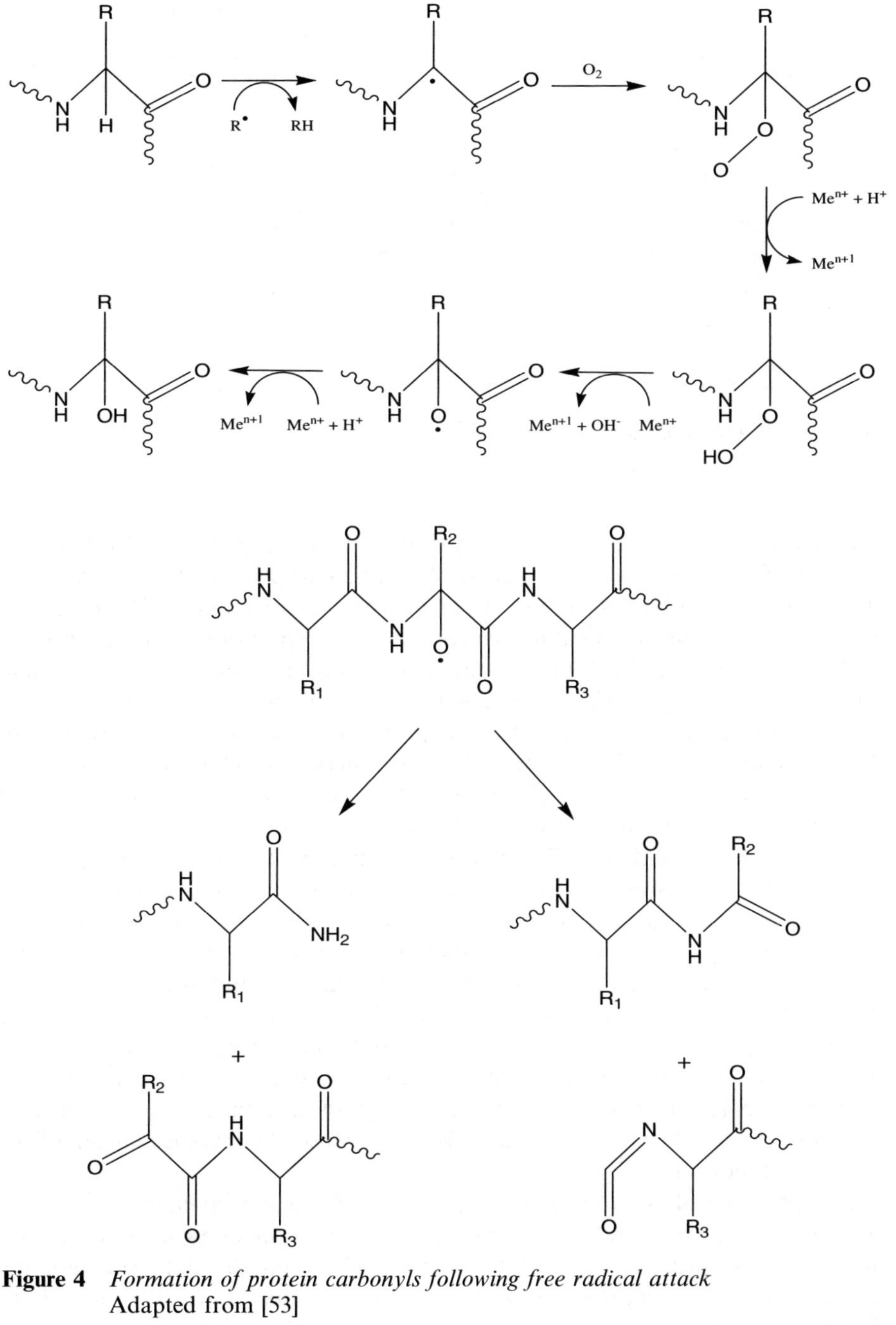

Figure 4 *Formation of protein carbonyls following free radical attack*
Adapted from [53]

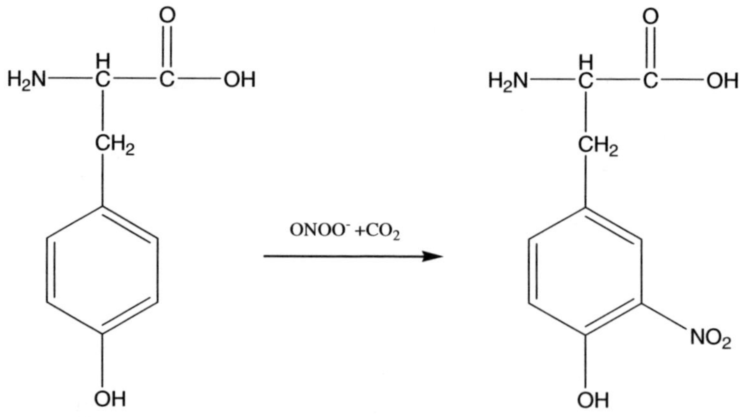

Figure 5   *Nitration of the tyrosine by peroxynitrite*

Increased levels of the reactive carbonyls were identified in AD as the product of protein oxidative damage.[73–76] The carbonyl-related damage was evident in senile plaques as well as in neurofibrillary tangles. The oxidative alterations to proteins critical for neurodegenerative pathologies like β-amyloid in AD or α-synuclein in PD may contribute to protein misfolding and protein impaired degradation. The latter events may result in toxic accumulation of soluble protofibrils and insoluble protein aggregates that may contribute to development of neurodegeneration. The products of oxidative stress increase with age (*vide supra*) and the ability of cells to repair the oxidative damage of proteins seems to decline with age. Thus, aging may contribute critically to aberrant protein buildup. The relations between oxidative stress and the specific protein misfolding besides general characteristics might have some very specific features that should be considered individually. Many issues are still very controversial and needs further studies. β-Amyloid peptide itself was proposed to generate the reactive oxygen,[77] while later studies were suggesting that peptide alone may play a role as modulator, able to increase or decrease reactive oxygen production.[78] However, β-amyloid can bind in some conditions the redox-active iron or copper and then can act as catalyst able to produce the oxygen radicals.[79] It was suggested that β-amyloid may have a direct impact on the production of H_2O_2 in cultured cells.[80] However, the real mechanism of the H_2O_2 production is not clear.

The presence of mutant genes, *e.g.* human A53T α-synuclein[81] or human ALS-SOD1[82] in cultured cells increases the oxidation of proteins or causes increased susceptibility to oxidative stress. Although the data showing the direct impact of oxidative stress are very convincing, it is still difficult to determine whether oxidative stress or protein aggregation is the initiating event in neurodegeneration process, although these two processes have a critical impact on each other.

3.5   Peroxidation of Lipids in Aging Brain

Peroxidation processes involving lipids yield large number of products including most importantly, from medical point of view, the active aldehydes and the isoprostanes. The level of the F_2-isoprostanes, the product of the non-enzymatic peroxidation of arachidonic acid is considered as the most valuable marker to index the oxidative damage to lipids *in vivo*.[83] The unique isoprostane-like compounds formed only in brain from peroxidation of docosohexonic acid were recently discovered and they were named neuroprostanes.[84,85] The level of this compounds in cerebral spinal fluid (CSF) of AD subjects was found to be distinctly higher than in normal age-matched controls. F_2-isoprostanes were also enhanced in the CSF of AD subjects although the total amount of the latter compounds was lower than that of neuroprostanes. Thus, it seems to be rather accepted that lipid peroxidation increases in the case of AD subjects when compared to controls. The AD animal model study indicated that brain trauma increases the amount of β-amyloid deposition and it also enhances the formation of isoprostanes long before the plaques are deposited.[86] This indicates the early involvement of the oxidative reactions in the development of the neurological disorders.

One of the most studied active aldehyde seems to be 4-hydroxy-2-nonenal (HNE) a major product of the peroxidation of omega-6-conjugated fatty acids including arachidonic acid and linoleic acid.[58,87] HNE is biologically very active. It reacts with amino acids in proteins (*e.g.* histidine[88]) and bases in DNA. It can also react with amino groups of lipids.[89] HNE is able to inactivate several enzymes like aldose reductase[90] (which metabolizes HNE[91]), glutathione peroxidase[92] and Na/K-ATPase.[93] The inactivation of the latter ion pump may be critical for its neurotoxic activities in neurons. The increased amount of the HNE adducts have been shown to be present in the PD and AD brains, however, not much is known about levels of HNE in aging brain.[94] Also impact of melanodialdehyde (MDA) is carefully studied, although in the case of this marker its relations with aging seem to be more complicated.[95]

3.6   Impact of Oxidative Stress on DNA

Damage to DNA is the most important factor for aging, especially for post-mitotic cells such as neurons. Oxidative damage to DNA results in large number of compounds derived from oxidized bases. The major oxidized product is 8-hydroxy-dGuanine (8-hydroxydG or 8-oxodG), a mutagenic compound, whose level is controlled by a set of proteins. The amount of 8-oxodG is increasing in the brain in exponential pattern with age.[96,97]

The interesting relations between aging, ROS and oxidative damage of DNA are observed in mitochondria. Aging is characterized well by the accumulation of mitochondrial DNA mutations especially in neurons.[98] Mitochondrial DNA (mtDNA) is inherited maternally and does not recombine. Its mutations accumulate through maternal lineages. One mitochondrion may contain from 2 to 10 double-stranded DNA encoding only 13 proteins, basic for mitochondrial functioning. There are multiple mitochondria inside one cell. Thus, normal and mutant mtDNA may coexist within the same cell including lethal mutations.[99] Nuclear DNA (nDNA) is more resistant to mutations as it is protected partly by histones and mtDNA mutation rate is around 17 times higher than that of nDNA.[100] mtDNA is located close to the site of mitochondrial ROS production, which seems to be mainly complex I involving FeS clusters. ROS generator is also centered at mitochondrial complex III but its products are directed rather toward cytosol.[101]

Some environmental agents like herbicide rotenone may induce the neuropathological disorder similar to PD.[102] This effect may result from the fact that rotenone is a selective complex I inhibitor causing the increase of the oxidation stress in mitochondrion, but the systemic administration of rotenone does not lead to serious mitochondrial dysfunction.

A reduced form of glutathione (GSH), a very effective reducing agent, plays an important role during the oxidation process. Its level can be affected, *e.g.* by dopamine quinines (DAQ) which form conjugates GSHDAQ. The latter species converted by proteases to 5CysDAQ, which is able to inhibit complex 1.[103,104] Elevated levels of 5CysDAQ and reduced amount of GSH are found in the *substantia nigra* of PD patients. The GSH deficit seems to precede the complex I activity loss.[105] The inhibition of the complex I activity due to GSH loss may result from the oxidation of thiol groups within the complex I. The depletion of GSH may also lead to GSH reductase (GSHRd) inhibition as a result of two active-site Cys thiols.[106] The inhibition of GSHRd results in further GSH loss and decrease of the GSH to oxidized glutathione (GSSG) ratio resulting in the total increase of the oxidative damage within the cells.

3.7   ROS and Cell Death

One of the characteristic features of oxidative stress (specifically the activity of ROS) is its ability to induce the death of neurons[107,108] and astrocytes[109] by apoptosis, known also as programmed cell death, or necrosis mechanisms. In most neurological pathologies apoptotic cell death predominates. Mitochondria may cause the excitotoxic neuron death *via* Ca^{2+}-glutamate mechanism.[107,110] Excessive activation of glutamate receptors may mediate neuronal injury or even death owing to excessive influx of calcium ions through the ionic channels into the cell.[111] The intracellular reaction to this event is overproduction of proteolitic enzymes, lipid peroxidation, production of ROS and RNS[112] and, as a result, programmed cell death.[113] Cell death is also the major event in neurodegenerative diseases. Thus the direct impact of ROS on neurodegenerative disorders is very likely even if it is still not known whether oxidative stress is a primary initiating cause of such pathologies like AD, PD or ALS. There is, however, very strong evidence that ROS production is crucial for propagation of the cell injury resulting in neurodegeneration.

References

1. J.K. Andersen, *Nat. Rev. Neurosci.*, 2004, **5**, S18.
2. S.I. Liochev, in *Metal Ions in Biological Systems*, A. Sigel and H. Sigel (eds), Marcel Dekker, New York, 1999, 1.
3. W.G. Barb *et al.*, *Trans. Faraday Soc.*1951, **47**, 462.
4. J.D. Rush and H.J. Bielski, *J. Chem. Phys.*, 1985, **89**, 5062.
5. S.I. Liochev and I. Fridovich, *Arch. Biochem. Biophys.*, 1991, **291**, 379.
6. B.P. Branchaud, in *Metal Ions in Biological Systems*, A. Sigel and H. Sigel (eds), Marcel Dekker, New York, 1999, 79.
7. O. Sergent, I. Morel and J. Cillard, in *Metal Ions in Biological Systems*, A. Sigel and H. Sigel (eds), Marcel Dekker, New York, 1999, 251.
8. K. Keyer and J.A. Imlay, *J. Biol. Chem.*, 1997, **272**, 27652.
9. K. Keyer and J.A. Imley, *Proc. Natl. Acad. Sci., USA*, 1996, **93**, 13635.
10. C.E. Cooper and J.B. Porter, *Biochem. Soc. Trans.*, 1996, **25**, 75.
11. R. Böhnke and B.F. Matzanke, *Biometals*, 1995, **8**, 223.
12. S.N. Wai *et al.*, *Mol. Microbiol.*, 1996, **20**, 1127.
13. S.L. Baader *et al.*, *FEBS Lett.*, 1996, **381**, 131.
14. P.R. Gardner, *Biosci. Rep.*, 1997, **17**, 33.
15. L. Castro, M. Rodriguez and R. Radi, *J. Biol. Chem.*, 1994, **269**, 29409.
16. M.C. Kennedy, W.E. Antholine and H. Beinert, *J. Biol. Chem.*, 1997, **272**, 20340.
17. A. Boveris and E. Cadenas, in *Lung biology in health and disease*, Oxygen Gene Expression and Cellular Function, , L.B. Clerch and D.J. Massaro (eds), vol 105. Marcell Dekker, New York, 1997, 1.
18. I. Fridovich, *J. Biol. Chem.*, 1970, **245**, 4053.
19. H.M. Hassan, in *Lung biology in health and disease*, Oxygen Gene Expression and Cellular Function, L.B. Clerch and D.J. Massaro (eds), vol 105. Marcell Dekker, New York, 1997, 27.
20. B. González-Flecha and B. Demple, *J. Biol. Chem.*, 1995, **270**, 13681.
21. S.I. Liochev, *Frez Rad. Res.*, 1996, **25**, 369.
22. A. Carlioz and D. Toutati, *EMBO J.*, 1986, **5**, 623.
23. R.S. Sohal, *FASEB*, 1997, **11**, 1269.
24. P.R. Gardnem *et al.*, *J. Biol. Chem.*, 1995, **270**, 13399.
25. Y. Li *et al.*, *Nat. Genet.*, 1995, **11**, 376.
26. R.M. Lebovitz *et al.*, *Proc. Natl. Acad. Sci. USA*, 1996, **93**, 9782.
27. D.J. Stuehr, *Annu. Rev. Pharmacol. Toxicol.*, 1997, **37**, 339.
28. J.M. Fukuto, *Adv. Pharmacol.*, 1995, **34**, 1.
29. J.F. Erwin Jr., J.R. Lancaster Jr. and P.L. Feldman, *J. Med. Chem.*, 1995, **38**, 4343.
30. J.M. Fukuto and D.A. Wink, in *Metal Ions In Biological Systems*, A. Sigel and H. Sigel (eds), Marcel Dekker, New York, 1999, 547.
31. S. Goldstein and G. Czapski, *Free Rad. Biol. Med.*, 1995, **19**, 505.
32. R.E. Huie and S. Padmaja, *Free Rad. Res. Commun.*, 1993, **18**, 195.
33. A.J. Hobby, *Trends Pharmacol. Sci.*, 1997, **18**, 484.
34. D.A. Wink *et al.*, *Arch. Biochem. Biophys.*, 1993, **300**, 115.

35. J. Stadler *et al.*, *Proc. Natl. Acad. Sci. USA*, 1994, **91**, 3559.
36. R. Farias-Eisner *et al.*, *J. Biol. Chem.*, 1996, **271**, 6144.
37. J.B. Hibbs Jr. *et al.*, *Biochem. Biophys. Res. Commun.*, 1988, **157**, 87.
38. Y.M. Kim, H. Begonia and J.R. Lancaster Jr., *FEBS Lett.*, 1995, **374**, 228.
39. W.H. Koppenol, in *Metal Ions in Biological Systems*, A. Sigel and H. Sigel (eds), Marcel Dekker, New York, 1999, 597.
40. M.S. Ramezanian, S. Padmaja and W.H. Koppenol, *Chem. Res. Toxicol.*, 1996, **9**, 232.
41. S. Golstein and G. Czapki, *Inorg. Chem.*, 1995, **34**, 4041.
42. L.M. Landino *et al.*, *Proc. Natl. Acad. Sci. USA*, 1996, **93**, 15069.
43. J.S. Beckman, in *The Neurobiology of NO· and OH·*, C.C. Chiueh, D.L. Gilbert and C.A. Colton (eds), NY Academy of Sciences, New York, 1995, 69.
44. J.B. Sampson, Y. Ye, H. Rosen and J.S. Beckman, *Arch. Biochem. Biophys.*, 1998, **356**, 207.
45. R.A. Floyd and K. Hensley, *Neurobiol. Aging*, 2002, **23**, 795.
46. R.A. Floyd and F.M. Carney, *Arch. Gerontol. Geriatr.*, 1991, **12**, 155.
47. R.A. Floyd, *Proc. Soc. Ex. Biol. Med.*, 1999, **222**, 236.
48. K. Hensley *et al.*, in *Understanding the Processes of Aging, the Roles of Mitochondria*, E. Cadenas and L. Packer (eds) New York, Marcel Dekker, 1999, 311.
49. K. Hensley, Q.N. Pye and M.L. Maidt *et al.*, *J. Neurochem.*, 1998, **71**, 2549.
50. S. Papa and V.P. Skulachev, *Mol. Cell Biochem.*, 1997, **174**, 305.
51. R.A. Floyd, F. West and K. Hensley, *Exp. Gerontol.*, 2001, **36**, 619.
52. D.A. Butterfield and E.R. Stadtman, *Adv. Cell Aging Gerontol.*, 1997, **2**, 161.
53. D.A. Butterfield and J. Kanski, *Mech. Aging Develop.*, 2001, **122**, 945.
54. G. Perry *et al.*, *Free Rad. Biol. Med.*, 2002, **33**, 1475.
55. J.S. Beckman, *Chem. Res. Toxicol.*, 1996, **9**, 836.
56. L. Levine *et al.*, *Proc. Natl. Acad. Sci. USA*, 1996, **93**, 15036.
57. E.R. Stadtman and B.S. Berlett, *Drug Metab. Rev.*, 1998, **30**, 225.
58. H. Esterbauer *et al.*, *Frez Rad. Biol. Med.*, 1991, **11**, 81.
59. R.J. Mark *et al.*, *J. Neurochem.*, 1997, **68**, 255.
60. J.N. Keller and M.P. Mattson, *Rev. Neurosci.*, 1998, **9**, 105.
61. R. Subramaniam *et al.*, *J. Neurochem.*, 1997, **69**, 1161.
62. S.M. Yatin *et al.*, *Neurochem. Res.*, 1999, **24**, 427.
63. D.A. Batterfield *et al.*, *Meth. Enzymol.*, 1999, **309**, 746.
64. K. Hensley *et al.*, *J. Neurosci.*, 1998, **18**, 8126.
65. J.S. Althaus *et al.*, *Free Rad. Biol. Med.*, 2000, **29**, 1085.
66. X. Huang *et al.*, *Biochim. Biophys. Acta, 2000, J. Nutr.*, 2000, **130**, 1488S.
67. E.R. Stadtman, *Science*, 1992, **257**, 1220.
68. C.D. Smith *et al.*, *Proc. Natl. Acad. Sci. USA*, 1991, **88**, 10540.
69. C. Leeuwenburgh *et al.*, *Arch. Biochem. Biophys.*, 1997, **346**, 74.

70. U. Catakay *et al.*, *Exp. Gerontol.*, 2001, **36**, 221.
71. I. Petropoulos *et al.*, *Biochem. J.*, 2001, **355**, 819.
72. W.R. Markesbery, *Free Rad. Biol. Med.*, 1997, **23**, 134.
73. C.D. Smith *et al.*, *Ann. NY Acad. Sci.*, 1992, **663**, 110.
74. M.P. Vitek *et al.*, *Proc. Natl. Acad. Sci. USA*, 1994, **91**, 4766.
75. S.D. Yan *et al.*, *Proc. Natl Acad. Sci. USA*, 1994, **91**, 7787.
76. M.D. Ledesma *et al.*, *J. Biol. Chem.*, 1994, **269**, 21614.
77. K. Hensley *et al.*, *Proc. Natl. Acad. Sci.USA*, 1994, **91**, 3270.
78. J. Joseph *et al.*, *Neurobiol. Aging*, 2001, **22**, 131.
79. C.A. Rottkamp *et al.*, *Free Radic. Biol. Med.*, 2001, **30**, 447.
80. C. Bel *et al.*, *Cell*, 1994, **77**, 817.
81. N. Ostrerova-Golts *et al.*, *J. Neurosci.*, 2000, **20**, 401.
82. M. Lee *et al.*, *J. Neurochem.*, 2001, **78**, 209.
83. L.J. Roberts II and J.D. Morrow, *Free Radic. Biol. Med.*, 2000, **28**, 505.
84. E.E. Reich *et al.*, *Biochemistry*, 2000, **39**, 2376.
85. L.J. Roberts II *et al.*, *J. Biol. Chem.*, 1998, **273**, 13605.
86. K. Uryn *et al.*, *J. Neurosci.*, 2002, **22**, 446.
87. W.G. Siems *et al.*, *EXS*, 1992, **62**, 124.
88. K. Uchida and E.R. Stadtman, *Proc. NATO. Acad. Sci. USA*, 1992, **89**, 4544.
89. M. Guichardant, P. Taili-Tronche, L.B. Fay and M. Lagarde, *Free Radic. Biol. Med.*, 1998, **25**, 1049.
90. A. Del Corso *et al.*, *Arch. Biochem. Biophys.*, 1998, **350**, 245.
91. S. Srivastava *et al.*, *Biochem. Biophys. Res. Commun.*, 1995, **217**, 741.
92. S. Boschi-Muller *et al.*, *J. Biol. Chem.*, 2000, **275**, 35908.
93. W.G. Siems, S.J. Hapner and F.J.G.M. Kuijk, *Free Radic. Biol. Med.*, 1996, **20**, 215.
94. A. Yaritaka *et al.*, *Proc. NATO. Acad. Sci. USA*, 93, 2696.
95. S. Block *et al.*, in *Critical Reviews of Oxidative Stress and Aging: Advances in Basic Sciences Diagnostics and Intervention*, R.D. Cutler and H. Rodriguez (eds), World Scientific Publishing, Singapore, 2003, 870.
96. M.L. Hamilton *et al.*, *Proc. Natl. Acad. Sci. USA*, 2001, **98**, 10469.
97. D. Nakae *et al.*, *Lab. Inst.*, 2000, **80**, 249.
98. G. Barja, *Trends In Neurosci.*, 2004, **27**, 595.
99. D.R. Johns, *N. Engl. J. Med.*, 1995, **333**, 638.
100. B. Bandy and A.J. Davidson, *Free Radic. Biol. Med.*, 1990, **8**, 523.
101. J. ST-Pierre *et al.*, *J. Biol. Chem.*, 2002, **277**, 44784.
102. R. Betarbet *et al.*, *Nat. Neurosci.*, 2000, **3**, 1301.
103. J.P. Spencer *et al.*, *J. Neurochem.*, 1998, **71**, 2112.
104. S.B. Berman and T.G. Hastings, *J. Neurochem.*, 1999, **73**, 1127.
105. D.T. Dexter *et al.*, *Ann. Neurol.*, 1994, **35**, 38.
106. J.E. Barker *et al.*, *Brain Res.*, 1996, **716**, 118.
107. J. Emeryt, M. Edeas and F. Bricaire, *Biomed. Pharmacotherapy*, 2004, **58**, 39.
108. M. Sonet *et al.*, *Neurotoxicology*, 2003, **24**, 443.

109. N. Rouach *et al.*, *Glia*, 2004, **45**, 28.
110. F. Hattori *et al.*, *J. Neurochem.*, 2003, **86**, 860.
111. S.A. Lipton and P.A. Rosenberg, *N. Engl. J. Med.*, 1994, **330**, 613.
112. S.A. Lipton *et al.*, *Nature*, 1993, **364**, 626.
113. R.M. Friedlander, *N. Engl. J. Med.*, 2003, **348**, 1365.

Copper Metabolism in the Brain

4.1 Introduction

The essentiality of Cu for mammals was firmly established by Hart *et al.*[1] who reported that Cu was required for erythropoiesis. The current US–Canadian Recommended Dietary Allowance (RDA) for Cu is 9 mg d^{-1} for adult men and women, with a tolerable upper intake level (UL) of 10 mg d^{-1} for adults.[2] The redox-cycling ability of Cu enables it to play an important role in electron transfer reactions. A deficit of Cu can result in[3,4]

- impaired energy production;
- abnormal glucose and cholesterol metabolism;
- increased oxidative damage;
- increased tissue iron accumulation;
- altered structure and function of circulating blood and immune cells;
- abnormal neuropeptide synthesis and processing;
- aberrant cardiac electrophysiology;
- impaired myocardial contractility; and
- persistent effects on neurobehavior and the immune system.

On the other hand, an excess of Cu is associated with oxidative stress and can be toxic to cells, tissues and organisms. Redox cycling of copper may generate reactive oxygen species (ROS), such as hydroxyl radicals, which are responsible for cellular damage that includes protein oxidation, lipid peroxidation in membranes and DNA damage.

In recent years, research on the role of nitric oxide (NO) in neurons has led to the identification of NO stress derivatives. The reaction between superoxide and NO generates peroxynitrite, a strong oxidant species able to produce highly reactive intermediates by the Fenton reaction plus hydrogen peroxide.[5] ROS generation is a normal cellular process but its imbalance is directly associated with the appearance of diverse pathologies. In this context, copper, as well as other metals, becomes crucial in the direct modulation of ROS generation and participation in pathological processes. It is important to emphasize that metals not only modulate the appearance of ROS, but also participate in other processes that indirectly affect ROS appearance.

Two well-characterized human genetic disorders that serve as examples of the consequences of Cu imbalance are Menkes' and Wilson's disease.[6] In Menkes' disease, the transport of dietary Cu from intestinal cells is impaired, leading to low serum Cu levels, while in Wilson's disease there is a defect in cellular Cu export which leads to the accumulation of high levels of copper. The genes for Menkes' and Wilson's disease share 55% amino acid homology, but their tissue-specific expression differs. Menkes' disease affects 1/50,000 to 1/250,000 live births and is a serious disorder caused by the dysfunction of the ATPase ATP7A (*vide infra*). Due to its essential function, ATP7A is found in many cell types including neurons and glia, as well as in several tumor cell lines including the neuronal cell lines PC12 and C6. As a consequence, people affected by Menkes' disease suffer neuronal disorders. The enzymatic activities that are most affected are those corresponding to Cu dependent-Superoxide Dismutase (SOD) and Cytochtome c Oxygenase (COX), two enzymes that play significant roles in controlling the production of ROS during active neuronal metabolism. A decrease in activities of these two enzymes likely accounts for the majority of the neurological symptoms.

Wilson's disease affects 1/100,000 live births and is an autosomal recessive disorder resulting in Cu poisoning. It affects the liver, kidneys, eyes and brain and can lead to death. This occurs as a result of mutations in the ATP7B Cu transporter gene involved in Cu excretion (*vide infra*). Contrary to what occurs in Menkes' disease, in Wilson's disease the blood concentration of Cu becomes unusually high when the liver, damaged by Cu accumulation, releases the metal directly into the bloodstream. Thus, Cu ions are carried throughout the body, damaging other organs. Wilson's patients with neurological symptoms show elevated Cu levels in cerebrospinal fluid (CSF).

One of the most significant advances in recent years has been the linking of a series of neurodegenerative diseases to alterations in the metabolism of copper. More than any other metal, it seems that control of Cu metabolism is more critical to the maintenance of healthy neurons. Besides Menkes' and Wilson's diseases, three other major neurodegenerative diseases have now been linked to Cu metabolism by virtue of the fact that the proteins central to these diseases all bind Cu. These proteins are the Amyloid Precursor Protein (APP), prion protein (PrP) and alpha-synuclein. These diseases will be dealt with in separate chapters but the complexity of Cu metabolism necessitates a more thorough examination of the issue and the potential of altered Cu metabolism to result in neurodegeneration.

4.2   Models of Copper Metabolism

Advances in understanding the mechanism of copper uptake and utilization by cells have come though the use of both bacterial and yeast system. The findings from these systems have been significant to understanding the mechanisms of copper utilization by mammalian cells because most of the proteins isolated from the model systems have homologues with very high homology in mammalian cells. The implication is that the fundamental basis of copper metabolism has not changed much during the course of evolution. The two main bacterial models used are the gram-positive *Enterococcus hirae* and the gram-negative *Escherichia coli*.[7,8] In bacteria genes related to a particular activity are usually grouped together on a single chromosome as an operon. In *E.* hirae the *copYZAB* operon contains the four main Cu regulating genes. These are CopY, CopZ, CopA and CopB. Both CopA and CopB are homologue of P-type ATPases.[9] CopA is involved in copper uptake in Cu deficient states while CopB is involved in excretion of copper when intracellular copper is too high. CopZ is an intracellular copper chaperone probably delivering copper to CopY and CopB.[10] CopY is a repressor protein that responds to low levels of copper by inhibiting expression of proteins by the *cop* operon.[11] When copper levels increase in the cell, CopZ becomes charged with copper. Two Cu-charged CopZ molecules then deliver Cu to CopY causing the displacement of zinc. The CopY protein is then released from the operon and gene expression increases (Figure 1).

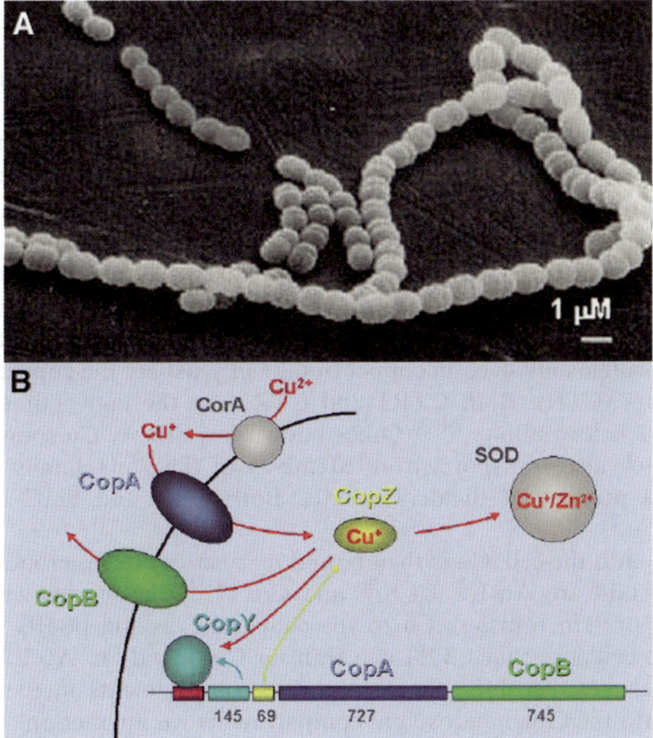

Figure 1 *Copper metabolism in E. hirae: (A) An electron microscope image of E. hirae form a line of cells. (B) The Cop operon of E. hirae. Shown are the transporters CopA and CopB, the inhibitor CopY and the chaperone CopZ that provides Cu for the SOD*

In *E. coli* four operons have been identified associated with copper metabolism. The *cue* operon regulates Cu efflux from the cell through expression of the CueR protein which has homology to the MerR protein.[12,13] The gene responds to levels of Cu^+ in the cytoplasm by upregulating CupA and CueO when copper levels are high. CopA is a P-type ATPase that expels excess Cu from the cytoplasm,[13] and CueO is a copper dependent oxidase that protects cells from Cu-induced toxicity.[14] The *cus* locus has two operons involved in copper sensing and is also involved in extrusion of copper from the cell. The *cusRS* operon is separately transcribed to the *cusCFBA* operon.[15,16] The gene products of *cusRS* sense copper levels and regulate the activity of *cusCFBA*, which constitutes a multicomponent Cu transport system. This system acts as a way of detoxifying Cu and exporting it across the outer membrane of the cell from the periplasm.[17] The fourth operon is only expressed in strains of *E.coli* that are resistant to high levels of extracellular Cu. The genes involved are plasmid encoded. The *pco* operon contains seven genes that confer resistance to copper toxicity. The mechanism of action of these genes remains unknown.[18,19] The mechanism of uptake of Cu in *E. coli* and the intracellular Cu chaperones have yet to be identified.

Somewhat more relevant to the mammalian system is the yeast system using *Saccharomyces cerevisiae*.[20] This yeast system has a proven advantage over the bacterial system in that genes of yeast can be mutated out and complementation with mammalian homologues has verified that the homologues from the mammals have the same function in yeast.[21] This system has been so effective that in only a few years the major copper transport proteins of both yeast and mammals have been identified (Figure 2). The Cu uptake system in yeast occurs through both a low and high affinity pathway. Low affinity systems are generally high capacity and therefore there are quite a number of low affinity transporters in yeast. High affinity uptake is limited and mediated by two closely related proteins. The first step for Cu entry into yeast cells involves the activity of two metalloreductases (Fre1 and Fre2) that reduce Cu^{2+} to Cu^+.[22,23] The main proteins involved in copper uptake in yeast are the copper transporting receptors (CTR). Both CTR1 and CTR3 are the high affinity receptors while CTR2 is low affinity.[24–26] Other known low affinity Cu transporters are Fet4,[27] which also transports iron. Members of the Smf1 family of proteins transport a number of divalent metals. Both Smf1 and Smf2 are able to transport Cu.[28]

On entry into the cell, Cu is then bound by a series of chaperones. The three most important are Atx1,[29] CCS,[30] and Cox17.[31] The mechanism by which the uptake protein transfers Cu to the chaperones is still poorly understood but there is evidence that CTR1 can transfer Cu directly to Atx1.[32] The three main chaperones distribute Cu to separate compartments in the cell. Cu is delivered into the Golgi/endosomal compartments *via* interaction of Atx1 with Ccc2, a P-type ATPase.[29,33] There Cu is utilized by proteins including Fet3, a Cu-dependent oxidase.[34] This incorporation requires the activity of Gef1, a chloride channel.[35] Fet3 is then able to form part of a iron transport complex at the cell membrane.[36]

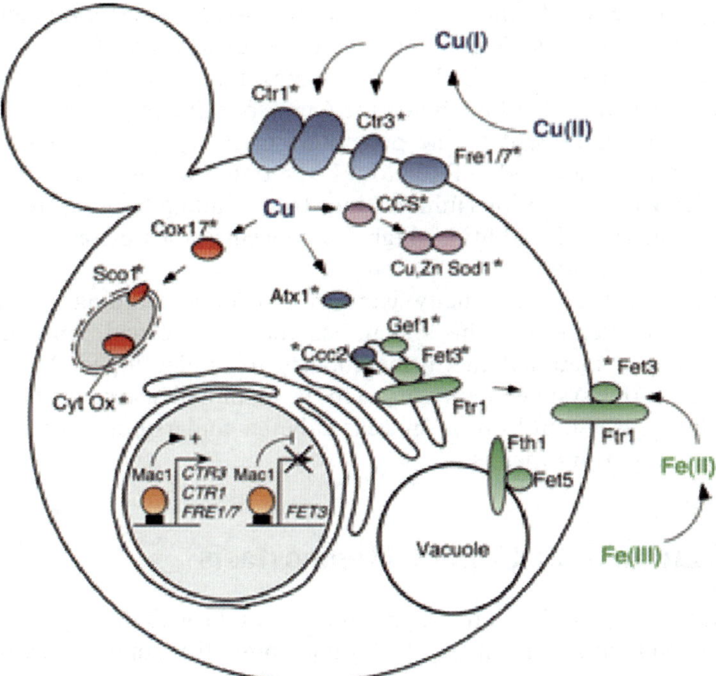

Figure 2 *Copper metabolism in yeast: This diagram shows the main proteins in yeast Cu metabolism and their interactions. This scheme is similar to one that could be created for mammalian Cu metabolism. CTR1 and CTR3 allow Cu entry into the cell. The chaperones cox17, CCS and Atx1 shuttle Cu to various compartments. Cu binding to Ccc2 alters the localization of Fet3 which influences Fe metabolism. Mac1 is one of the main nuclear regulators influencing expression of Cu metabolism genes*

Cox17 delivers Cu to the mitochondrion where Cu is essential for respiration as a co-factor of Cytochrome c oxidase.[31] Two proteins are possibly involved in the transport of Cu away from Cox17 into the mitochondrion. These are Sco1 and Cox11.[37,38] The chaperone CCS remains in the cytosol and is currently only known to be necessary for incorporation into Cu/Zn SOD.[30] The exchange follows transient formation of a heterodimer between the two proteins.[39]

These transport pathways normally result from uptake *via* high affinity transporters. Low affinity uptake can result in high concentrations of Cu in the cytoplasm. In response to high concentrations of Cu yeast cells express Cu-sequestering proteins, Cup1 and Crs5.[40–42] In particular, Cup1 allows yeast to survive in media containing high Cu. In addition, high Cu results in increased expression of Cu/Zn SOD. The very high affinity of Cu for this protein allows it to sequester excess Cu and inhibit its ability to generate oxidative radicals *via* the Fenton reaction.[43]

Regulation of Cu by cells requires tight control of protein transcription particularly of those proteins directly involved in the regulation of Cu transport and storage. In yeast cells there are two Cu responsive transcription factors that have opposing activities. Ace1 allows increased expression of Cu detoxifying genes when Cu levels are high.[44] The genes that bind Ace1 have metal regulatory elements.[45,46] Mac1 binds the Cu responsive sequence (TTTGC(T/G)C(A/G)) that is present in the promoters of many genes including those encoding Fre1, CTR1 and CTR3. Mac1 responds to low concentrations of Cu.[47] High Cu concentrations inhibit Mac1 from binding to promoter domains of its target genes.[48] In addition, high Cu concentrations increase the rate at which Mac1 is degraded.[49]

Although there are still many gaps in our understanding of yeast Cu metabolism, the richness of this system has allowed a logical approach to the understanding of mammalian Cu metabolism to be developed. Many of the mechanisms in this system are preserved in the mammalian system. This implies that regulation of Cu utilization by cells is better understood for Cu than for most of the important metals.

4.3 Mammalian Copper Homeostasis

The human body absorbs all the copper it requires through dietary sources. The recommended intake is *ca.* 10 mg per day and comes from many sources such as legumes, beef and shell-fish.[50,51] Absorption of copper is *via* the mucosal of the stomach and small intestines.[52] It is not clear if this process occurs by diffusion or active transport. However, two possible candidate proteins for absorption are the divalent metal transporter (DMT-1) also known as NRAM-2 (natural resistance associated macrophage protein) or the human homologue of CTR1.[53,54] Once Cu is transported into the mucosal cells of the intestine it is excreted into the blood in the portal region of the circulation. This energy-dependent process requires the MNK ATPase. Mutations in this protein cause Menkes' disease.[55] Cu in the blood is absorbed by the liver, which is the primary organ of copper homeostasis. The liver determines the fate of all absorbed Cu. Cu within the body is associated with proteins or other ligands such as amino acids but it is too redox-active to exist in an unbound form:[56] an upper limit of 10^{-18} M has been evaluated for the free concentration of Cu^{2+} in unstressed cells.[56,57] Most proteins have histidine or cysteine residues that have high affinity for Cu. However, there are specific proteins that play an important role in carrying Cu in the blood, to and from the liver. Once in the liver the Cu is either utilized by enzymes that require it as a co-factor, bound into transport proteins or excreted in the bile. The main proteins that bind Cu and are thought to play a role in its transport are albumin,[58] trascuperin,[59] and ceruloplasmin.[60] Greater than 95% of all blood serum Cu is associated with ceruloplasmin. This protein plays a major role in its transport to tissues such as the brain, lung and kidney. There are specific receptors for ceruloplasmin expressed in different organs.[61–65] The neurodegenerative disorder aceruloplasminemia is associated with loss of ceruloplasmin expression. However, this disorder shows no disturbance to copper homeostasis but instead suggests that this protein has more to do with iron metabolism than Cu, despite the large amount of Cu the protein binds.[66–68] In the absence of ceruloplasmin many other proteins can substitute and aid copper transport.[60]

Transport of Cu across the blood brain barrier is thought to require the protein MNK P-type ATPase. This process is blocked in Menkes' disease.[69] Once in the brain, uptake and utilization of Cu is thought to be similar to other organs such as the liver. It would therefore utilize similar proteins. However, the brain expresses large amounts of other Cu-binding proteins such as the APP and the prion protein. It is currently unclear how expression of these proteins could alter Cu metabolism but there is strong evidence that this is the case.[70,71] Nevertheless, the wide expression of most of the main Cu-binding proteins in organs that include the brain suggests that a basic description of Cu utilization by cells is relevant to the brain as well as to cells such as hepatocytes.

4.4 Mammalian Cellular Copper Metabolism

As indicated above, DMT-1 and CTR-1 are the two proteins that are associated with entry of Cu into cells. Initial studies of Cu uptake by mammalian cells have suggested that there are two uptake pathways. One of these is low affinity and high capacity and the other is high affinity.[72] The use of radioactive copper allowed an assessment of uptake by brain tissue indicating that the high affinity process had a K_m of 6 μM and a V_{max} of 23 pmol min^{-1} mg^{-1} protein. The low affinity process had a K_m of 40 μM and a V_{max} of 425 pmol min^{-1} mg^{-1} protein. Further study showed that the Cu taken up by brain tissue *via* the high affinity process was rapidly released by depolarization of neurons with veratridine.[73] Although DMT-1 is able to transport Cu, it also plays a role in the uptake of Fe, Zn and Mn.[53] Also research shows that DMT-1 is more likely to be relevant for intestinal absorption of Cu than entry into neurons.[74] In comparison the kinetics of Cu uptake mediated by CTR-1 are very similar to those observed using radioactive Cu in brain slices.[75,76] CTR-1 knockout mice show severe developmental problems and greatly altered tissue Cu level implying that CTR-1 plays a major role in Cu metabolism.[77,78] CTR-1 is expressed in the plasma membrane but localization experiments show that it is also expressed in intracellular vesicles suggesting a role in movement of Cu within the cell.[79] CTR-1 forms a multimeric complex creating a membrane spanning channel for the entry of Cu into the cell.[80] Other proteins have been suggested to play a role in Cu uptake. CTR-1 independent Cu uptake processes occur.[76] In particular, it has been suggested that the prion protein can aid cellular uptake of Cu.[81] The low affinity, high capacity process of Cu uptake has not been fully elucidated but could potentially be a result of the activity of a number of proteins such as DTM-1 and PrP.

Once within the cell Cu has one of four possible fates. Binding to Atox-1 the human homologue of Atx-1 can occur.[82] Atox-1 can transport Cu to the WND P-type ATPase that allows entry into the *trans*-Golgi network.[83,84] Mutations in this protein are associated with Wilson's disease. This pathway is associated with excretion of Cu from cells. The second pathway involves binding of Cu to the chaperone Cox17 and transport to the mitochondria in a process similar to that described for yeast.[85] The third pathway involves binding to the chaperone CCS and incorporation into Cu/Zn SOD.[86] The final pathway involves interaction of Cu with other proteins in the cell and, in particular, binding to glutathione and storage in the metallothioneins.[87] Therefore, as stated above, there is a strong similarity between Cu utilization and transport in both mammals and yeast. However, when considering the brain, this similarity ends abruptly. Mutations or alterations in a whole range of Cu-binding proteins are associated with neurodegenerative diseases.

4.5 Neurodegenerative Diseases and Copper

Copper is also associated with redox activity which can either lead to the formation of ROS or convert ROS into other forms. Alternatively, copper, harnessed to specific proteins such as SOD, can detoxify radicals. However, free-copper ions are dangerous for the cell because of its oxidizing potential. Neurons, in particular, are very sensitive to oxidative stress. The implication of this is that close regulation of copper-binding to proteins and of its concentration in the neuron is necessary. The likely consequence of this is that all copper-binding proteins in a neuron are co-regulated. The original identification of a link between neurodegeneration and Cu relates to two inherited disorders that were clearly linked to Cu metabolism initially. These are Menkes' disease and Wilson's disease. Association between Cu and other neurodegenerative diseases was not so easy to recognize. Most of these are not necessarily related to inherited mutations and the diseases and changes to Cu metabolism were not necessarily life threatening or even apparent. Four proteins associated with these diseases have now been shown to be Cu-binding proteins. Some cases of Amyotrophic Lateral Sclerosis (ALS) are associated with mutations in the Cu/Zn SOD. The APP and its brain-damaging derivative, β-amyloid (Aβ) have both been shown to bind Cu and Aβ is associated with Alzheimer's disease.[88,89] Transmissible spongiform encephalopathies or prion diseases are associated with a misfolded form of the prion protein, now accepted as a Cu-binding protein.[90] Changes in brain Cu levels occur in prion disease.[91] More recently, it has been found that the protein α-synuclein associated with a variety of diseases including Parkinson's disease can also bind Cu.[92] This could potentially play a role in aggregation of the protein. Alzheimer's disease, prion disease and Parkinson's disease will all be dealt with in separate chapters.

4.6 Wilson's and Menkes' Diseases

Wilson's disease is an autosomal recessive disorder characterized by liver cirrhosis and neuronal degeneration resulting from severe impairment of biliary copper excretion.[93] Liver disease is the most common presentation in children. Neuropsychiatric changes begin in the third and fourth decade of life.[94] There is copper deposition in the basal ganglia which results in neuronal loss, vacuolation of the affected brain region, gliosis and neurological features including tremors and dystonia. Behavioral changes include depression and schizophrenia. Copper also forms deposits in the cornea of the eye and can be seen as Kayser–Fleischer rings. Almost any organ can accumulate Cu in Wilson's disease but such changes do not occur in every case. Other commonly affected organs are the heart and musculature.[95] Diagnostic features include high Cu in the urine. One of the main biochemical changes in the diseases is a low level of ceruloplasmin. However, the genetic cause of the disease is mutation of the gene for the WND P-type ATPase.[96] The main way Wilson's disease is treated is to decrease Cu in the liver and other organs. This is largely done with chelation therapy (see Chapter 12). Various chelators have been used to increase excretion of Cu. The compounds used include D-penicillamine, trientine, tetrathiolmolybdate or zinc salts. D-penicillamine is used preferentially but other treatments are used when problems with D-penicillamine results in unwanted side effects.[95] Such treatments are often effective in preventing life threatening symptoms. In some more severe cases, liver transplant is used.

Further studies of the expression of the WND protein have shown directly that WND protein functions as a copper-translocating P-type ATPase in mammalian cells. Importantly, it has also been shown that the mutation of the conserved amino acid residue methionine at position 1386 to a valine (associated with disease) causes a loss of Cu-translocating activity.[97] Recent hybridization studies have shown that the WND has a number of splice variants. One of these is found in the pineal gland and the retina.[98] This variant (PINA) is expressed in a dramatic diurnal fashion with a 100-fold higher expression level at night. This finding suggests that Cu homeostasis and the circadian rhythm are somehow linked. Several transcripts with skipped exons are brain specific.[99] One of these is a truncated protein and another lacks the cysteine–proline–cysteine (CPC) motif, thought to be essential for Cu transport. It is unclear why the alternative transcripts are produced but they could be a result of brain-specific regulation indicating that brain-specific isoforms are required for specific activities in the brain that may or may not be related to Cu metabolism.

Menkes' disease was first described in 1962.[100] It is an x-linked recessive disorder. It is a disease of children sometimes termed kinky hair syndrome. This is because one of the immediately obvious symptoms is abnormal hair that is colorless and brittle. The children begin to have seizures early in life. Gradually patients with Menkes' disease develop mental retardation, hypothermia, feeding difficulties and decreased muscle tone. Magnetic resonance imaging has shown that the disease causes abnormalities in myelination and atrophy. Neuronal loss is predominant in the cerebellum. Purkinje cells are the most-affected cell type with abnormal dendritic arborization and focal axonal swelling. The fundamental cause of the disease is a failure of Cu absorption from food.[101] As well as changes to absorption, there is also evidence that intracellular transport of copper is also disturbed.[102] It has been found that intravenous injections of copper (as opposed to increased oral doses) are able to correct the Cu deficiency but do not protect against further neurodegeneration.[103] Some treatments with chelated Cu do produce some improvement.[104] Copper metabolism remains impaired in patients even when injected with Cu-histidine as cellular copper utilization remains impaired. Some changes are present at birth in Menkes' patients and it has been suggested that, as yet, unidentified cuproenzymes are necessary for the normal development of the nervous system.[105] The genetic cause of this disease is similar to that in Wilson's disease in that the mutations that cause the disease result in the inactivity of a P-type ATPase.[106] Mutations in the MNK ATPase result in failure in copper transport across the placenta, gastrointestinal tract and blood–brain barrier. There is therefore a failure of copper incorporation in certain enzymes such as ceruloplasmin, cytochrome C and lysyl oxidase. Loss of lysyl oxidase activity results in connective tissue problems that are central to the disease.

Evidence that the MNK ATPase plays a role in copper efflux from cells came from experiments with cells that are resistant to copper toxicity. These cells were found to have an over expression of the MNK protein.[107] In normal cells the MNK protein is found in the *trans*-Golgi network.[108] Upon transfer of cells to medium with high Cu levels the MNK protein is rapidly relocalized to the plasma membrane in an ATP-dependent process that does not require new protein expression.[108] Additionally, copper transport in fibroblast cells isolated from Menkes' patients show high intracellular levels of Cu. Transfecting these cells to overexpress normal MNK protein results in restoration of normal Cu levels and excretion from the cell of excess Cu.[109] Overexpression of the WND ATPase has a similar effect, eliminating Cu accumulation, further emphasizing the similarity in function of these two proteins.[110] Overexpression of the MNK protein has also been shown to have effects on other Cu-binding proteins. Human fibroblasts overexpressing the MNK protein have significantly reduced APP levels and down-regulated APP gene expression.[111] This implies that the reduction in Cu levels induced by MNK causes decreased expression of proteins such as APP that might also have a role in Cu efflux.

The MNK and WND proteins share considerable homology (55%) and are expressed from two related genes (*atp7a* and *atp7b*).[6] The two proteins contain an MXCXXC motif in the N-terminals which functions to bind Cu,[112] and a transmembrane CPC component involved in translocation of Cu across the membrane.[83] In addition, they both have phosphorylation sites and ATP-binding domains (Figure 3).[112] Both proteins are synthesized as single-polypeptide chains. Mutations that cause the diseases are therefore found in similar regions in both proteins.[105]

ATP7A and ATP7B possess a long N-terminal cytosolic tail containing six putative metal-binding domains whose sequences are significantly similar to one another. The number of metal-binding domains in ATP7A and ATP7B homologues is variable, ranging from one to six, with proteins from higher eukaryotic organisms, *e.g.*, mammals, having a higher number of such domains than prokaryotic (typically one or two) or yeast (two) homologues.[113]

As far as ATP7A is concerned, the structure of the fourth domain (MNK4) has been determined in solution in the apo and Ag(I) forms,[114] and a structure of the apo form of the second domain (MNK2) is just available.[115]

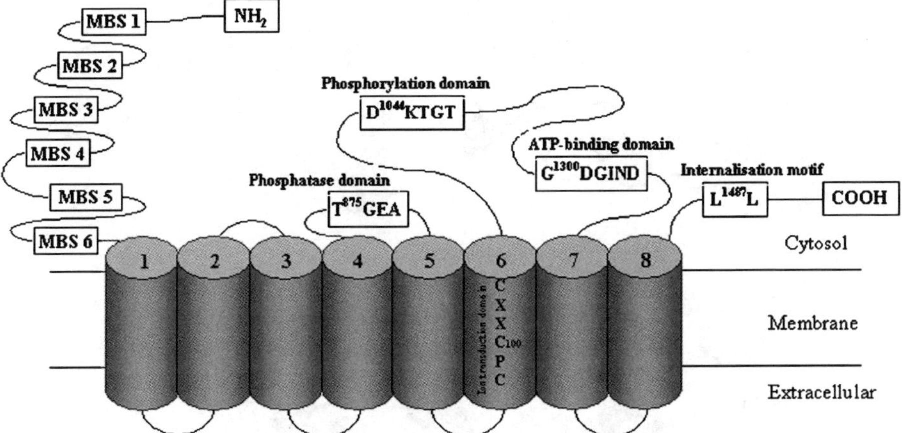

Figure 3 *A representation of the protein mutated in Menkes' disease, the Cu transporting MNK ATPase. It shows the eight transmembrane domains and the folds forming the metal binding site (MBS). Also shown are sites binding ATP, the internalisation motif, the site of phosphorylation and the site of phosphatase activity*

The structure of the second domain of the human Menkes' protein (MNK2), formed by 72 residues, was recently determined by NMR in both the apo- and copper-loaded forms (Figure 4).[116] The loop involved in copper binding is part of a hydrophobic patch, which is maintained in both forms.

MNK2 adopts the classical βαββαβ fold regardless of the presence of the metal ion and, when metalated, the copper ion is close to the protein surface and exposed to the solvent. Cu^+ binding mainly affects the Cys-containing loop (loop 1) and the N-terminal region of the first α-helix. The superposition of the two structures (Figure 4) highlights that this is the region where structural rearrangement occurs upon metal binding, while the remainder of the poly-peptide chain experiences negligible structural variations.

The structure of MNK2 is similar both to that of the first metal-binding domain of the homologous yeast protein Ccc2,[117] and to that of MNK4 (Figure 5). With respect to these two systems, the most interesting difference is probably that the system presented here shows variations in the structure of the metal-binding loop in the apo- and holo-protein larger than those observed for the other two proteins. This rearrangement is also accompanied by variations in protein dynamics, with the apo-protein featuring conformational exchange processes that are absent in holo-MNK2.

In all proteins shown in Figure 5, residues Met12, Cys17 (loop 1, *i.e.*, the copper-binding loop), Leu38 (loop 3) and Phe66 (loop 5) form a single-compact core, through a network of hydrophobic contacts, which is present in both the apo- and holo-proteins. Met12 is conserved in all eukaryotic ATPases; Leu38 is conserved in all mammal ATPases and replaced with another hydrophobic residue in only 6% of the eukaryotic ATPase domains, while Phe66 is always replaced with a Pro in the third domain of mammal ATPases (in both Menkes'

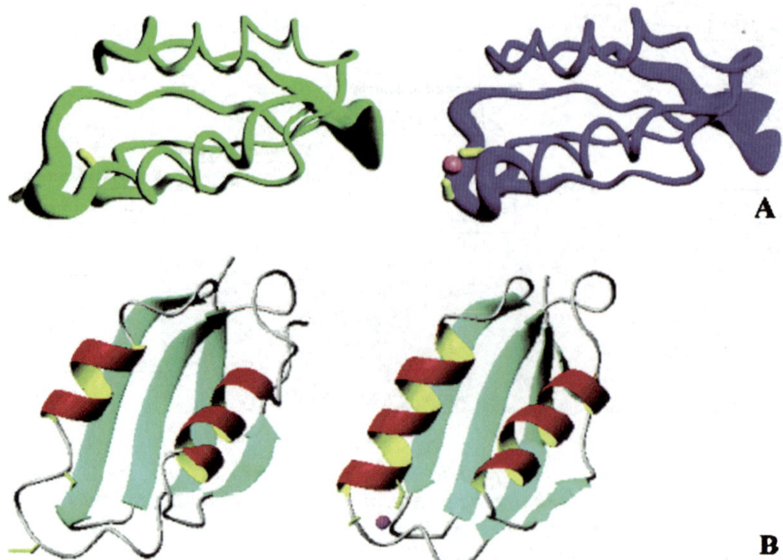

Figure 4   *Menkes' protein: Comparison of the solution structures of apo- (left) and copper(I)-bound MNK2 (right). In panel A, the two families are compared using a representation as a tube with variable radius, proportional to the local backbone rmsd. In panel B, the two energy-minimized average structures are compared, using a ribbon representation. The side chains of Cys14 and Cys17 are shown in yellow; the copper(I) ion is shown as a pink sphere*
(Reprinted with permission from ref. 116.)

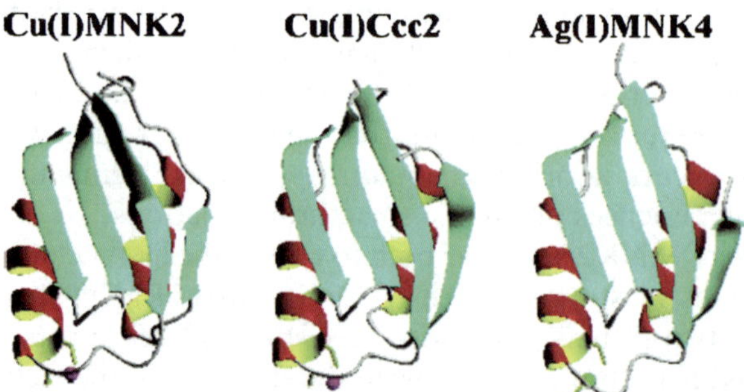

Figure 5   *Ribbon representations of the structures of (from left to right) copper(I)-bound MNK2, copper(I)-bound Ccc2,[117] and silver(I)-bound MNK4.[115] The side chains of the metal-binding cysteines are shown in yellow; the metal ions are shown as pink (copper(I)) or green (silver(I)) spheres*
(Reprinted with permission from ref. 116.)

and Wilson's protein homologues). Since the variation in side-chain contacts and solvent accessibility caused by the Phe66Pro substitution is likely to affect the stability of both the protein and metal site structures, it was concluded that the strict conservation of the substitution pattern for residue 66 in mammals is functionally relevant.[116]

4.7   Familial Amyotrophic Lateral Sclerosis

Amyotrophic lateral sclerosis (ALS) is a neurodegenerative disease arising from the loss of motor neurons, leading to progressive muscle weakness, atrophy and death often within three years of the onset of symptoms. ALS approximately affects one in 200,000 people all over the world. Around 10% of all cases of this disease are familial ones. In some cases, (*ca.* 20%), Familial amyotrophic lateral sclerosis (FALS) has been associated with point mutations scattered throughout the gene for the protein, Cu/Zn SOD.[118,119] There are more than 100 known mutations, most of which are missense mutations.[120] Neuropathology of ALS is characterized by degeneration and loss of motor neurons and gliosis. Intracellular inclusions are found in degenerating neurons and glia. Familial ALS (FALS) neuropathology is characterized by neuronal Lewy body-like inclusions (*vide infra*) and astrocytic inclusions composed largely of mutant SOD. The effects of mutations on the enzyme's structure and function are not clearly identified and they were hypothesized either to reduce SOD activity or to induce new functions.[121,122] SOD catalyzes the conversion of the superoxide radical to hydrogen peroxide, which can be converted to water by catalase or glutathione peroxidase (Figure 6). Details of the structure and function of wild type and mutated Cu/Zn SOD will be discussed in the specific chapter.

   Since reactive-free radicals can be toxic to cells and cause neuronal injury, the disease might be a result of reduced SOD activity, which would produce an increased contribution of superoxide radicals or, as it seems more likely, be due to the new ability in catalyzing other toxic reactions, such as formation of hydroxyl radical (OH$^{\bullet}$) from hydrogen peroxide and the nitration of proteins on tyrosine residues by peroxynitrite (ONOO^{-}). The hypothesis of a gain-of-function in the FALS mutants, proposed by several authors,[123–125] was recently supported by Crow *et al.*[126] who demonstrated that SOD mutants catalyze the nitration of tyrosines by peroxynitrite in the rod and head of neurofilament-L. Neurofilaments are major structural proteins expressed in motor neurons and are important for their survival *in vivo*; nitration inhibits the assembly of neurofilament subunits. This new function seems to be related to different affinity of SOD mutants for the metal ions. Since different sets of FALS mutations lead to similar disruption of metal-binding properties, structural changes that are associated with these altered properties could be reasonably responsible of the toxic property gained by FALS mutants.[127] The alteration in metal-binding behavior has effects, as far as the structural and redox properties are concerned, which resemble those observed in the reduced Cu^{+} and in the zinc-free wild-type SOD. The links between the disease and structural and dynamics modifications related to FALS point mutations and to the metal-binding properties occurring in the FALS mutants can be elucidated by the investigation of the solution structure of these mutants and of single-metallated proteins, *i.e.*, Cu^{+}, Zn-free SOD and Cu-free, Zn^{2+}SOD, and by the comparison with the properties of native proteins. The results of such studies suggest that the mutant proteins show altered Cu incorporation and protein

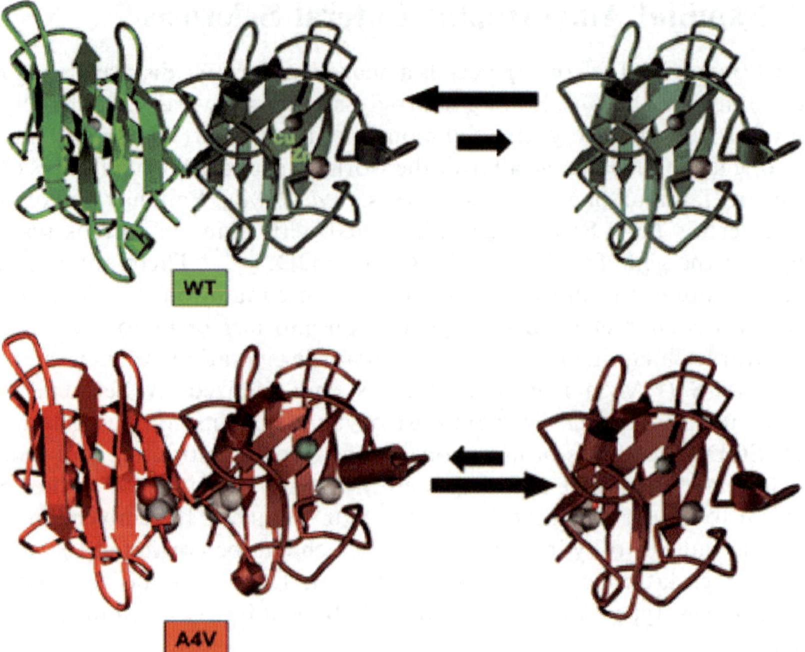

Figure 6   *SOD and ALS: In familial amyotrophic lateral sclerosis mutations in Cu/Zn SOD potential are causal to the disease. The mechanism of action of the mutations is unclear but one possibility is that the mutations alter the potential of the protein to dimerise. The molecule is shown as a ribbon diagram in this figure. Monomeric and dimeric forms are in equilibrium. In the wild-type protein this equilibrium favors the dimer but with mutations such as A4V, the equilibrium would favor the monomer. This could alter the activity of the mutant SOD*

destabilization.[128] The damaging activity of the mutants is therefore likely to be due to the Cu bound to SOD mutants. Additional support for this hypothesis comes from studies of the interaction of mutant SOD with the Cu chaperone CCS. Disease associated mutants of SOD can also interact with CCS and acquire Cu.[118,129] Therefore a possible strategy to treat ALS could be reduce or prevent Cu delivery to mutant SOD.

4.8   Conclusions

Copper metabolism is well studied and researched. The mechanism of transport and utilization of Cu by cells has been shown to have an elaborate series of pathways that ensure that Cu in cells is kept in check. This is because of the possible damage that redox active Cu can have in terms of oxidation of cellular components. It is also clear that when handing of Cu is altered through mutation or disease it has dire consequences for cells and for survival. Other redox active metals are not as well understood. This understanding has come from the utilization of yeast and bacterial cell lines and the tight evolutionary conservation of the proteins involved. Understanding Cu provides insight into the ways other metals might be regulated by cells. As many neurodegenerative diseases have been found to be linked in some way to Cu, this broad base of understanding provides an excellent platform to begin investigation of the roles of other key proteins in both Cu metabolism and how mutations or disease might be explained by changes in Cu metabolism.

References

1. E.B. Hart *et al.*, *J. Biol. Chem.*, 1928, **77**, 797.
2. Institute of Medicine, Copper. Dietary reference intakes: vitamin A, vitamin K, arsenic, boron, chromium, copper, iodine, iron, manganese, molybdenum, nickel, silicon, vanadium, and zinc, Food and Nutrition Board, National Academy Press, Washington, DC, 2002, 224.
3. E.D. Harris, *Crit. Rev. Clin. Lab. Sci.*, 2003, **40**, 547.
4. C.L. Keen *et al.*, *J. Nutr.*, 2003, **133**, 1477S.
5. J.Y. Wang *et al.*, *J. Neurosci. Res.*, 2003, **72**, 508.
6. P.C. Bull *et al.*, *Nat. Genet.*, 1993, **5**, 327.
7. M. Solioz and J.V. Stoyanov, *FEMS Microbiol. Rev.*, 2003, **27**, 183.
8. C. Rensing and G. Grass, *FEMS Microbiol. Rev.*, 2003, **27**, 197.
9. A. Odermatt *et al.*, *J. Biol. Chem.*, 1993, **268**, 12775.
10. P. Cobine *et al.*, *FEBS Lett.*, 1999, **445**, 27.
11. D. Strausak and M. Solioz, *J. Biol. Chem.*, 1997, **272**, 8932.
12. F.W. Outten *et al.*, *J. Biol. Chem.*, 2000, **275**, 31024.
13. C. Petersen and L.B. Moller, *Gene*, 2000, **261**, 289.
14. G. Grass and C. Rensing, *Biochem. Biophys. Res. Commun.*, 2001, **286**, 902.
15. G.P. Munson *et al.*, *J. Bacteriol.*, 2000, **182**, 5864.
16. T. Oshima *et al.*, *Mol. Microbiol.*, 2002, **46**, 281.
17. S. Franke *et al.*, *J. Bacteriol.*, 2003, **185**, 3804.
18. N.L. Brown *et al.*, *Plasmid*, 1992, **27**, 41.
19. N.L. Brown *et al.*, *Mol. Microbiol.*, 1995, **17**, 1153.
20. E.M. Rees and D.J. Thiele, *Curr. Opin. Microbiol.*, 2004, **7**, 175.
21. J. Lee *et al.*, *Gene*, 2000, **254**, 87.
22. A. Dancis *et al.*, *Proc. Natl. Acad. Sci. USA*, 1992, **89**, 3869.
23. L.J. Martins *et al.*, *J. Biol. Chem.*, 1998, **273**, 23716.
24. K. Kampfenkel *et al.*, *J. Biol. Chem.*, 1995, **270**, 28479.
25. A. Dancis *et al.*, *J. Biol. Chem.*, 1994, **269**, 25660.
26. S.A. Knight *et al.*, *Genes Dev.*, 1996, **10**, 1917.
27. D.R. Dix *et al.*, *J. Biol. Chem.*, 1994, **269**, 26092.
28. X.F. Liu *et al.*, *J. Biol. Chem.*, 1999, **274**, 4863.
29. S.J. Lin *et al.*, *J. Biol. Chem.*, 1997, **272**, 9215.
30. V.C. Culotta *et al.*, *J. Biol. Chem.*, 1997, **272**, 23469.
31. D.M. Glerum *et al.*, *J. Biol. Chem.*, 1996, **271**, 14504.

32. Z. Xiao *et al.*, *Chem. Commun.*, 2002, **6**, 588.
33. D.S. Yuan *et al.*, *Proc. Natl. Acad. Sci. USA*, 1995, **92**, 2632.
34. C. Askwith *et al.*, *Cell*, 1994, **76**, 403.
35. S.R. Davis-Kaplan *et al.*, *Proc. Natl. Acad. Sci. USA*, 1998, **95**, 13641.
36. R. Stearman *et al.*, *Science*, 1996, **271**, 1552.
37. H.S. Carr *et al.*, *J. Biol. Chem.*, 2002, **277**, 31237.
38. Y.C. Horng *et al.*, *J. Biol. Chem.*, 2004, **279**, 35334.
39. A.L. Lamb *et al.*, *Biochemistry*, 2000, **39**, 14720.
40. S. Fogel and J.W. Welch, *Proc. Natl. Acad. Sci. USA*, 1982, **79**, 5342.
41. D.H. Hamer *et al.*, *Science*, 1985, **228**, 685.
42. V.C. Culotta *et al.*, *J. Biol. Chem.*, 1994, **269**, 25295.
43. V.C. Culotta *et al.*, *J. Biol. Chem.*, 1995, **270**, 29991.
44. D.J. Thiele, *Mol. Cell. Biol.*, 1988, **8**, 2745.
45. P. Furst *et al.*, *Cell*, 1988, **55**, 705.
46. C. Buchman *et al.*, *Mol. Cell. Biol.*, 1990, **10**, 4778.
47. S. Labbe *et al.*, *J. Biol. Chem.*, 1997, **272**, 15951.
48. J.A. Graden and D.R. Winge, *Proc. Natl. Acad. Sci. USA*, 1997, **94**, 5550.
49. Z. Zhu *et al.*, *J. Biol. Chem.*, 1998, **273**, 1277.
50. B. Lonnerdal, *Am. J. Clin. Nutr.*, 1996, **63**, 821S.
51. M.C. Linder and M. Hazegh-Azam, *Am. J. Clin. Nutr.*, 1996, **63**, 797S.
52. R.F. Crampton *et al.*, *J. Physiol.*, 1965, **178**, 111.
53. H. Gunshin *et al.*, *Nature*, 1997, **388**, 482.
54. B. Zhou and J. Gitschier, *Proc. Natl. Acad. Sci. USA*, 1997, **94**, 7481.
55. Y. Murata *et al.*, *Pediatr. Res.*, 1997, **42**, 436.
56. T.D. Rae *et al.*, *Science*, 1999, **284**, 805.
57. S.J. Lippard, *Science*, 1999, **284**, 748.
58. C.A. Owen Jr., *Am. J. Physiol.*, 1965, **209**, 900.
59. K.C. Weiss and M.C. Linder, *Am. J. Physiol.*, 1985, **249**, E77.
60. Z.L. Harris and J.D. Gitlin, *Am. J. Clin. Nutr.*, 1996, **63**, 836S.
61. S.J. Orena *et al.*, *Biochem. Biophys. Res. Commun.*, 1986, **139**, 822.
62. L. Dini *et al.*, *Eur. J. Cell Biol.*, 1990, **52**, 207.
63. R.V. Stern and E. Frieden, *Anal. Biochem.*, 1990, **190**, 48.
64. R.V. Stern and E. Frieden, *Anal. Biochem.*, 1993, **212**, 221.
65. E.D. Harris, *Ann. Rev. Nutr.*, 2000, **20**, 291.
66. J.I. Logan *et al.*, *Quart. J. Med.*, 1994, **87**, 663.
67. Z.L. Harris *et al.*, *Proc. Natl. Acad. Sci. USA*, 1995, **92**, 2539.
68. L.A. Meyer *et al.*, *J. Biol. Chem.*, 2001, **276**, 36857.
69. H. Kodama, *J. Inherit. Metab. Dis.*, 1993, **16**, 791.
70. A.R. White *et al.*, *Brain Res.*, 1999, **842**, 439.
71. D.R. Brown, *J. Neurochem.*, 2003, **87**, 377.
72. D.E. Hartter and A. Barnea, *J. Biol. Chem.*, 1988, **263**, 799.
73. A. Barnea *et al.*, *Am. J. Physiol.*, 1989, **257**, C315.
74. M. Arredondo *et al.*, *Am. J. Physiol.*, 2003, **284**, C1525.
75. L.B. Moller *et al.*, *Gene*, 2000, **257**, 13.
76. J. Lee *et al.*, *J. Biol. Chem.*, 2002, **277**, 4380.
77. J. Lee *et al.*, *Proc. Natl. Acad. Sci. USA*, 2001, **98**, 6842.

78. Y.M. Kuo *et al.*, *Proc. Natl. Acad. Sci. USA*, 2001, **98**, 6836.
79. A.E. Klomp *et al.*, *Biochem. J.*, 2002, **364**, 497.
80. M.J. Petris *et al.*, *J. Biol. Chem.*, 2003, **278**, 9639.
81. D.R. Brown, *J. Neurosci. Res.*, 1999, **58**, 717–725.
82. L.W. Klomp *et al.*, *J. Biol. Chem.*, 1997, **272**, 9221.
83. I.H. Hung *et al.*, *J. Biol. Chem.*, 1997, **272**, 21461.
84. X.L. Yang *et al.*, *Biochem. J.*, 1997, **326**, 897.
85. R. Amaravadi *et al.*, *Hum. Genet.*, 1997, **99**, 329.
86. P.C. Wong *et al.*, *Proc. Natl. Acad. Sci. USA*, 2000, **97**, 2886.
87. R.J. Cousins, *Physiol. Rev.*, 1985, **65**, 238.
88. L. Hesse *et al.*, *FEBS Lett.*, 1994, **349**, 109.
89. C.S. Atwood *et al.*, *J. Neurochem.*, 2000, **75**, 1219.
90. D.R. Brown *et al.*, *Nature*, 1997, **390**, 684.
91. B.-S. Wong *et al.*, *J. Neurochem.*, 2001, **78**, 1400.
92. R.M. Rasia *et al.*, *Proc. Natl. Acad. Sci. USA*, 2005, **102**, 4294.
93. S.A.K. Wilson, *Brain*, 1912, **34**, 295.
94. J.M. Walshe, *Quart. J. Med.*, 1989, **70**, 253.
95. J.A. Cuthbert, *J. Investig. Med.*, 1995, **43**, 323.
96. R.E. Tanzi *et al.*, *Nat. Genet.*, 1993, **5**, 344.
97. I. Voskoboinik *et al.*, *Biochem. Biophys. Res. Commun.*, 2001, **281**, 966.
98. J. Borjigin *et al.*, *J. Neurosci.*, 1999, **19**, 1018.
99. K. Petrukhin *et al.*, *Hum. Mol. Genet.*, 1994, **3**, 1647.
100. J.H. Menkes *et al.*, *Pediatrics*, 1962, **29**, 764.
101. D.M. Danks *et al.*, *Lancet*, 1972, **1**, 1100.
102. D.M. Danks *et al.*, *Lancet*, 1973, **1**, 891.
103. A.D. Garnica, *Eur. J. Pediatr.*, 1984, **142**, 98.
104. J.F. Mercer *et al.*, *Nat. Genet.*, 1994, **6**, 374.
105. D.J. Waggoner *et al.*, *Neurobiol. Dis.*, 1999, **6**, 221.
106. C. Vulpe *et al.*, *Nat. Genet.*, 1993, **3**, 7.
107. J. Camakaris *et al.*, *Hum. Mol. Genet.*, 1995, **4**, 2117.
108. M.J. Petris *et al.*, *EMBO J.*, 1996, **15**, 6084.
109. S. La Fontaine *et al.*, *J. Biol. Chem.*, 1998, **273**, 31375.
110. S. La Fontaine *et al.*, *Hum. Mol. Genet.*, 1998, **7**, 1293.
111. S.A. Bellingham *et al.*, *J. Biol. Chem.*, 2004, **279**, 20378.
112. S. Lutsenko *et al.*, *J. Biol. Chem.*, 1997, **272**, 18939.
113. P.C. Bull and D.W. Cox, *Trends Genet.*, 1994, **10**, 246.
114. J. Gitschier *et al.*, *Nat. Struct. Biol.*, 1998, **5**, 47.
115. C.E. Jones *et al.*, *J. Struct. Biol.*, 2003, **143**, 209.
116. L. Banci *et al.*, *Biochemistry*, 2004, **43**, 3396.
117. L. Banci *et al.*, *J. Biol. Chem.*, 2001, **276**, 8415.
118. D.R. Rosen *et al.*, *Nature*, 1993, **362**, 59.
119. H.X. Deng *et al.*, *Science*, 1993, **261**, 1047.
120. M.E. Cudkowicz *et al.*, *Ann. Neurol.*, 1997, **41**, 210.
121. S.I. Liochev *et al.*, *Arch. Biochem. Biophys.*, 1998, **352**, 237.
122. J.J. Goto *et al.*, *J. Biol. Chem.*, 2000, **275**, 1007.
123. A. Okado-Matsumoto *et al.*, *Free Radical Res.*, 2000, **33**, 65.

124. C. Café *et al.*, *J. Mol. Neurosci.*, 2000, **15**, 71.
125. L.J. Hayward *et al.*, *J. Biol. Chem.*, 2002, **277**, 15923.
126. J.P. Crow *et al.*, *J. Neurochem.*, 1997, **69**, 1936.
127. L.I. Bruijn *et al.*, *Science*, 1998, **281**, 1851.
128. M.A. Hough *et al.*, *Proc. Natl. Acad. Sci. USA*, 2004, **101**, 5976.
129. R.L. Casareno *et al.*, *J. Biol. Chem.*, 1998, **273**, 23625.

Lithium, A Neuroprotective Element

Lithium salts have been known as mood stabilizers since the nineteenth century and have been reintroduced in 1949 by Cade.[1] Believing that mania might be caused by intoxication by the normal body products, Cade's experiments led him to focus on lithium urate. The observation of lethargy induced on guinea pigs and the use of careful controls revealed that lithium had a calming effect by itself. Clinical trials of lithium with manic patients finally led to its employment for the control of manic excitement in bipolar disorder (BD).

BD, also known as manic-depressive illness (MDI) is a severe, chronic disease characterized by two apparently opposite mood states, mania and depression. It is a devastating disease affecting 1–3% of population. BD is a brain disorder that causes unusual shifts in a person's mood, energy and ability to function. Different from the normal ups and downs that everyone goes through, the symptoms of BD are severe and can result in damaged relationships, poor job or school performance and even suicide. In most patients, the outcome is poor and characterized by cognitive and functional impairment and psychosocial disability.[2] BD has been listed as the 6th leading cause of disability worldwide;[3] it has been evaluated that disability and the high incidence of premature death (mainly due to suicide)[4] are associated with a very high economic cost (tens of billions of dollars in the U.S. alone).[5] BD typically develops in late adolescence or early adulthood. However, some people have their first symptoms during childhood, and some develop them late in life. It is often not recognized as an illness, and people may suffer for years before it is properly diagnosed and treated. Like diabetes or heart disease, BD is a long-term illness that must be carefully managed throughout a person's life.

It may be helpful to think of the various mood states in BD as a spectrum or continuous range. At one end is severe depression, above which is moderate depression and then mild low-mood, which many people call "the blues" when it is short-lived but is termed "dysthymia" when it is chronic. Then there is normal or balanced mood, above which comes hypomania (mild to moderate mania), and then severe mania (Figure 1).

In some people, symptoms of mania and depression may occur together in what is called a mixed bipolar state. Symptoms of a mixed state often include

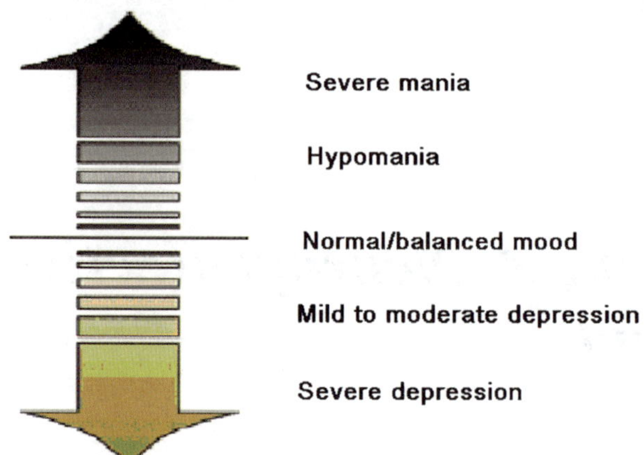

Figure 1   *Mood states in dipolar disorder*

agitation, trouble sleeping, significant change in appetite, psychosis and sui-
cidal thinking. A person may have a very sad, hopeless mood, while at the same
time feeling extremely energized.

It is generally agreed that there is no single cause for BD and that, rather,
many factors act together to produce this illness.

BD tends to run in families and, consequently, specific genes that may
increase a person's chance of developing the illness have been sought. But genes
are not the whole story. Studies of identical twins, who share all the same genes,
indicate that not only genes but also other factors play a role in BD. If BD were
caused entirely by genes, then the identical twin of someone with the illness
would *always* develop the illness, and research has shown that this is not the
case. But if one twin has BD, the other twin is more likely to develop the illness
than another sibling.

In addition, findings from gene research suggest that BD, like other mental
illnesses, does not occur because of a single gene.[6] It appears likely that many
different genes act together, and in combination with other factors of the
person or the person's environment cause the BD. Finding these genes, each of
which contributes only a small amount toward the vulnerability to BD, is of
course extremely difficult.

Brain-imaging studies are helpful in delineating what goes wrong in the brain
to produce BD and other mental illnesses.[7,8] There is evidence from imaging
studies that the brains of people with BD may differ from the brains of healthy
individuals. As the differences are more clearly identified and defined through
research, a better understanding will be gained of the underlying causes of the
illness such that the types of treatment most effectively working will eventually
become predictable.

Most people with BD – even those with the most severe forms – can achieve
substantial stabilization of their mood swings and related symptoms with

proper treatment.[9-11] Because BD is a recurrent illness, long-term preventive treatment is strongly recommended and almost always indicated. A strategy that combines medication and psychosocial treatment is optimal for managing the disorder over time.

In most cases, BD is much better controlled if treatment is continuous than if it is on and off. But even when there are no breaks in treatment, mood changes can occur and adjustments are to be made to the treatment plan.

Several different types of mood stabilizers are available. In general, people with BD continue treatment with mood stabilizers for extended periods of time (years). Other medications are added when necessary, typically for shorter periods, to treat episodes of mania or depression that break through despite the mood stabilizer.

- Lithium, the first mood-stabilizing medication approved by the FDA for treatment of mania, is often very effective in controlling mania and preventing the recurrence of both manic and depressive episodes.
- Anticonvulsant medications, such as valproate (Formula 1) or carbamazepine, (Formula 2) also can have mood-stabilizing effects and may be especially useful for difficult-to-treat bipolar episodes.
- Newer anticonvulsant medications, including lamotrigine (Formula 3), gabapentin (Formula 4) and topiramate (Formula 5), are being studied to determine how well they work in stabilizing mood cycles.
- Anticonvulsant medications may be combined with lithium, or with each other, for maximum effect.

Sodium valproate

Carbamazepine

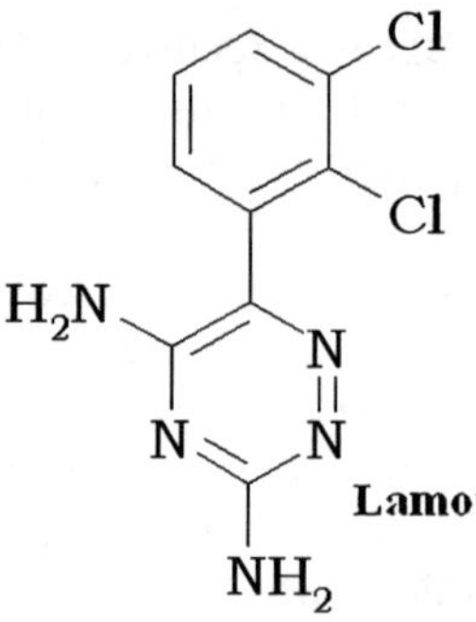

Lamotrigine

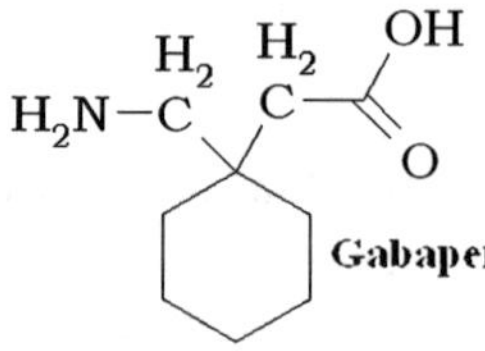

Gabapentin

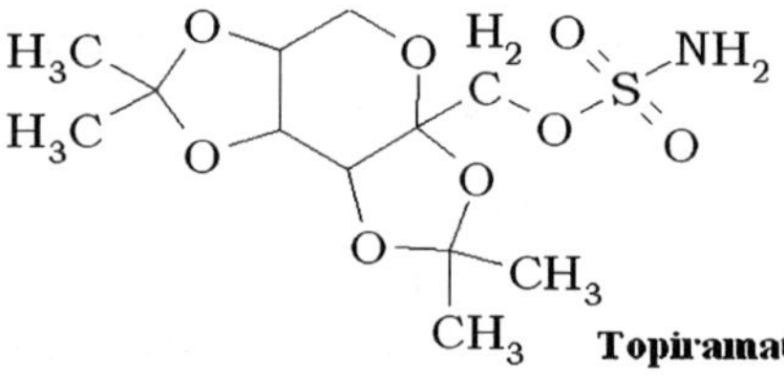

Topiramate

Lithium is an alkali metal similar to magnesium and sodium in its properties. Lithium compounds are found in natural waters and in some food. The average dietary intake is estimated to be about 2 mg per day.[12] The salt used as a therapeutic agent is lithium carbonate. Soluble lithium compounds are readily absorbed through the gastrointestinal tract but not the skin; distribution is rapid to the liver and kidneys but slower to other organ systems.[13] Lithium crosses the human placenta and can also be taken up by infants through breast milk. Lithium is not metabolized and is excreted primarily in the urine.

The oral toxicity of most lithium compounds is relatively low; oral LD_{50} values for several compounds and animal species range from 422 to 1165 mg kg^{-1}. Case histories described by Gosselin *et al.*,[14] indicate that doses of 12–60 g (171–857 mg kg^{-1} per day for a 70 kg person) can result in coma, respiratory and cardiac complications and death in humans. For chronic therapeutic use, the standard dose of lithium carbonate is 1–2 g per day (14–28 mg kg^{-1} per day).

Signs and symptoms of lithium toxicity include anorexia; nausea; diarrhea; alopecia; weight gain; thirst; pre-tibial edema (sodium retention); polyuria; glycosuria; aplastic anemia, tremors, acne, muscle spasm, and rarely dysarthria, ataxia, impaired cognition and pseudo-tumor cerebri.[15] Toxic effects that may appear after prolonged therapeutic use may include neurological symptoms, changes in kidney function, hypothyroidism and leukocytosis.

The nervous system is the primary target organ of lithium toxicity. Neurologic effects occurring during prolonged therapy often include minor effects on memory, motor activity and associative productivity.[16] Movement disorders (myoclonus, choreoathetosis), proximal muscle weakness, fasciculations, gait disturbances, incontinence, corticospinal tract signs and a Parkinsonian syndrome (cogwheel rigidity, tremor) have been reported. Cases of severe lithium neurotoxicity, which may occur during chronic therapy as a result of increased lithium retention, may be characterized by disorientation, incoherence, paralysis, stupor, seizure and coma. Permanent brain damage has occurred in several patients on long-term lithium therapy.[10]

During chronic lithium therapy, changes in kidney function may appear as transient natriuresis, polydipsia/polyuria, nephrogenic diabetes insipidus, partial renal tubular acidosis, minimal change disease and nephrotic syndrome. Degenerative changes may occur in the glomeruli or in the distal convoluted tubules or collecting ducts. In rare cases, acute renal failure may occur.

Cohort studies indicate that the risk of major congenital malformations among women receiving lithium during early pregnancy is slightly higher (4–12%) than that among control groups (2–4%).[17] Evidence also suggests that women on lithium therapy may have a higher risk of pre-mature birth.

'Although not completely devoid of side effects and neurotoxicity, the efficacy of lithium has provided the starting point for reshaping the concept of mental illness and has led to gain valuable information on signal transduction pathways involved in BD.

The similarity of the ionic radii of Li^+ and Mg^{2+} makes lithium effective in inhibiting magnesium-requiring enzymes,[18,19] such as phosphoglucomutase (PGM), glycogen synthase kinase-3 (GSK-3), inositol monophosphatase (IMPase) *etc.* Interestingly, lithium ions inhibit a group of at least four related phosphomonoesterases including, in mammals, besides IMPase, also inositol polyphosphate 1-phosphatase (IPPase), fructose 1,6-bisphosphatase (FBPase) and bisphosphate nucleotidase (BPNase).[20] All members of the group share a common core structure that binds metal ions and relates to the catalytic function.[20]

Among all the enzymes inhibited by lithium, IMPase and GSK-3 play relevant roles in CNS functions,[21] and, consequently, have been given prominent attention as possible relevant targets of lithium action. A simplified picture of related intracellular pathways is provided in Figure 2.[5,22]

5.1 IMPase

IMPase (together with IPPase) is involved in either recycling or *de novo* synthesis of inositol, a necessary component of the phosphoinositol (PI) signaling pathway. Many extracellular receptors, such as the serotonin $(5\text{-}HT)_2$, α_1, and muscarinic (M)1, 3 and 5 receptors are coupled to the G protein $G_{q/11}$, in a way that, through activation of phospholipase C (PLC), mediates the hydrolysis of a cellular membrane phospholipid, phosphoinositide 4,5-bisphosphate (PIP_2) (Formula 6), to form the second messengers diacylglycerol (DAG) and inositol-1,4,5-triphosphate (IP_3).[23] DAG and IP_3 subsequently modulate the activity of a multitude of intracellular events.

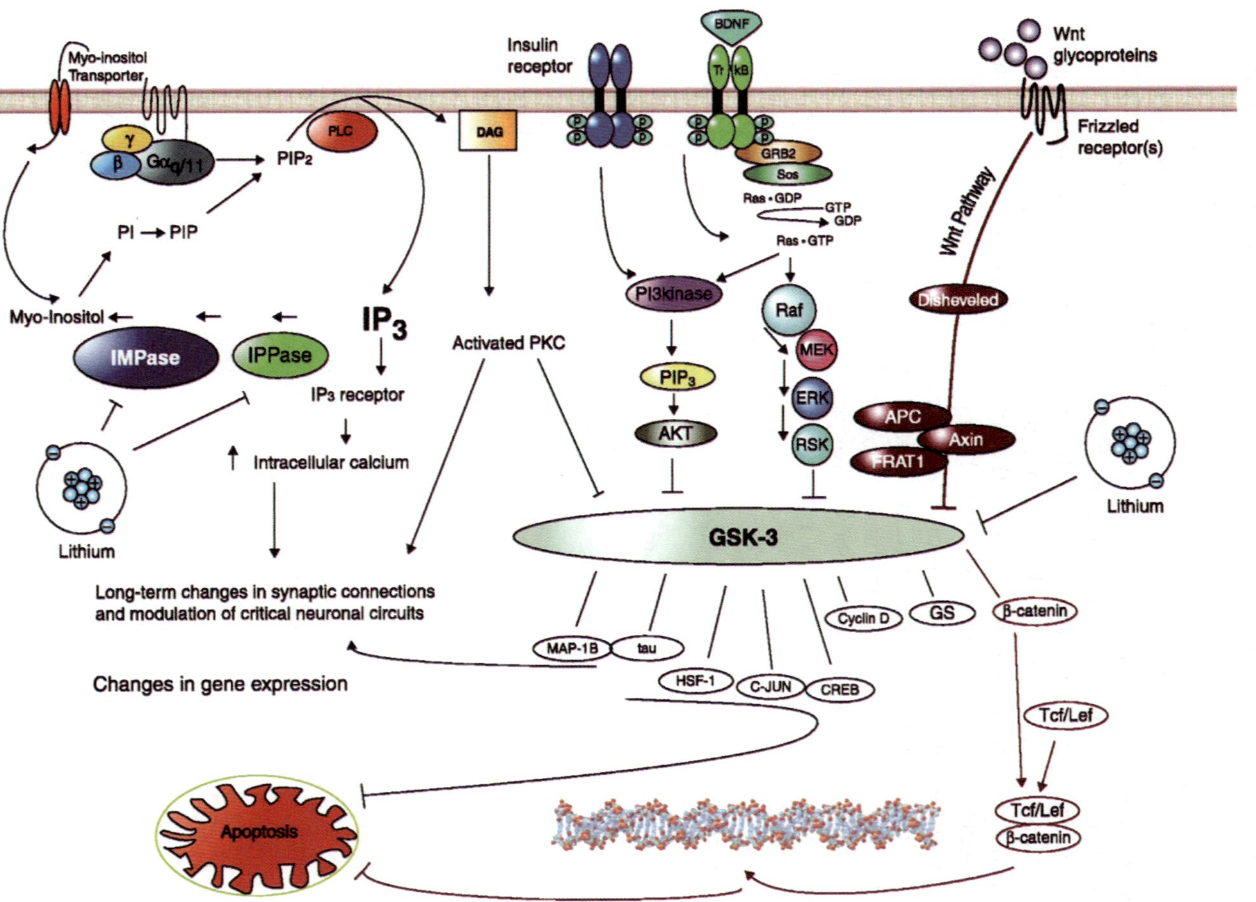

Myo-inositol Transporter
Insulin receptor
BDNF
TrkB
Wnt glycoproteins
γ
β
Gα q/11
PIP₂
PLC
DAG
GRB2
Sos
Ras · GDP
GTP
GDP
Ras · GTP
Frizzled receptor(s)
Wnt Pathway
PI → PIP
PI3kinase
Raf
MEK
ERK
RSK
Disheveled
Myo-Inositol
IP₃
IMPase
IPPase
Activated PKC
PIP₃
AKT
APC
Axin
FRAT1
IP₃ receptor
Intracellular calcium
Lithium
Lithium
GSK-3
Long-term changes in synaptic connections and modulation of critical neuronal circuits
Changes in gene expression
MAP-1B
tau
HSF-1
C-JUN
CREB
Cyclin D
GS
β-catenin
Tcf/Lef
Tcf/Lef
β-catenin
Apoptosis

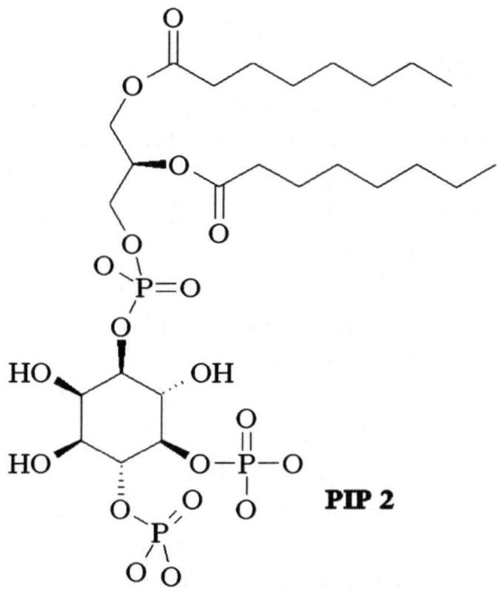

PIP 2

A number of IPPase enzymes are involved in the dephosphorylation (recycling) of IP_3 to inositol, a precursor of membrane PIP_2. This recycling is necessary to maintain PI-mediated signaling in cell types where inositol is not freely available. The enzyme IMPase catalyzes the final (and rate-limiting) step in the conversion of IP_3 into inositol. IPPase removes a phosphate from inositol-1,4-bisphosphate, at the point just prior to where IMPase participates. Both appear to be critical steps in the maintenance of inositol levels and continuation of PI-mediated signaling.[24]

Lithium's direct effect on IMPase,[25] and secondary on IPPase,[26] leads to the inositol depletion hypothesis of lithium's action (Figure 2).[27,28] The inositol depletion hypothesis suggests that lithium exerts its mood-stabilizing effect by inhibiting IMPase, decreasing inositol concentrations, and thus the amount of PIP_2 available for signaling cascades that rely upon this pathway, including but not limited to neurotrophin signaling pathways, receptor tyrosine kinase

---

Figure 2 *GSK-3 and IMPase are direct targets of lithium. This simplified figure highlights relevant interactions among intracellular pathways related to lithium's action. GSK-3 functions as an intermediary in a number of signaling pathways, including neurotrophic signaling pathways, the insulin–phosphatidylinositol 3 kinase (PI3K) pathway and the Wnt pathway – activation of these pathways inhibits GSK-3. The upper-left portion of the figure depicts lithium's actions on the PI signaling pathway. Activation of some G proteins induces phopsholipase C hydrolysis of phosphoinositide-4,5-bisphosphate (PIP_2) to DAG and inositol-1,4,5-triphosphate (IP_3). DAG activates protein kinase C (PKC). IP_3 binds to the IP_3 receptor, which also functions as a calcium channel in the cell. IP_3 is recycled back to PIP_2 by IMPase and inositol polyphosphatase phosphatase (IPPase); both of which are inhibited by lithium. The inositol depletion hypothesis suggests that lithium exerts its therapeutic actions by depleting free inositol and thus dampening the activation of downstream signaling pathways in neurons*

(Reproduced with permission from ref. 5.)

pathways, and some G protein-mediated signaling. It is hypothesized that the brain is particularly sensitive to lithium because of inositol's relatively poor penetration across the blood–brain barrier or of a reduced ability of specific neuronal populations to transport inositol across their cell membranes. Furthermore, based on the noncompetitve inhibition profile of lithium, more active cells and brain regions may be affected to a greater degree;[29] however, a recent study suggests that depletion of inositol may not have major effects on PI-mediated signaling. Specifically, Berry and colleagues found that the reduction of intracellular inositol in the brain sodium–myoinositol transporter (SMIT1) knockout in mice has no effect on PI levels.[30]

Although the data are not entirely consistent, lithium does decrease free inositol levels in brain sections and in the brains of rodents treated with lithium.[31] Thus, the inositol depletion hypothesis remains a viable one for the mechanism of action of lithium. However, no clinically approved inhibitors of either IPPase or IMPase are available, and therefore it remains difficult to test the inositol depletion hypothesis in patients with BD. Past pharmaceutical industry efforts have attempted to develop a brain-penetrating IMPase inhibitor by altering the primary substrate of IMPase – inositol monophosphate.[32] Compounds with sufficient inhibition properties were developed, but have thus far failed to advance through clinical trials because they are too highly charged, or extremely lipophilic, both properties that limit bioavailability in the brain.[32]

IMPase (EC 3.1.3.25) is a magnesium-activated dimer of identical 277-amino acid subunits (30 kDa),[33] the crystal structure of which was first resolved at 2.1 Å resolution.[34] It was found that each subunit is folded into a five-layered sandwich of three pairs of α-helices and two β-sheets (Figure 3),[34] forming a penta-layered $\alpha\beta\alpha\beta\alpha$ core structure. Sulfate and an inhibitory Gd^{3+} ion are bound at identical sites on each subunit and establish the positions of the active sites (Figure 4).[34] Each site is located in a large hydrophilic gorge at the base of the two central helices where several segments of secondary structure intersect. Comparison of the phosphatase-aligned sequences of several diverse genes suggested that the products of these genes and the phosphatase form a structural family with a conserved metal-binding site.

The structures of ternary complexes of human IMPase with inhibitory Gd^{3+} and either D- or L-myoinositol 1-phosphate were then determined to 2.2–2.3 Å resolution using X-ray crystallography.[35] Substrate and metal are bound identically in each active site of the phosphatase dimer. The substrate is present at full occupancy, while the metal is present at only 35% occupancy, suggesting that Li^+ from the crystallization solvent partially replaces Gd^{3+} upon substrate binding. The phosphate groups of both substrates interact with the phosphatase in the same manner with one phosphate oxygen bound to the octahedrally co-ordinated metal ion in the active site and another oxygen forming hydrogen bonds with the amide groups of residues 94 and 95 (Figure 5).[35] The active-site orientations of the inositol rings of D- and L-myoinositol 1-phosphate differ by rotation of nearly 60° about the phosphate ester bond. Each substrate utilizes the same key residues (Asp-93, Ala-196, Glu-213 and Asp-220) to form the same number of hydrogen bonds with the enzyme. Mutagenesis experiments

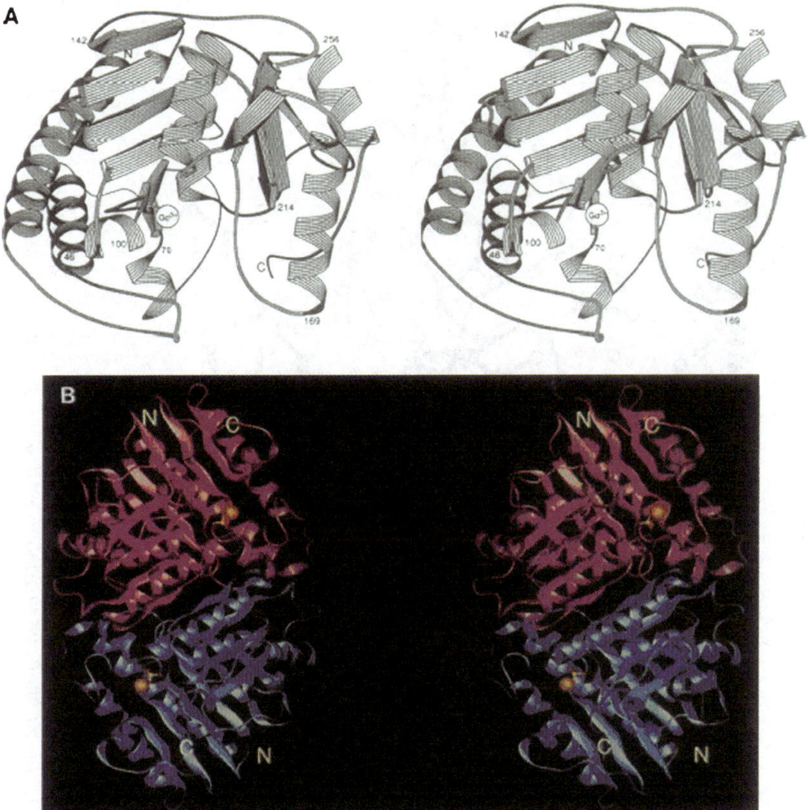

Figure 3 *Stereo drawings of ribbon representations of the secondary structure of the IMPase monomer (A) and dimer (B). The view of the monomer is into the subunit interface of the molecule and the active site is identified by Gd^{3+}. The view of the dimer is approximately down the dimer twofold axis and shows the locations of sulphate (ball and stick) and Gd^{3+} (orange sphere)* (Reproduced with permission from ref. 34.)

confirm the interaction of Glu-213 with the inositol ring and suggest that interactions with Ser-165 may develop during the transition state. The structural data suggest that the active-site nucleophile is a metal-bound water that is activated by interaction with Glu-70 and Thr-95. Expulsion of the ester oxygen appears to be promoted by three aspartate residues acting together (90, 93 and 220), either to donate a proton to the leaving group or to form another metal-binding site from which a second Mg^{2+} co-ordinates the leaving group during the transition state.

Modeling, kinetic and mutagenesis studies on IMPase revealed the requirement for two metal ions in the catalytic mechanism.[36] While the first metal ion may activate water for nucleophilic attack, a second metal ion, co-ordinated by three aspartate residues, appears to act as a Lewis acid, stabilizing the leaving inositol oxyanion. In this model, the 6-OH group of substrate acts as a ligand

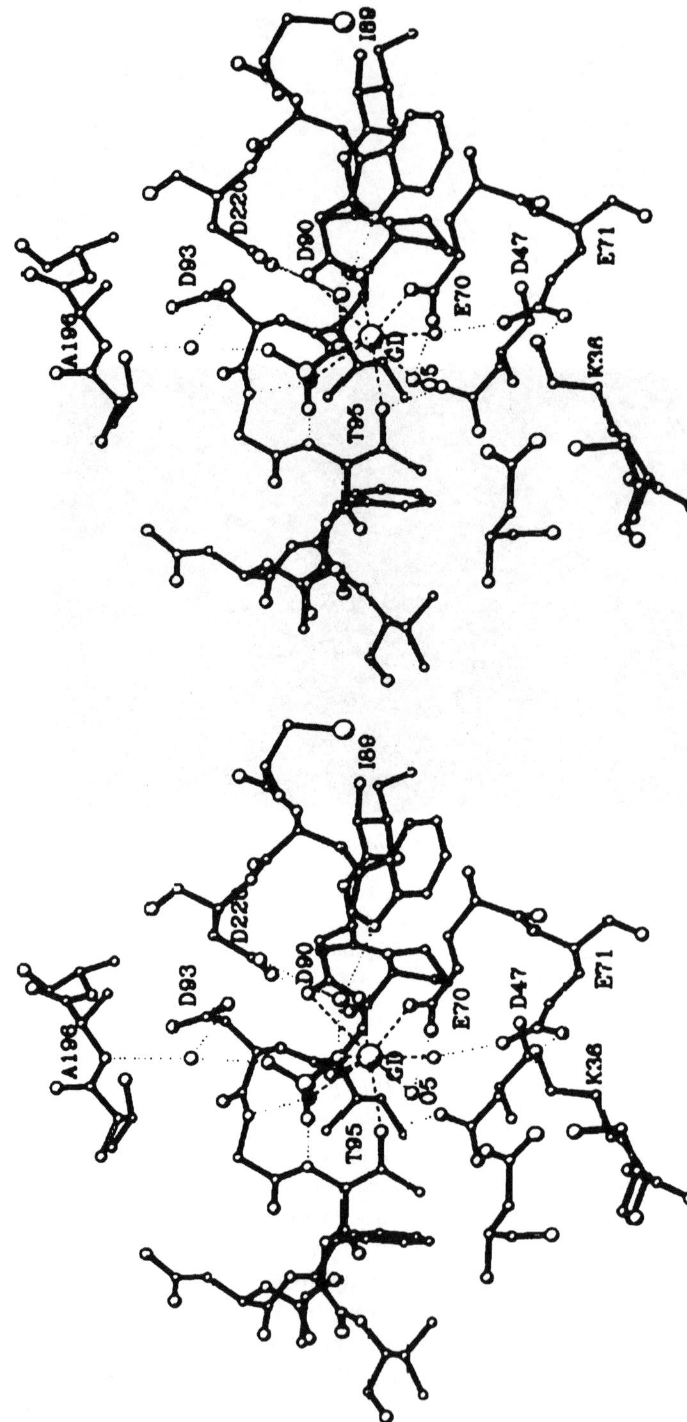

Figure 4 *The IMPase active site. Stereo drawing of the IMPase active site showing the residues involved in Gd^{3+} and sulphate (solid bonds) binding. Possible intermolecular hydrogen bonds are indicated by dotted lines and possible coordinate bonds are indicated by dashed lines (Reproduced with permission from ref. 34.)*

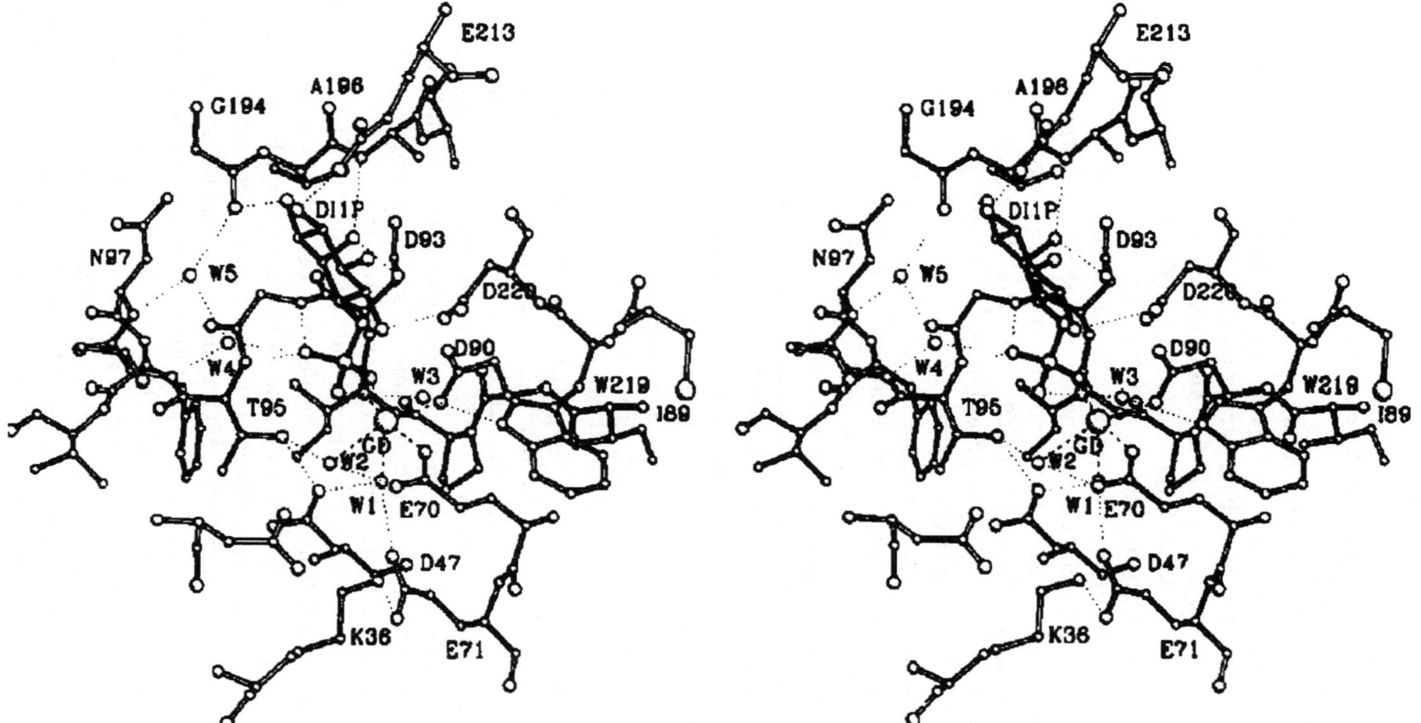

Figure 5 *The complex of L-Ins(1)P with IMPase and Gd^{3+}. Stereodrawing of the active site of the structure of the complex of L-Ins(1)P with IMPase and Gd^{3+}. Enzyme residues are shown in open bonds and D-Ins(1)P with filled bonds. Possible hydrogen bonds are shown with dotted lines, and coordinate bonds are shown with dashed lines. For comparison, a stick figure of D-Ins(1)P from the D-Ins(1)P complex with the phosphatase has been superimposed on the structure. The superposition was accomplished using the Cα atoms of subunit A of each structure*
(Reproduced with permission from ref. 35.)

for this second metal ion, consistent with the reduced catalytic activity observed with substrate analogues lacking the 6-OH. Evidence from Tb^{3+} fluorescence quenching and the two-metal kinetic titration curves suggests that Li^+ binds at the site of this second metal ion.

Phosphate ^{18}O-ligand exchange occurs only in the presence of substrate, indicating that IMPase operates *via* a sequential ternary complex mechanism and that water attacks the phosphate ester bond directly.[37] As a matter of fact, site-directed mutagenesis studies together with kinetic and molecular-modeling data identified a water molecule co-ordinated by metal 1 and activated by Thr-95 as the nucleophile.[36] The mechanism proceeds *via* inline displacement with inversion of configuration at phosphate and the metal ion at site 2 was proposed to assist in phosphate co-ordination and charge stabilization during the transition state. In contrast, Gani and co-workers suggested that although metal 1 co-ordinates the phosphate moiety, it is metal 2 that activates a water molecule for nucleophilic attack on phosphorus.[38] In this proposal, the substrate 6-OH group forms the H-bond to the nucleophile so that it is positioned for a non-inline attack on the phosphate P atom with adjacent displacement of the inositol moiety. Such mechanism would proceed *via* an adjacent attack involving pseudo-rotation with retention of configuration. The recent determination of the stereochemical course of the reaction demonstrated that hydrolysis occurs with inversion at phosphorus,[39] indicating that the water nucleophile is indeed associated with metal 1. However, this interpretation excluded a role for the substrate 6-OH group, which is essential for catalysis. A new unified mechanism was proposed by Gani and co-workers in which the metal 2-bound water molecule serves as a proton donor for the inositolate leaving group and the 6-OH substrate group hydrogen bonds to this water to lower its pK_a value further for inline attack of the 1-inositolate anion.[40]

Because Li^+ inhibits uncompetitively with respect to the first Mg^{2+} ion required for high-affinity substrate binding, uncompetitively with respect to substrate but competitively with respect to the second Mg^{2+} ion, it was proposed that Li^+ binds to the Mg^{2+}-IMPase ternary complex at metal site 2 and prevents the dissociation of inorganic phosphate after hydrolysis. The H217Q and C218A mutations in the proximity of site 2 eliminate not only the uncompetitive inhibition by Li^+ but also by excess of Mg^{2+}.

However, the retention of a third metal ion (Ca^{2+} or Mn^{2+}) and certain kinetic data were found difficult to reconcile with a two-metal mechanism.[41] Moreover, the proposed two-metal ion site mechanisms for FBPase and the archaeal FBPase/IMPases were amended to three-metal mechanisms.[42]

The crystal structure of bovine IMPase has been determined at 1.4 Å resolution in complex with the physiological magnesium ion ligands, showing that three magnesium ions are octahedrally co-ordinated at the active site of each of the two subunits of the dimer.[42] This structure led to postulate a detailed three-metal mechanism.[42] Ligands to the three metals include the side chains of Glu-70, Asp-90, Asp-93 and Asp-220, the backbone carbonyl group of Ile-92 and several solvent molecules, including the proposed nucleophilic water molecule (W1) ligated by both Mg-1 and Mg-3 (Figure 6).[42]

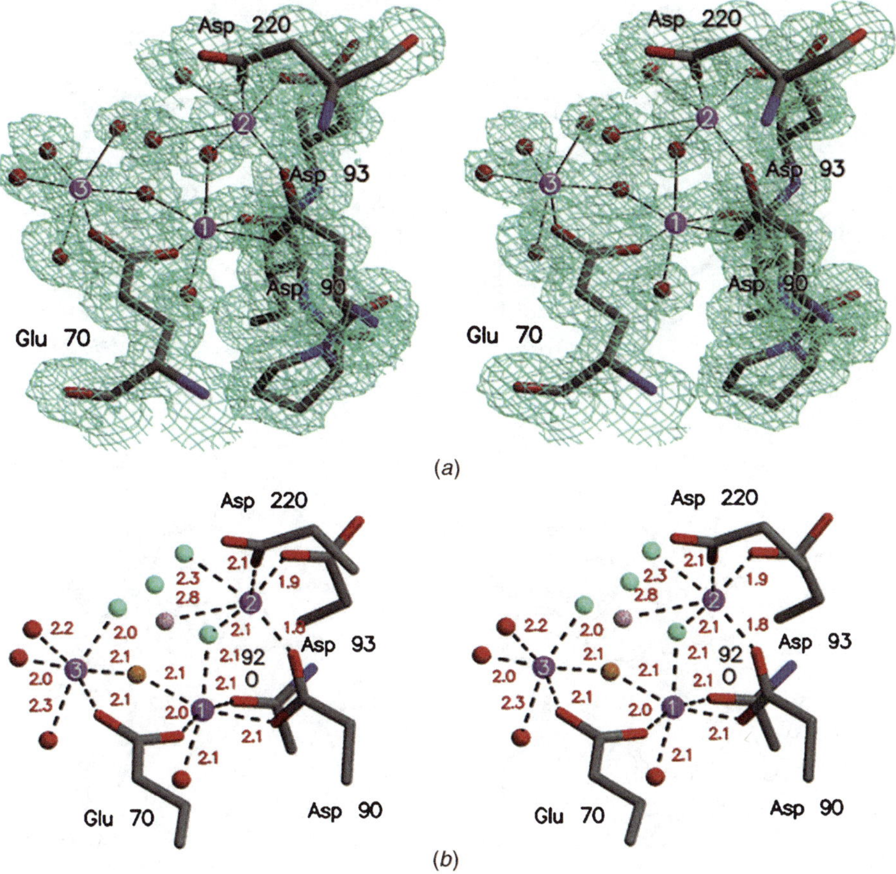

(a)

(b)

Figure 6 *The bovine IMPase. (a) An example of the quality of the electron density at the active site allowing unambiguous modeling of the magnesium and water structure. The map was contoured at 1.5σ. (b) The active site of bovine IMPase depicting the octahedral coordination of each magnesium ion (distances in Å). Magnesium ions are depicted in purple and water molecules are depicted in red, light blue, orange and pink*
(Reproduced with permission from ref. 42.)

Modeling of the phosphate moiety of inositol monophosphate to superpose the axial phosphate O atoms onto three active-site water molecules orientates the phosphoester bond for in-line attack by the nucleophilic water which is activated by Thr-95. Modeling of the pentacoordinate transition state suggests that the 6-OH group of the inositol moiety stabilizes the developing negative charge by hydrogen bonding to a phosphate O atom. Modeling of the post-reaction complex suggests a role for a second water molecule (W2) ligated by Mg-2 and Asp-220 in protonating the departing inositolate (Figure 7).[42] This second water molecule is absent in related structures in which lithium is bound at site 2, providing a rationale for enzyme inhibition by this simple monovalent cation (Figure 8).[42]

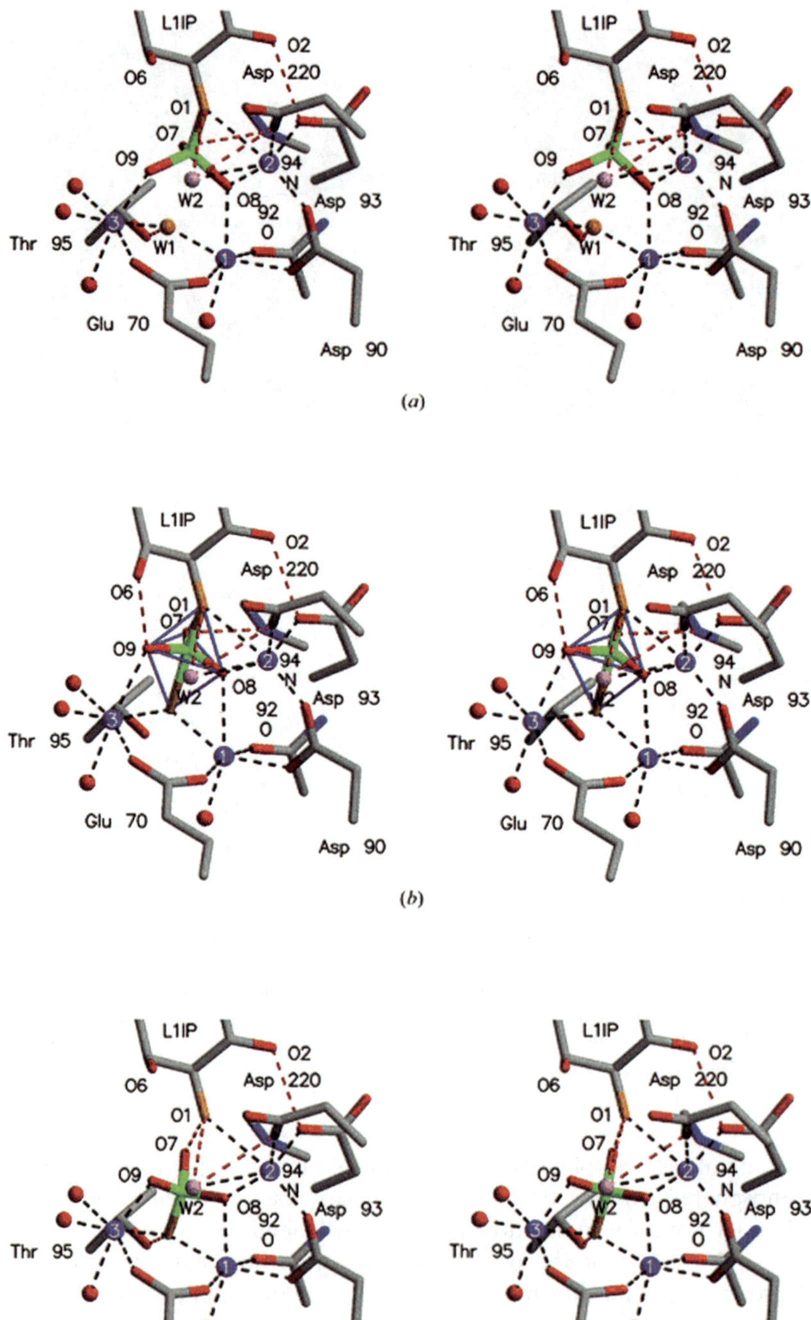

(*a*)

(*b*)

(*c*)

The higher resolution structural information on the active site of IMPase will hopefully facilitate the design of substrate-based inhibitors and aid in the development of better therapeutic agents for BD (manic depression). Although X-ray methods are unable to observe Li^+ directly, metal site 2 is clearly formed and most likely occupied by Li^+ in crystals grown in high concentrations of Li_2SO_4.[43] A nuclear magnetic resonance study of ^7Li binding to IMPase provided the first direct evidence of Li^+ ion binding to IMPase,[44] the dissociation constant of 1.0 mM being in excellent agreement with the kinetic K_i value and consistent with binding at site 2. Remarkably, in contrast to IMPase, the bifunctional archaeal IMPase/FBPase enzymes are not inhibited by Li^+ in the submillimolar range despite possessing a similar Mg^{2+}-coordination geometry at site 2. A structural comparison implicates the conformation of the catalytic loop (β-hairpin residues 32–43) involved in positioning the third metal ion.[45] It has been proposed that Li^+ binds at site 3 but that in the three-metal mechanism its smaller charge would be unable to create the water nucleophile, effectively inhibiting the enzyme. Since five of the six Mg^{2+} ligands at site 3 are water molecules, the change from octahedral to tetrahedral co-ordination could be easily accommodated. In support of this hypothesis, Li^+ inhibition of K36Q human IMPase is greatly reduced and noncompetitive and the mutant enzyme is no longer inhibited at high Mg^{2+} concentration.

5.2 GSK-3

GSK-3 was one of the first kinases to be identified and studied and is part of the machinery that controls protein phosphorylation, the most common post-translational mechanism used by cells to regulate enzymes and structural proteins, together with *ca.* 520 protein kinases and *ca.* 80 protein phosphatases. It was initially studied for its function in the regulation of glycogen

---

Figure 7 *The essential features of the catalytic mechanism of IMPase. (a) Modeling of the phosphate moiety of L-Ins(1)P in the pre-reaction state. A slight rotation of the phosphate moiety about O1 superposes the axial phosphate O atoms onto the positions of three active-site waters and orientates the phosphoester bond for a direct inline attack by the putative water nucleophile (W1), which is depicted in orange. Mg-1 and Mg-3 coordinate W1, thus lowering its pK_a and facilitating proton removal by the Thr95/Asp49 dyad. (b) Modeling of the trigonal bipyramidal transition state based upon the structure of the pentavalent phosphorus intermediate of phosphorylated β-PGM. The O6 hydroxyl of L-Ins(1)P is within hydrogen-bonding distance of O9 of the phosphate moiety. Whether the species represents a classical transition state as implied here or a trappable pentavalent intermediate is not the focus of the present work. (c) Modeling of the post-reaction structure based upon the yeast Hal2p PAPase-3Mg^{2+}-AMP-P_i end-product complex. The collapse of the transition state yields inositolate complexed to Mg-2 and the cleaved phosphate, formed by an inversion of configuration. Inositol is generated through protonation by W2 (depicted in pink) and released as the first product*
(Reproduced with permission from ref. 42.)

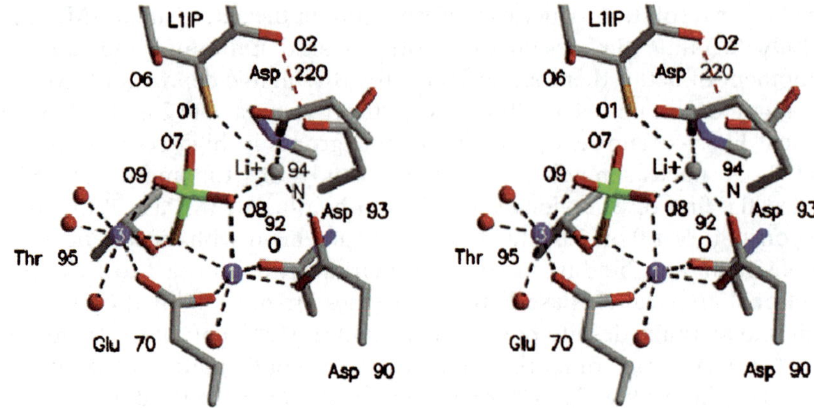

Figure 8   *Modelling of the lithium-inhibited IMPase structure. The model is based upon that of the yeast Hal2p PAPase-2Mg^{2+}-AMP anion-Li^{+} complex, in which the tetrahedral coordination of Li^{+} at site 2 has been inferred. Replacement of Mg^{2+}-2 by Li^{+}-2 precludes the coordination of W2 and prevents the protonation of the inositolate group after phosphoester hydrolysis, trapping inositolate and P$_i$ in the active site and effectively inhibiting the enzyme* (Reproduced with permission from ref. 42.)

synthase.[46–48] Interest in GSK-3 has grown far beyond glycogen metabolism during the past decade and GSK-3 is now known to occupy a central stage in many cellular and physiological events, including Wnt,[†49] and Hedgehog signaling, transcription, insulin action, cell-division cycle, response to DNA damage, cell death, cell survival, patterning and axial orientation during development, differentiation, neuronal functions, circadian rhythm and others. The name GSK-3, thus, appears to be a rather limited tribute to the large diversity of its physiological effects (Figure 9).

In fact, GSK-3 is a serine–threonine kinase that is normally highly active in cells. Cellular targets of GSK-3 are numerous and often depend on the signaling pathway that is acting upon it (due to cellular localization and regional sequestration). For example, Wnt pathway-mediated inhibition of

---

[†] The name Wnt is an amalgam of the two founding members: int-1 in the mouse (now called Wnt-1; int stands for integration site, and wingless in Drosophila. Other Wnt genes have been numbered according to their order of discovery. A simple outline of the current model of Wnt signal transduction is presented in Figure 9. Wnt proteins released from or presented on the surface of signaling cells act on target cells by binding to the Frizzled (Fz)/low-density lipoprotein (LDL) receptor-related protein (LRP) complex at the cell surface. These receptors transduce a signal to several intracellular proteins that include, Dishevelled (Dsh), glycogen synthase kinase-3β (GSK-3), Axin, Adenomatous Polyposis Coli (APC) and the transcriptional regulator, β-catenin (Figure 9). Cytoplasmic β-catenin levels are normally kept low through continuous proteasome-mediated degradation, which is controlled by a complex containing GSK-3/APC/Axin. When cells receive Wnt signals, the degradation pathway is inhibited, and consequently β-catenin accumulates in the cytoplasm and nucleus. Nuclear β-catenin interacts with transcription factors such as lymphoid enhancer-binding factor 1/Tcell-specific transcription factor (LEF/TCF) to affect transcription. A large number of Wnt targets have been identified that include members of the Wnt signal transduction pathway itself, which provide feedback control during Wnt signaling.

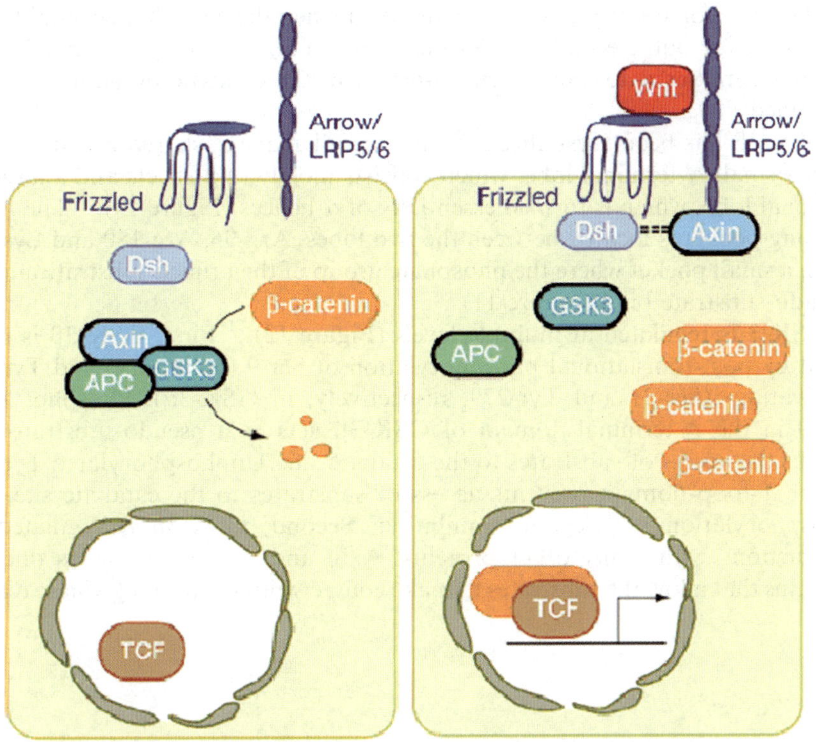

Figure 9 *The canonical Wnt signaling pathway. In cells not exposed to a Wnt signal (left panel), β-catenin is degraded through interactions with Axin, APC and GSK-3. Wnt proteins (right panel) bind to the Frizzled/LRP receptor complex at the cell surface. These receptors transduce a signal to Disheveled (Dsh) and to Axin, which may directly interact (dashed lines). As a consequence, the degradation of β-catenin is inhibited, and this protein accumulates in the cytoplasm and nucleus. β-catenin then interacts with TCF to control transcription. Negative regulators are outlined in black. Positively acting components are outlined in color*
(Reproduced with permission from ref. 49.)

GSK-3 activates the transcription factor β-catenin, whereas in the insulin–PI3K signaling pathway, inhibition of GSK-3 results in activation of the enzyme glycogen synthase. Targets of GSK-3 include, among others, transcription factors, proteins bound to microtubules, cell-cycle mediators, and regulators of metabolism (Figure 2).

The GSK-3 kinase family is highly conserved throughout evolution. In humans, two genes, which map to 19q13.2 and 3q13.3, encode two distinct but closely related GSK-3 forms, GSK-3α (51 kDa) and GSK-3β (47 kDa). They display 84% overall identity (98% within their catalytic domains) with the main difference being an extra Gly-rich stretch in the *N*-terminal domain of GSK-3α. However, they are not interchangeable functionally, as demonstrated by the embryonic-lethal phenotype observed when the gene that encodes GSK-3β is knocked out. Recently, GSK-3β2, an alternative-splicing variant of GSK-3β that contains a 13-amino-acid insertion in the catalytic domain, has been identified.

GSK-3β has been crystallized.[50] The overall shape is shared by all kinases, with a small *N*-terminal lobe, which consists mostly of β-sheets and a large *C*-terminal lobe, which is formed essentially of α-helices (Figure 10).[50] The ATP-binding pocket is located between the two lobes. Arg-96, Arg-180 and Lys-205 form a small pocket where the phosphate group of the primed substrate and the pseudo-substrate bind (Figure 11).[50]

GSK-3 is regulated at multiple levels (Figure 12).[51] First, GSK-3β is regulated by post-translational phosphorylation of Ser-9 (inhibitory) and Tyr-216 (activating) (Ser-21 and Tyr-279, respectively, in GSK-3α). Phosphorylated Ser-9 in the *N*-terminal domain of GSK-3β acts as a pseudo-substrate that blocks the access of substrates to the catalytic site. Unphosphorylated Tyr-216 in the T-loop domain prevents access of substrates to the catalytic site, and phosphorylation releases this inhibition. Second, GSK-3β is regulated by interactions with many other proteins. Axin and pre-senilin act as docking proteins that allow the substrates to make contact with the priming kinase (casein

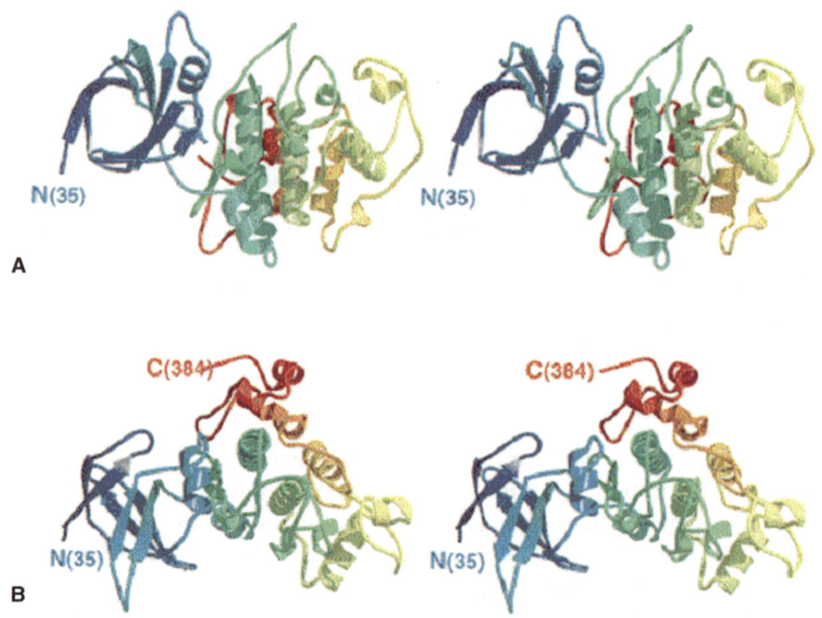

Figure 10 *Structure of human GSK3β. (A) Stereo pair secondary structure cartoon of human GSK-3β colored blue-red from the visible N-terminus at residue 35 to the visible C-terminus at residue 384. The orthogonal β barrel formed by the N-terminal domain is on the left. (B) As (A), but with the view rotated by 90° around the horizontal*
(Reproduced with permission from ref. 50.)

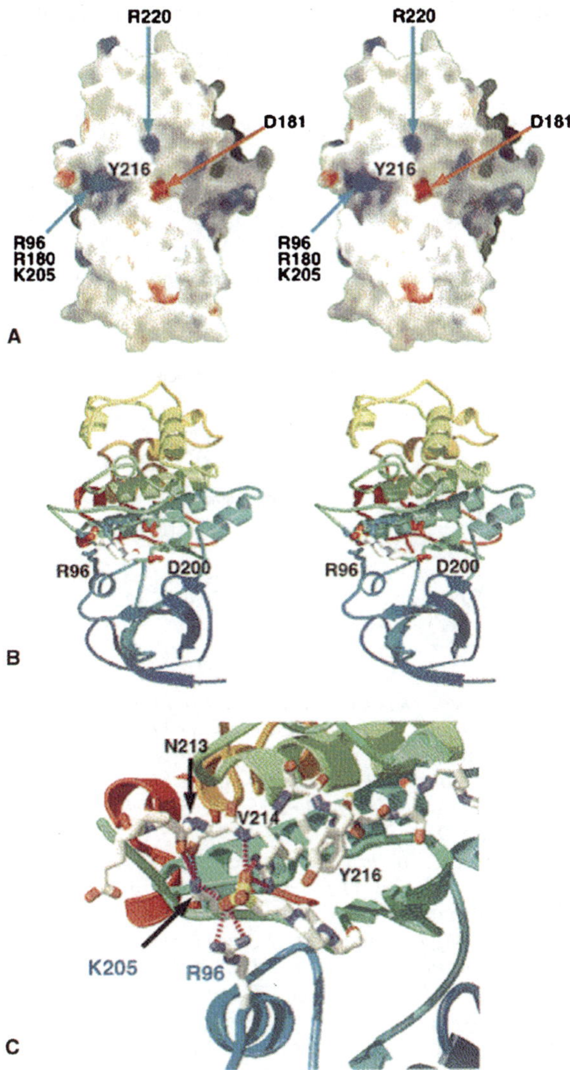

Figure 11 *Oxyanion binding site. (A) Stereo pair of solvent-accessible surface of GSK3β colored according to the electrostatic potential: red, negative; blue, positive. The intense positive patch generated by the basic side chains of Arg96, Arg180, and Lys205 is indicated, as is the location of the catalytic Asp181 and Arg220, which could interact with a phosphorylated Tyr216. (B) Position of the bound HEPES molecule relative to the overall structure of GSK3β. The negatively charged sulfonate group of HEPES is close to the positive patch (see [A]) generated by Arg180 and Lys205 from the large domain and Arg96 on the C helix of the N-terminal domain (blue side chains). The side chains of the catalytic residues Asp181 and Asp200 are shown in red. (C) Detail of the hydrogen-bonding network involving the sulfonate of HEPES, the side chains of Arg96, Arg180 and Lys205, and the main chain of Asn213 and Val214 in the activation segment*
(Reproduced with permission from ref. 50.)

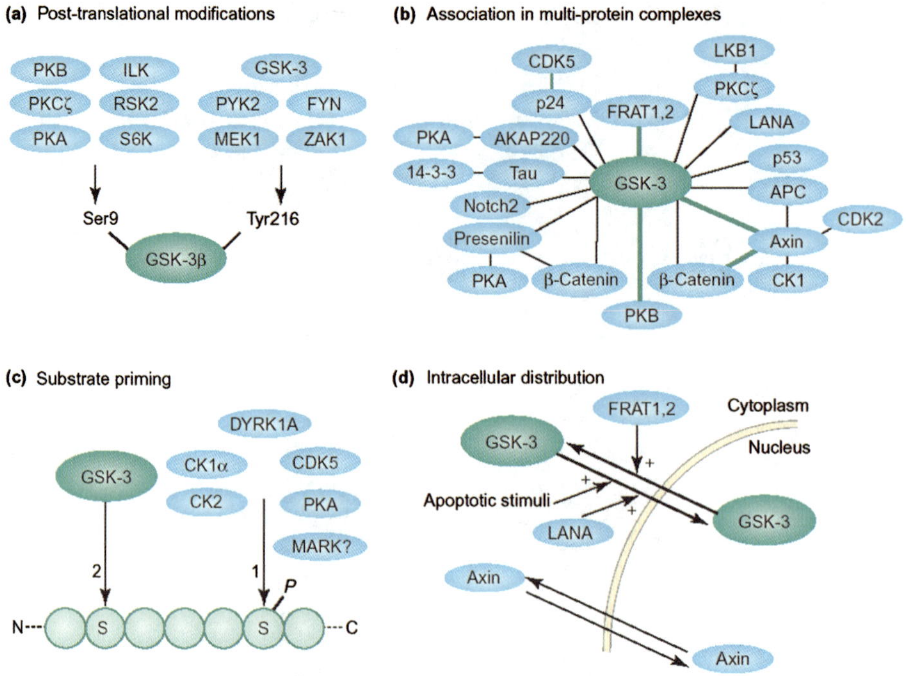

Figure 12 *GSK-3 regulation. (a) Post-translational modifications. GSK-3β is regulated by post-translational phosphorylation of Ser9 and Tyr216. Phosphorylation of Ser9 can be carried out by $p70^{S6K}$, $p90^{rsk}$, protein kinase A (PKA), PKB (AKT), PKC isoforms and integrin-linked kinase (ILK). The src-like FYN kinase, the Ca^{2+}-sensitive praline-rich tyrosine kinase 2 (PYK2), a putative homolog of the Dictyostelium zaphod kinase 1 (ZAK1) tyrosine kinase and mitogen-activated protein kinase kinase 1 (MEK1) might be responsible for Tyr216 phosphorylation. (b) Association with partners. GSK-3β is regulated by interactions with many proteins. Interactions that have been characterized by crystallography are indicated by green lines. (c) Priming of substrates. The substrate-recognition site of GSK-3 is S–X–X–X–S–P, where S–P is a priming, pre-phosphorylated serine residue. The priming kinase can be casein kinase 2 (CK2) (for glycogen synthase), CK1α (for β-catenin), dual-specificity tyrosine-phosphorylation regulated kinase 1A (DYRK1A) (for eukaryotic protein synthesis initiation factor 2B), PKA (for Cubitus interruptus) or cyclin-dependent kinase 5 (CDK5)/p25 (for Tau). (d) Intracellular distribution. GSK-3β is predominantly cytoplasmic, but it is also present in the nucleus and in mitochondria. The activity of nuclear GSK-3 is higher than cytoplasmic GSK-3, and the nuclear pool is stimulated further in apoptotic cells or by p53. GSK-3β is localized predominantly in the neuronal soma and processes, whereas GSK-3β2 is localized in the soma. FRAT1,2 (frequently rearranged in advanced T-cell lymphomas 1,2) promotes the nuclear efflux of GSK-3β, whereas latent nuclear antigen (LANA), a Kaposi virus protein, sequesters GSK-3β in the nucleus and mimics Wnt signaling. Abbreviations: AKAP220, A-kinase anchoring protein; APC, adenomatous polyposis coli; LKB1, gene denomination of serine/threonine kinase 11; MARK, microtubule affinity-regulating kinase; RSK, ribosomal S6 kinase; S6K S6 kinase*
(Reproduced with permission from ref. 50.)

kinase (CK1) and protein kinase A, respectively) and GSK-3. Docking proteins might, thus, specify different GSK-3 functions in the cell. Third, GSK-3 action requires the priming phosphorylation of its substrates by another kinase on a serine residue located four amino acids *C*-terminal to the GSK-3 phosphorylation site. Fourth, GSK-3 is regulated through its intracellular distribution.

Both GSK-3β and GSK-3β of vertebrates appear to be inhibited by Li^+ *in vitro* and *in vivo*.[52,53] Enzyme kinetic experiments suggest that this is through competition for Mg^{2+} binding at a site distinct from the ATP-binding site.[54,55] The K_i for GSK-3β is 2.0 mM, whereas it appears to be a little higher for GSK-3α. However, reducing the Mg^{2+} concentration to 0.75 mM, approximately that of free Mg^{2+} in cells, decreases the IC_{50} to 0.8 mM, well within the therapeutic range of 0.6–1.2 mM.

The main effect of GSK-3 inhibition reflects on the growth cones at the ends of developing axons that are "endowed with amoeboid movements" that respond to chemical signals. These motile growth cones are responsible for the pattern of axonal and dendritic processes as they extend from the neuronal cell body. Through their response to guidance cues, growth cones steer the developing axons and dendrites to their targets, where they develop into the synapse.

Growth cone motility requires both F-actin (filamentous actin) and microtubule cytoskeleton.[56] Actin polymerization at the periphery drives protrusion of the leading edge of the growth cone, and traction is achieved by points of adhesion between actin filaments and the extracellular matrix. Attractive guidance signals promote local actin polymerization, and repellent signals induce its local collapse. As the growth cone moves forward, F-actin flows from the periphery toward the center of the growth cone, where it disassembles to be recycled to the leading edge for further growth.

Stable microtubules form the major structural component of the axon, but as they project from the end of the axon into the growth cone they become much more dynamic, and exhibit a highly complex behavior, splaying, bending and projecting in different directions. Changes in microtubule structure appear to be responsible for branching along the axon, although this also involves changes in the actin cytoskeleton. Li^+ and Wnt proteins alter growth cone morphology, axonal branching and microtubule organization,[57] an effect also seen with the specific GSK-3 inhibitors SB216763 (Formula 7) and SB415286 (Formula 8).[58] These results suggest that Wnt signaling may regulate the microtubule cytoskeleton of developing neurons through inhibition of GSK-3 activity.

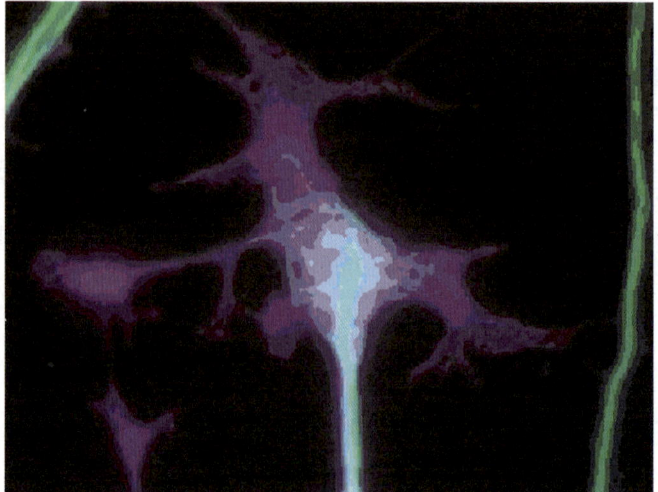

GSK-3 affects growth cone collapse and microtubule dynamics, as shown in Figure 13.[59]

Early studies suggested that peripheral administration of lithium inhibited brain GSK-3 in the 7-day-old rat brain.[60] More recently, studies suggest that this enzyme is significantly inhibited in the rodent brain in the presence of therapeutic serum lithium concentrations during long-term treatment. For example, it was demonstrated that nine days of lithium treatment (at a mean serum concentration of 0.8 mM) increased cytosolic protein levels of β-catenin in rats, one of the transcription factors regulated directly by GSK-3.[61,62] Several other pre-clinical (*i.e.*, animal or cellular) studies have clearly suggested that therapeutic serum concentrations of lithium produce a biologically signif-icant inhibition of GSK-3 in the mammalian brain and that GSK-3 represents a *therapeutic* target of lithium. GSK-3 represents a strong candidate as a medi-ator of lithium's neuroprotective effects most likely because GSK-3 in the brain is significantly inhibited by therapeutic lithium concentrations. Additionally, recent evidence suggests that the behavioral effects of lithium, at least in rodent models, may also be due to inhibition of GSK-3. It has been in fact found that

Figure 13 *A neuronal growth cone treated with lithium. Lithium induces both an alter-ation in microtubule dynamics (stained green with anti-acetyl tubulin antibod-ies) and an expansion of the growth cone (stained blue with calcein)*

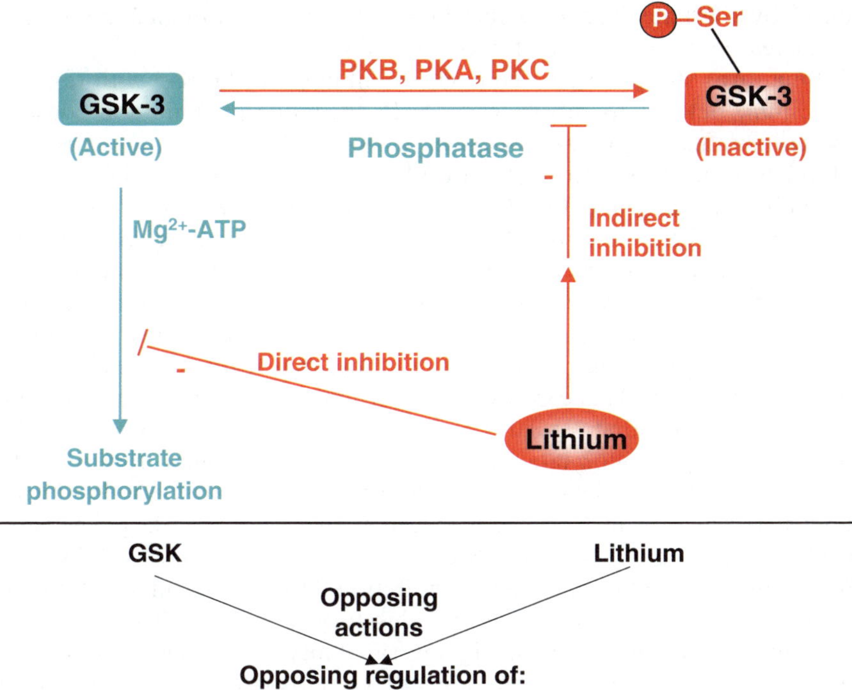

Figure 14 *Interactions between lithium and GSK-3. GSK-3 catalyzes the phosphorylation of many protein substrates in the presence of Mg^{2+}–ATP. Lithium is a competitive inhibitor of the Mg^{2+}, which results in inhibition of the activity of GSK-3 (indicated as direct inhibition of GSK-3). GSK-3 is inactivated by phosphorylation on a serine in the N-terminal domain: Ser9 in GSK-3β and Ser21 in GSK-3α. This modification can be catalyzed by several different kinases, such as protein kinase B (PKB; Akt), protein kinase A (PKA), and protein kinase C (PKC). The inactive phospho-serine-GSK-3 can be reactivated by phosphatase activity that removes the phosphate from the serine. In the presence of lithium, the action of this phosphatase is reduced, leaving more of the GSK-3 in the phosphorylated, inactive form (indicated as indirect inhibition of GSK-3). Therefore, following lithium exposure, GSK-3 is inhibited both by a direct action of lithium and indirectly by increased phosphorylation of serine. Consequently, GSK-3 and lithium have opposing influences on many important cellular processes. Blue indicates aspects of signaling that are associated with active GSK-3. Red indicates inhibition of GSK-3 activity (Reproduced with permission from ref. 62.)*

administration of GSK-3 inhibitors results in antidepressant-like effects in the forced swim test paradigm following either intracerebral ventricle injections in mice,[63] peripheral administration to rats[64] or lithium administration to mice.[62] Furthermore, the behavioral effects of knocking out a single copy of the GSK-3β gene were producing the same antidepressant-like behavior induced by alternate pharmacological inhibition and by lithium administration (*i.e.*, increased mobility in the forced swim test).[62]

Lithium was shown to be a competitive inhibitor of GSK-3 with respect to magnesium, but not to substrate or ATP,[54] a mode of inhibition conserved between mammalian and *Dictyostelium* GSK-3 isoforms, and not experienced with other group I metal ions (Figure 14).[65] As a consequence, the potency of Li^+ inhibition is dependent on Mg^{2+} concentration. GSK-3 was also found to be sensitive to chelation of free Mg^{2+} by ATP and to be progressively inhibited when ATP concentrations exceed that of Mg^{2+}.[65] Given the cellular concentrations of ATP and Mg^{2+}, these results indicate that Li^+ is likely to have a greater effect on GSK-3 activity *in vivo* than expected from *in vitro* studies.

Another category of inhibitors includes those that regulate an enzyme by an indirect mechanism. This can involve controlling a post-translational modification of an enzyme that regulates its activity, such as phosphorylation. GSK-3 is inhibited by phosphorylation of a serine in its *N*-terminal region: Ser9 in GSK-3β, the most widely studied subtype, or Ser21 in GSK-3α. Lithium administered chronically to mice has been shown to cause a large increase in the phosphorylation of Ser9 of GSK-3β *in vivo*,[66] and of Ser21 of GSK-3α *in vitro*.[67] Thus, in addition to its direct inhibition, lithium also inhibits GSK-3 by increasing the phosphorylation of the inhibitory serine of GSK-3.

Lithium has, therefore, the intriguing capability of leading to inhibition of GSK-3 in two ways: directly and indirectly by phosphorylation of the enzyme. These two inhibitory actions following exposure to lithium probably act concomitantly *in vivo* to achieve a cumulative inhibition of GSK-3. Identification of these dual inhibitory effects answers the criticisms that the direct inhibitory effect of lithium is too small and too fast to be relevant therapeutically because the phosphorylation mechanism adds to the direct inhibitory effect and occurs more slowly.

References

1. J.F.J. Cade, *Med. J. Aust.*, 1949, **36**, 349.
2. F.K. Godwin and S.N. Ghaemi, *Am. J. Psychiatry*, 2003, **160**, 2077.
3. C.J.L. Murray and A.D. Lopez, *The Global Burden of Disease Summary*, Harvard School of Public Health, Cambridge, MA, 1996.
4. J.A. Quiroz *et al.*, *Mol. Interv.*, 2004, **4**, 259.
5. S.E. Hyman, *Biological Psychiatry*, 1999, **45**, 518.
6. J.C. Soares and J.J. Mann, *Biological Psychiatry*, 1997, **41**, 86.
7. J.C. Soares and J.J. Mann, *J. Psychiatr. Res.*, 1997, **31**, 393.
8. G.S. Sachs and M.E. Thase, *Biological Psychiatry*, 2000, **48**, 573.
9. N.A. Huxley *et al.*, *Harvard Review of Psychiatry*, 2000, **8**, 126.
10. R.P. Beliles. Lithium in: *Patty's Industrial Hygiene and Toxicology*, 4th edn, vol II, Part C, G.D. Clayton and F.E. Clayton (eds), Wiley, New York, 1994.
11. A. Jaeger *et al.*, *J. Toxicol. Clin. Toxicol.*, 1985–86, **23**, 501.

12. R.E. Gosselin *et al.*, *Clinical Toxicology of Commercial Products*, 5th edn, Williams and Wilkins, Baltimore, MD, 1984.
13. J.M. Arena, (ed), *Poisoning: Toxicology, Symptoms, Treatments*, Charles C. Thomas, Springfield, IL, 1986.
14. J.H. Kocsis *et al.*, *J. Clin. Psychopharmacol.*, 1993, **13**, 268.
15. L.S. Cohen *et al.*, *J. Am. Med. Assoc.*, 1994, **271**, 146.
16. L. Amari *et al.*, *Anal. Biochem.*, 1999, **272**, 1.
17. S.P. Davies *et al.*, *Biochem. J.*, 2000, **351**, 95.
18. J.D. York *et al.*, *Proc. Natl. Acad. Sci. USA*, 1995, **92**, 5149.
19. T.D. Gould *et al.*, *J. Clin. Psychiatry*, 2004, **65**, 10.
20. T.D. Gould *et al.*, *Mol. Psychiatry*, 2004, **9**, 734.
21. P.W. Majerus, *Annu. Rev. Biochem.*, 1992, **61**, 225.
22. D. Gani *et al.*, *Biochim. Biophys. Acta*, 1993, **1177**, 253.
23. L.M. Hallcher and W.R. Sherman, *J. Biol. Chem.*, 1980, **255**, 10896.
24. R.C. Inhorn and P.W. Majerus, *J. Biol. Chem.*, 1988, **263**, 14559.
25. M.J. Berridge *et al.*, *Biochem. J.*, 1982, **206**, 587.
26. M.J. Berridge *et al.*, *Cell*, 1989, **59**, 411.
27. S.R. Nahorski *et al.*, *Trends Pharmacol. Sci.*, 1991, **12**, 297.
28. G.T. Berry *et al.*, *Mol. Genet. Metab.*, 2004, **82**, 87.
29. J.H. Allison and M.A. Stewart, *Nat. New Biol.*, 1971, **233**, 267.
30. J.R. Atack, *Med. Res. Rev.*, 1997, **17**, 215.
31. G. McAllister *et al.*, *Biochem. J.*, 1992, **284**, 749.
32. R. Bone *et al.*, *Proc. Natl. Acad. Sci. USA*, 1992, **89**, 10031.
33. R. Bone *et al.*, *Biochemistry*, 1994, **33**, 9460.
34. S.J. Pollack *et al.*, *Proc. Natl. Acad. Sci. USA*, 1994, **91**, 5766.
35. A.P. Leech *et al.*, *Eur. J. Biochem.*, 1993, **212**, 693.
36. C.M.J. Fauroux *et al.*, *J. Am. Chem. Soc.*, 1999, **121**, 8385.
37. D.J. Miller *et al.*, *ChemBioChem.*, 2000, **1**, 262.
38. A.J. Ganzhorn *et al.*, *Biochemistry*, 1996, **35**, 10957.
39. R. Gill *et al.*, *Acta Crystallogr. D Biol. Crystallogr.*, 2005, **61**, 545.
40. S. Patel *et al.*, *J. Mol. Biol.*, 2002, **315**, 677.
41. V. Saudek *et al.*, *Eur. J. Biochem.*, 1996, **240**, 288.
42. K.A. Stieglitz *et al.*, *J. Biol. Chem.*, 2002, **277**, 22863.
43. S. Frame and P. Cohen, *Biochem. J.*, 2001, **359**, 1.
44. B.W. Doble and J.R. Woodgett, *J. Cell Sci.*, 2003, **116**, 1175.
45. R.S. Jope and G.V.W. Johnson, *Trends Biochem. Sci.*, 2004, **29**, 95.
46. C.Y. Logan and R. Nusse, *Annu. Rev. Cell Dev. Biol.*, 2004, **20**, 781.
47. R. Dajani *et al.*, *Cell*, 2001, **105**, 721.
48. L. Meijer *et al.*, *Trends Pharmacol. Sci.*, 2004, **25**, 471.
49. P.S. Klein and D.A. Melton, *Proc. Natl. Acad. Sci. USA*, 1996, **93**, 8455.
50. V. Stambolic *et al.*, *Curr. Biol.*, 1996, **6**, 1664.
51. W.J. Ryves and A.J. Harwood, *Biochem. Biophys. Res. Commun.*, 2001, **280**, 720.
52. W.J. Ryves *et al.*, *Biochem. Biophys. Res. Commun.*, 2002, **290**, 967.
53. E.W. Dent and F.B. Gertler, *Neuron*, 2003, **40**, 209.
54. F.R. Lucas and P.C. Salinas, *Dev. Biol.*, 1997, **192**, 31.

55. R. Owen and P.R. Gordon-Weeks, *Mol. Cell Neurosci.*, 2003, **23**, 626.
56. R.S. Williams *et al.*, *Nature*, 2002, **417**, 292.
57. J.R. Munoz-Montano *et al.*, *FEBS Lett.*, 1997, **411**, 183.
58. T.D. Gould *et al.*, *Neuropsychopharmacology*, 2004, **29**, 32.
59. W.T. O'Brien *et al.*, *J. Neurosci.*, 2004, **24**, 6791.
60. O. Kaidanovich-Beilin *et al.*, *Biol. Psychiatry*, 2004, **55**, 781.
61. T.D. Gould *et al.*, *Int. J. Neuropsychopharmacol.*, 2004, **7**, 387.
62. R.S. Jope, *Trends Pharmacol. Sci.*, 2003, **24**, 441.
63. P. De Sarno *et al.*, *Neuropharmacology*, 2002, **43**, 1158.
64. E. Chalecka-Franaszek and D.M. Chuang, *Proc. Natl. Acad. Sci. USA*, 1999, **96**, 8745.

CHAPTER 6
Neurotoxicity of Aluminum

6.1   Neurochemistry of Aluminum

Al is a metal present in high amounts in the earth's crust, where it represents 8% of the total components. Although limited by the formation of Al-silicate complexes, the bioavailability of this metal can be raised by environmental and industrial factors, such as contamination, acid rain,[1] water purification treatment with Al compounds,[2] the use of Al as food additives, and certain pharmacological and therapeutic treatments such as buffered aspirin and Al-based antacids.[3] Al is present in high amounts in certain foods and beverages such as tea, herbs and spices, grain products, processed cheese and salt. Deodorants with Al-based formula have been claimed to be a risk for human health since Al might be rapidly absorbed in the nasal cavity to the perineural space that surrounds the olfactory neurons reaching the nervous system.

There are over 2000 references in the National Library of Medicine on adverse effects of aluminum. Its toxicity has been recognized in many settings where

- exposure is heavy or prolonged;
- where renal function is limited, such that ready excretion of absorbed Al into the urine is impaired; and
- where a previously accumulated bone burden is released in stress or illness.

Toxicity may include

- encephalopathy (stuttering, gait disturbance, myoclonic jerks, seizures, coma, abnormal EEG);
- osteomalacia or aplastic bone disease (associated with painful spontaneous fractures, hypercalcemia, tumorous calcinosis);
- proximal myopathy;
- increased risk of infection;
- increased left-ventricular mass and decreased myocardial function,
- microcytic anemia with very high levels; and
- sudden death.

The contributory role of Al in the dialysis encephalopathy syndrome has been shown. Avoidance of the major Al sources contributing to the syndrome has greatly reduced this problem, although occasional outbreaks still occur. There has been concern about the suggested role of Al in Alzheimer disease (AD) since the initial report of elevated brain Al in victims of this condition (*vide infra*). Some studies have shown a small positive correlation between drinking-water Al concentration and dementias, including AD.[4] As a result of continued concern about the neurotoxicity of Al, the U.S. Environmental Protection Agency has put Al on its contaminant candidate list,[†] the U.S. Food and Drug Administration recently implemented labeling requirements for Al in large- and small-volume parenterals,[‡] and Canada established operational guidance limits for drinking-water Al on the basis of the precautionary principle.[§]

The gastrointestinal tract is relatively impervious to aluminum, absorption normally being only about 2%. Aluminum is absorbed by a mechanism related to that for calcium. Gastric acidity and oral citrate favors absorption, and H2-blockers reduce absorption. As the significance of aluminum toxicity has become apparent, considerable attention has been given to defining the chemical speciation of the Al^{3+} ion in human serum. Several fractionation studies have reported the distribution of Al^{3+} ions between proteins and low-molecular-mass ligands. Most of the earlier studies reported that $\sim 80\%$ of the aluminum in serum was bound to proteins, but, as methodologies have improved, this figure has tended to increase to $\sim 90\%$.[5] This now appears that essentially all the high-molecular-mass aluminum in serum is bound to the iron transport protein transferrin.

Transferrin consists of two similar lobes. Each lobe is further divided into two domains, with a high-affinity metal-binding site located within a cleft between the two domains. The Fe^{3+} ion in each lobe is coordinated to four ligands from protein side chains: two tyrosine phenolic groups, an imidazole group from one histidine residue, and the carboxylate side chain of an aspartic acid. The fifth and sixth coordination sites on the Fe^{3+} ion are occupied by a bidentate carbonate anion derived from the buffer that is called the synergistic anion (Figure 1).[6]

No crystal structure has been reported for the Al-transferrin complex, but it is highly likely that an Al^{3+} ion binds in essentially the same manner as that of a Fe^{3+} ion. Apotransferrin (apoTf) binds two Al^{3+} ions that compete with Fe^{3+} ion for the same binding sites.[7] Different UV studies confirm that the Al^{3+} ion binds to tyrosine side chains,[8] and ^{13}C NMR studies have shown that the binding of an Al^{3+} ion involves the binding of a synergistic carbonate anion.[9] Last, small-angle X-ray scattering studies indicate that aluminum binding causes an overall protein conformational change that is similar to that caused by iron binding.[10]

[†] U.S. Environmental Protection Agency. Announcement of the drinking water contaminant candidate list. Fed Reg 63:10273–10287 (1998).
[‡] U.S. Food and Drug Administration. Aluminum in large and small volume parenterals used in total parenteral nutrition. Code of Federal Regulations, Vol. 21 CFR201.323, 2000.
[§] Health Canada. Aluminum. Environmental Health Program, 1998.

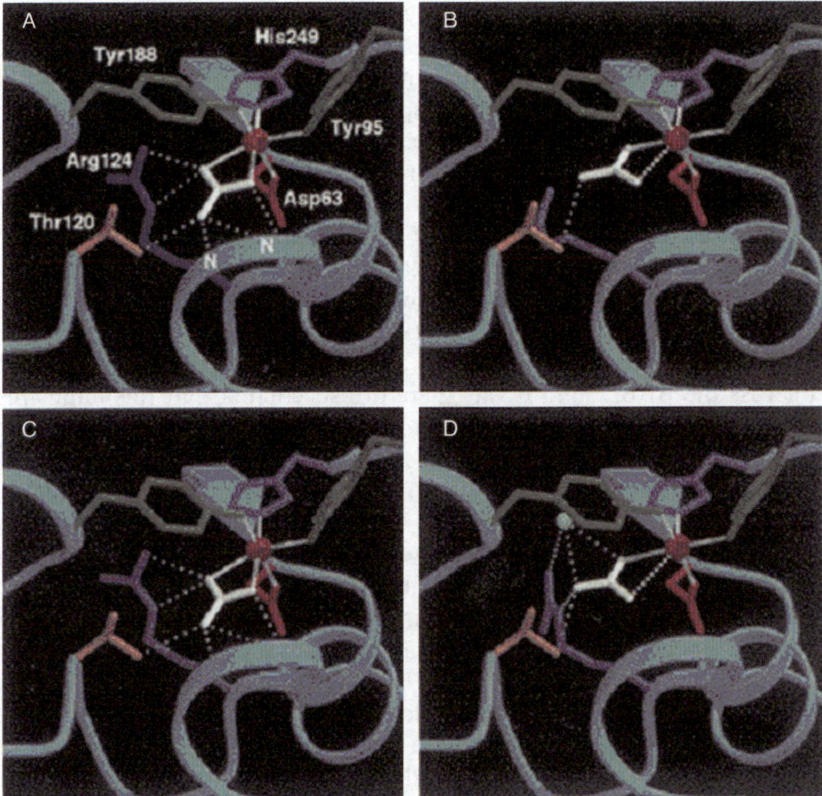

Figure 1 *Structures of the alternative conformations of the iron-binding sites in two different crystal forms of hTF/2N. (A) Orthorhombic crystal form of conformer A, (B) orthorhombic crystal form of conformer B, (C) tetragonal crystal form of conformer A, and (D) tetragonal crystal form of conformer B. The iron atom is indicated by the red sphere and the carbonate by the white triangular structure. Ligands to the iron from Asp-63, Tyr-95, Tyr-188, His-249 and the carbonate anion are indicated by thin white cylinders. Hydrogen bonds are indicated by dashed white lines. The side chain for Thr-120 is included; other side chains and water molecules have been omitted to simplify the figure. The main chain amide nitrogen atoms of Ala-126 and Gly-127 are indicated by N* (Reproduced with permission from ref. 6.)

Cells appear to take up aluminum from transferrin rather than from citrate. In fact, purified preparations of ferritin from brain and liver have been found to contain aluminum.

In order to protect the brain from the noxious effects of Al, there is an active efflux of the metal at the blood-brain barrier mediated by a monocarboxylate transporter.[11,12] However, exposure to high amounts of Al or an increased blood Al concentration due to a decreased renal functionality can lead to brain Al accumulation. Al brain accumulation occurs physiologically with aging, since it has been estimated that Al deposits in the brain at a rate of 6 µg per year of life.[13]

Al may enter the brain from blood, either through the choroid plexuses or the BBB (Figure 2).

There is a choroid plexus in each of the four cerebral ventricles of the brain. They synthesize most of the cerebrospinal fluid (CSF) that fills the brain ventricles and the subarachnoid space that surrounds the brain and spinal cord. The total surface area of the choroid plexuses is approximately 10 cm^2. Brain atlases of the rat and human show brain regions as far as 1 mm away from the nearest component of the CSF compartment.

An uncharacterized monocarboxylate transporter isoform and/or organic anion transporting protein family member expressed at the BBB was first suggested to be responsible for Al^{3+}-citrate transport:[14] L-glutamate, or L- or D-aspartate were interestingly reported to act as chelators, removing Al from the brain across the BBB,[15] and it was therefore speculated that this might result from an amino acid exchange between D-aspartate and endogenous Al^{3+}-L-glutamate. On the other hand, because Al^{3+}-citrate is the major species in the brain extracellular fluid and these amino acids are the ligands for the glutamate transporter family,[16] it is possible that their findings might be resulted from counter transport of these amino acids with Al citrate in the brain extracellular fluid *via* glutamate transporter(s) across the BBB. Recent results have suggested that the uptake of Al^{3+}-citrate into brain-cell lines is transporter-mediated and that the Na^+-independent glutamate transporter is a good candidate for such process.[17]

The brain is particularly susceptible to Al^{3+}-mediated damage, causing neurodegeneration and interfering with nerve myelinization.[18] Al^{3+} has been associated with the etiology of certain dementias, such as amyotrophic lateral sclerosis, Parkinsonism dementia and AD. In those pathologies, increased amounts of Al^{3+} in the brain of post-mortem patients have been reported. Al^{3+} content was found to be 19- and 5-times higher in Parkinsonism dementia,[19] and in AD,[20] compared to healthy individuals, respectively. On the other hand, in transgenic mice over expressing the human amyloid precursor peptide, the administration of Al^{3+} caused an increase in oxidative stress end-products

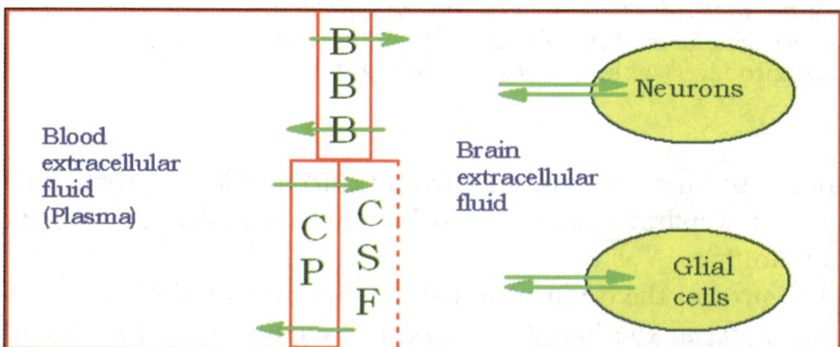

Figure 2 *The routes of distribution between plasma and brain: through the BBB and the choroid plexus (CP). Arrows show membranes through which distribution might occur by carrier-mediated transport in addition to diffusion. The dashed line depicts the absence of a membrane barrier to distribution between the CSF and brain ECF*

that correlated with the deposition of β-amyloid and the formation of senile plaques similar to that observed in Alzheimer's patients.[21] However, the participation of Al^{3+} in the etiology and/or in the development of AD is still a subject of discussion (*vide infra*).

Even though Al has no redox capacity in biological systems, it is well known that this metal can act as a pro-oxidant both *in vitro* and *in vivo*.[22] Aluminum has been in fact shown to facilitate (i) iron-induced lipid peroxidation,[23] (ii) non-iron-induced lipid peroxidation,[24] (iii) non-iron-mediated oxidation of NADH[25] and (iv) non-iron-mediated formation of the hydroxyl radical ($HO^•$).[26] A number of mechanisms have been proposed to account for these pro-oxidative effects. These include influences on

- the target substrate, *e.g.*, membrane lipids;
- other pro-oxidants, *e.g.*, iron; and
- the oxidant, *e.g.*, the superoxide radical anion ($O_2^{-•}$).

Although about 50 years have passed since oxygen toxicity was proposed to be due to free radical formation,[27] *ca.* 40 since the discovery of superoxide dismutase that limited a role for $O_2^{-•}$ in biological oxidation,[28] a consensus has not been reached as to the mechanism of toxicity of $O_2^{-•}$.[29] Oxidative damage has alternatively been ascribed to

- direct effects of $O_2^{-•}$ or the perhydroxyl radical ($HO_2^•$);
- $O_2^{-•}$-dependent formation of $HO^•$; and
- $O_2^{-•}$-dependent formation of the peroxynitrite anion ($OONO^-$).

The superoxide theory of oxygen toxicity is complicated by the observation that $O_2^{-•}$, like oxygen, is not a strong oxidant in that it does not easily damage biomolecules.[30] However, $O_2^{-•}$ will react in water to form its conjugate acid, $HO_2^•$, which is a much more powerful oxidant than $O_2^{-•}$:

$$O_2^{-•} + H^+ \rightarrow HO_2^{-•}$$

The pK of this reaction ensures that at physiological pH about 1% of all $O_2^{-•}$ formed will be protonated. The oxidative potential of $HO_2^•$ is significantly greater than that of $O_2^{-•}$ so that, for example, it will oxidize NADH 10^4 times faster than $O_2^{-•}$.[31] The potential of $O_2^{-•}$ to act as a Lewis base in forming $HO_2^•$ may be paralleled by an analogous interaction with other electrophilic atoms or molecules to form superoxide complexes of increasing oxidative power at physiological pH.

A number of non-redox-active metals have been shown by electron paramagnetic resonance (EPR) to form superoxide–metal complexes under anhydrous conditions.[32] The oxidative power of these complexes is linearly related to the binding energy between the metal and $O_2^{-•}$ which, in turn, reflects both the Lewis acidity of the metal and the ionic radii of metals within the same oxidation state (Figure 3).[33]

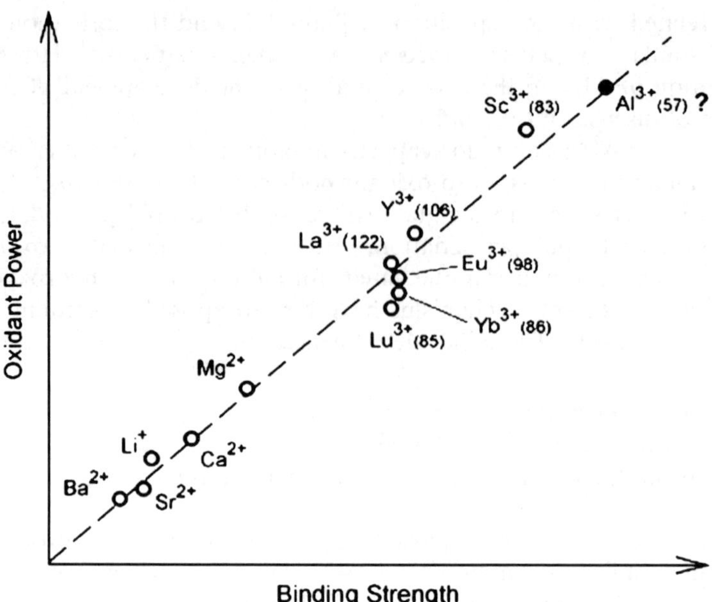

Figure 3 *Metal–superoxide complexes generated under anhydrous conditions and iden-tified by EPR observe a strikingly linear relationship between binding strength and oxidizing power of the complex. The putative $AlO_2^{\bullet 2+}$ is predicted to form the strongest complex with the highest oxidizing power*
(Reprinted with permission from ref. 33.)

Metal–superoxide complexes have long been known to be involved in bio-logical oxidation,[34] and a number of oxo-iron species have been implicated, though not definitely demonstrated, in classic superoxide-driven Fenton chemistry. In the same way, formation of a putative aluminum superoxide semi-reduced radical ion, $AlO_2^{\bullet 2+}$ might explain the pro-oxidant activity of aluminium, as well as bridge between the catalytic activity of this non-redox-active metal and both superoxide-driven and iron-driven biological oxidation.

Scarce but convincing evidence has been collected about the stimulation by aluminum of non-iron, superoxide-driven biological oxidations.[33,34] As an exam-ple, Fridovich *et al.* showed that the rate of oxidation of NADH by either photochemically or enzymatically generated $O_2^{-\bullet}$ was increased threefold in the presence of aluminum.[33] Superoxide dismutase inhibited only the aluminum-induced effect and not the initial rate of oxidation whereas catalase did not influence either reaction rate. The conclusion was that the catalytic influence of aluminum indicated *"the formation of an oxidizing complex from Al^{3+} plus $O_2^{-\bullet}$"*.

The efficiency of non-redox active trivalent metals to catalyze the oxidation of NADH *via* $O_2^{-\bullet}$ mirrored the Lewis acidity of their real and proposed super-oxide–metal complexes. Thus, for those metal–superoxide complexes that have already been identified under anhydrous conditions (La, Y and Sc), Sc^{3+}, which had the highest affinity for $O_2^{-\bullet}$, also showed the highest rate of oxidation of

NADH and the highest amount of melanin-induced lipid peroxidation. The superoxide complexes of Al^{3+} and Ga^{3+} have not yet been identified, but they may be predicted to be more stable than the complex with Sc^{3+} and they are both significantly more potent catalysts of $O_2^{-\bullet}$-mediated biological oxidation.

The mechanism by which aluminum increases the rate of oxidation of NADH by $O_2^{-\bullet}$ remains to be elucidated. In the absence of aluminum the oxidation of NADH was inhibited neither by SOD nor by catalase. Probable oxidants might include molecular oxygen, O_2, and the perhydroxyl radical $(HO_2^{\bullet})$.

Since oxidation of NADH was not evident in the absence of generated $O_2^{-\bullet}$ and its basal rate of oxidation in the presence of $O_2^{-\bullet}$ was independent of the presence of either SOD or catalase (the activity of both of which would generate O_2) this would seem to discount O_2 as the oxidant. However, the $HO_2^{\bullet}$ radical is 10^4 times more effective an oxidant of NADH, than superoxide. If the formation of this radical was sufficiently rapid with respect to the dismutation of $O_2^{-\bullet}$ by SOD then this could account for the SOD/catalase independent basal rate of oxidation of NADH. In the presence of aluminum the rate of oxidation of NADH was increased up to threefold and the increase in rate was inhibited by SOD. The following interaction could account for the promotion by aluminum:

$$2O_2^{-\bullet} + H^+/Al^{3+} \rightarrow HO_2^{\bullet} + AlO_2^{\bullet 2+}$$

As for iron-driven biological oxidation, there are many reports documenting the facilitation by aluminum. The chemistry involved in iron-driven biological oxidation can be summarized as follows:

$$Fe^{2+} + O_2 \rightleftharpoons Fe^{3+} + O_2^{-\bullet}$$
$$2O_2^{-\bullet} + 2H^+ \rightarrow H_2O_2 + O_2$$
$$Fe^{2+} + H_2O_2 \rightarrow OH^- + OH^\bullet + Fe^{3+}$$

The relatively large volume of research into this aspect of the pro-oxidant activity of aluminum has resulted in the elucidation of many different agonists and antagonists of this effect, including the physical and chemical properties of the substrate and the pH of the environment in which the oxidation event is taking place.[35] The suggested mechanisms of the pro-oxidant activity of aluminum can be divided into those which address an influence of aluminum on (1) the oxidizable substrate (*e.g.*, lipids), (2) another pro-oxidant metal (*e.g.*, iron) or (3) the oxidant (*e.g.*, $O_2^{-\bullet}$).

Since the first demonstration of the acceleration by aluminum of the stimulation of lipid peroxidation by iron salts,[31] it was suggested that aluminum produced "*an alteration in membrane structure that facilitates lipid peroxidation*". A number of non-redox active trivalent metals promote iron-stimulated lipid peroxidation, being the catalytic efficiency of these metals in order of their increasing oxidative power: $Ga < Al < Sc < Y < La$.[36]

Each of these trivalent metals is also capable of increasing lipid packing in both liposomal and membrane preparations in an order mirroring exactly the order of their catalytic efficiencies in promoting oxidative damage. It is evident that a direct relationship there occurs between the ability of a metal to induce lipid packing and its promotion of iron-stimulated lipid peroxidation. The suggestion that metal-induced increases in lipid packing might make a membrane more susceptible to damage *via* free radicals must also be open to debate as it is, for example, well known that a number of structural antioxidants, such as cholesterol, achieve their effect by increasing the rigidity of the membrane. While different agonists may induce changes in membrane structure *via* different mechanisms, it is clear that membrane rigidification *per se* is not sufficient to promote oxidative damage.

It is generally believed that the external charge on a membrane or liposome preparation is important to the efficiency of metal-promoted iron-stimulated lipid peroxidation, being the catalytic efficiency of aluminum limited to membranes bearing a negative external charge. The interactions of Al^{3+} with membrane phospholipid polar headgroups occur through the formation of both, *cis* and *trans* complexes. *Trans* interactions, that result in membrane aggregation and fusion, are not particularly involved in the stimulation of lipid oxidation. By contrast, due to *cis* interactions within the liposomes, Al^{3+} induces the rearrangement of negatively charged phospholipids in the lateral phase of the bilayer leading to the formation of microdomains enriched in acidic phospholipids.[37] In this process, phospholipids adopt an energetically favorable disposition in the bilayer, releasing water molecules from their headgroups.[38] This newly adopted disposition causes the immobilization of the phospholipids involved in metal binding, affecting not only the polar region of the membrane, but also the hydrophobic core. Given the fact that brain phosphatidylserine (the major acidic phospholipid) is highly enriched in poly-unsaturated fatty acids, Al^{3+} causes a local accumulation of oxidizable substrates. Both, the local enrichment of unsaturated fatty acids, as well as their immobilization, can lead to a higher rate of propagation of lipid oxidation (Figure 4).[39]

Another possible mechanism that could be involved in Al toxicity is the alteration of certain cell-signaling cascades. Studies performed both *in vitro* and *in vivo* reported that Al^{3+} causes a decrease in the hydrolysis of phosphatidylinositol biphosphate (PIP_2).[40–42] This metabolic pathway is mediated by a phosphatidylinositol-specific phospholipase C (PI-PLC) that renders inositol triphosphate (IP_3) and diacylglycerol. These two molecules are important intracellular-second messengers. While diacylglycerol activates enzymes such as protein kinase C (PKC), IP_3 is responsible for the intracellular mobilization of Ca^{2+} with the subsequent triggering of multiple signals.

The inhibitory effect of Al^{3+} on PIP_2 hydrolysis could be through the interaction between the metal and at least, one of the major components of this signaling system: the receptor, the enzyme PI-PLC and the substrates. Until now, there is no experimental evidence supporting a plausible inhibitory effect of Al^{3+} on the membrane receptor or on the G protein associated to the enzyme. Recently, it has been shown that (i) *in vitro* Al^{3+} binds to the polar headgroup of phosphoinositides, mainly to those containing one or more phosphate residues at the inositide molecule and (ii) the binding of Al to

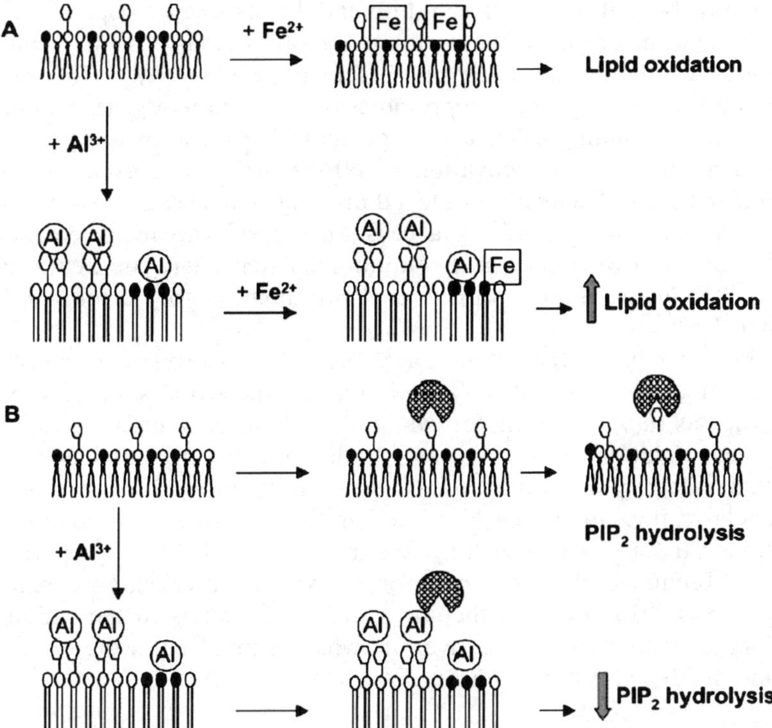

Figure 4 *Potential mechanisms involved in (A) Al-pro-oxidant action and (B) Al-induced alterations in PIP₂ hydrolysis. Negatively charged phospholipids, phosphatidyl inositol, phosphatidyl inositol-specific phospholipase C* (Reproduced with permission from ref. 39.)

PIP_2 results in a net increase in membrane surface potential.[43] In addition, through the formation of *cis* bondings with other PIP_2 molecules in the surroundings, Al creates microenvironments enriched in these phospholipids that limit the accessibility of the enzyme PI-PLC to its substrates (Figure 4). Interestingly, when membrane integrity is disrupted by the action of a non-polar detergent (Triton X-100), the activity of the enzyme is fully recovered, even when Al remains bound to the membrane. This finding argues against the hypothesis of a direct interaction between Al and the catalytic site of the enzyme. The inhibitory effect of Al^{3+} on PIP_2 hydrolysis by PI-PLC can alter several PIP_2-dependent processes. For example, the exocytosis of neurotransmitter-containing vesicles is mediated by CAPS, a calcium-dependent activated protein for secretion that depends on PIP_2 binding for its activity.[44] Al^{3+} could also interfere with phospholipase D activity, since this enzyme also depends on PIP_2 to be fully active,[45] and could also influence the cytoskeleton organization through the alteration of PIP_2 binding with actin-binding proteins.

The oxidant-responsive transcription factor NF-κB has been recently investigated in terms of its involvement in Al-mediated neurodegeneration. The activation of NF-κB by Al^{3+} has a differential consequence depending on the cell type (neurons or glial cells). In human glioblastoma cells, Al^{3+} caused an increase in the levels of tumor necrosis factor-α (TNF-α).[46] This cytokine activates NF-κB,[47] which in turn promotes the activation of genes that mediate inflammation, including other cytokines, inducible nitric oxide synthase and complement factors. The activation of NF-κB leads to a cycle of cytokine secretion and NF-κB activation which, if prolonged, may cause neuronal death, and the proliferation of reactive glial cells. These results are in agreement with a recent finding showing that in a rotation-mediated aggregate neural culture system, Al^{3+} increases the production of pro-inflammatory cytokines TNF-α and MIP-1α.[48]

On the other hand, the activation of NF-κB in neurons may constitute a cytoprotective-signaling pathway that induces the expression of protective genes, such as those encoding for calbindin and antioxidant enzymes.[49]

Cell accumulation of Al^{3+} affects both, mitochondria and endoplasmic reticulum, integrity and functionality. Al^{3+} enters into the neuron following cell depolarization, inhibiting Na^+/Ca^{2+} exchange, with the accumulation of mitochondrial Ca^{2+}. Consequently, the mitochondrial transition pores open, with the subsequent release of cytochrome c, which triggers caspase 9-mediated cell apoptosis.[50] In addition, the stress that Al^{3+} causes to the endoplasmic reticulum leads to caspase-12 activation, which is involved in the endoplasmic reticulum-mediated cell death by apoptosis.[51]

6.2 Aluminum and the Etiology of Alzheimer's Disease

The *Al hypothesis* for the etiology of AD originated when it was found that the injection of Al^{3+} salts into the brain of rabbits produced neurofibrillary degeneration,[52] and that the Al^{3+} concentration is increased in the brain of AD patients.[53] However, a controversy developed regarding the role of Al^{3+} in the etiology of this disease because subsequent experiments produced conflicting results.[54,55] The absence of aluminum in plaque cores, however, was not considered to contradict the Al hypothesis,[56] since Al was shown to be more often associated with the neurofibrillary tangles than with the plaques,[57] as demonstrated by employing the Laser Microprobe Mass Analysis (LAMMA) technique.[58]

Among the evidence in support of the Al hypothesis, minuscule insoluble Al^{3+} silicate granules, like tiny grains of sand, were found in the brain of patients with AD.[59] The granules were surrounded by equally insoluble amyloid protein plaques, leading to suggestion that the granules may represent an early or initiating factor in plaque formation. Binding of Al^{3+} to paired-helical filament (PHF) of tau protein in AD neurofibrillary degeneration was also demonstrated.[60]

The distribution of aluminum in the brain is not deeply known and limited studies have been done. However, it seems that Al mainly localizes in temporal cortex and in hippocampus, two regions that are also known to be significantly involved in AD.[61] At the subcellular level, the distribution of aluminum is more selective to lipofuscin, cytosolic, mitochondrial, lysosomal and nuclear compartments.[62] In the nucleus Al^{3+} is strongly bound to DNA.[63] Moreover, it has also been shown that astrocytes are more sensitive to Al^{3+} binding compared to neuronal cells.[64]

As already stated, the brain Al^{3+} content increases with age and AD patients display an almost threefold increased absorption of Al compared to healthy controls. The possible pathways leading to such Al^{3+} accumulation in AD brain are summarized in Figure 5.[65]

As previously stated, intraneuronal neurofibrillary tangles (NFT), made of phosphorylated proteins aggregated within the neuronal cytoplasm, are the most consistent post-mortem characteristic of AD. NFT are composed of PHFs of tau proteins. These are bound by microtubules in healthy neurons and are essential for the form and function of the neuronal cytoskeleton. Aluminum is believed to be involved in the formation of PHF.[66] Binding of Al^{3+} to PHF may involve the interaction with phosphate groups of hyper-phosphorylated tau proteins (Figure 6).[67]

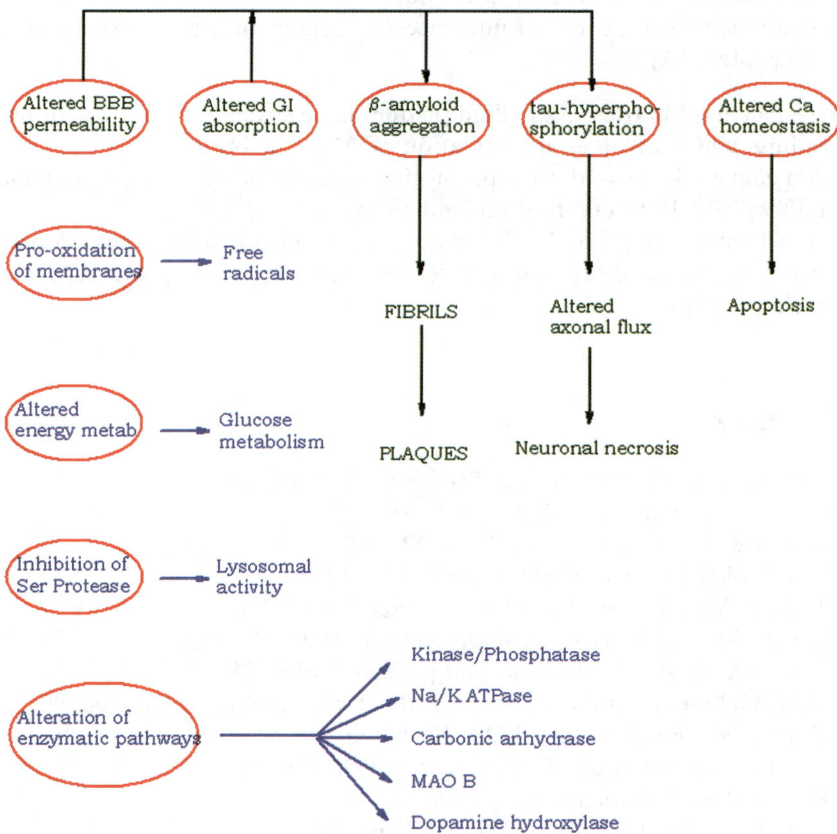

Figure 5 *Some biochemical and biophysical effects of aluminium* (Adapted with permission from ref. 67.)

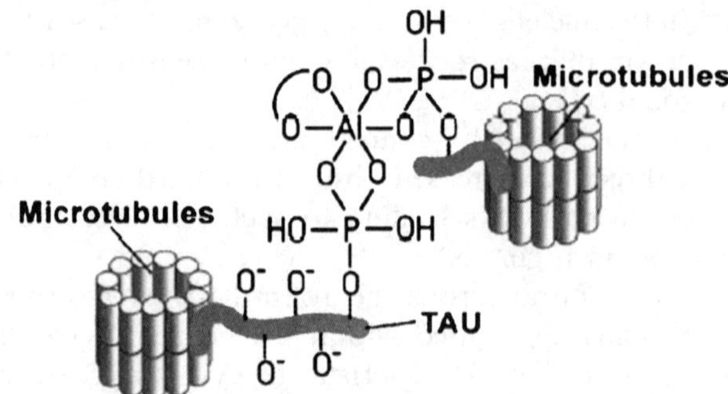

Figure 6 *Hypothetical interaction between Al^{3+} and phosphate groups of the hyper-phosphorylated tau proteins in AD*
(Reproduced with permission from ref. 67.)

By the way, the Al^{3+}-ATP complex mimics the glutamate in stimulating glutamate-receptor activity which determines an enhanced level of neuronal tau and self-assembly of free tau (*i.e.*, not bound to microtubules) into PHF.

Aluminum is also known to influence the aggregation and toxicity of the amyloid protein (Aβ).

- An increased burden of amyloid plaques is observed in patients with renal failure, which involves accumulation of Al in the brain.
- Al behaves like Fe and Zn in promoting aggregation of Aβ in physiological buffers with 100–1000 fold rate enhancements.
- Al accumulated in the AD brain accelerates the amyloidogenetic processing of the Amyloid Precursor Protein by activating serine proteases such as α-chymotrypsin.

References

1. P.P. Ganrot, *Environ. Health Perspect.*, 1986, **65**, 363.
2. T.P. Flaten, *Brain Res. Bull.*, 2001, **55**, 187.
3. A. Lione, *Pharmacol. Ther.*, 1985, **29**, 255.
4. R.A. Yokel, *Neurotoxicology*, 2000, **21**, 813.
5. A. Sanz-Medel *et al.*, *Coord. Chem. Rev.*, 2002, **228**, 373.
6. R.T.A. MacGillivray *et al.*, *Biochemistry*, 1998, **37**, 7919.
7. R.W. Harris and J. Sheldon, *Inorg. Chem.*, 1990, **29**, 119.
8. M.H. Gelb and D.C. Harris, *Arch. Biochem. Biophys.*, 1980, **200**, 93.
9. Y. Li *et al.*, *Biochemistry*, 1998, **37**, 14157.
10. J.G. Grossmann *et al.*, *J. Mol. Biol.*, 1993, **229**, 585.
11. R.A. Yokel, *Neurotoxicology*, 2000, **21**, 813.
12. R.A. Yokel *et al.*, *Toxicol. Sci.*, 2001, **64**, 77.
13. J. Edwardson, *Aluminum in Chemistry, Biology and Medicine*, Raven Press, New York, 1991.

14. R.A. Yokel *et al.*, *Brain Res.*, 2002, **930**, 101.
15. R. Deloncle *et al.*, *Brain Res.*, 2002, **946**, 247.
16. N.C. Danbolt, *Prog. Neurobiol.*, 2001, **65**, 1.
17. K. Nagasawa *et al.*, *Toxicol. Lett.*, 2005, **155**, 289.
18. M.S. Golub *et al.*, *Neurotoxicology*, 1999, **20**, 961.
19. M. Yasui *et al.*, *Rinsho Shinkeigaku*, 1991, **31**, 1095.
20. N. Ward and J. Mason, *J. Radioanal. Nucl. Chem.*, 1987, **113**, 515.
21. D. Pratico *et al.*, *Faseb J.*, 2002, **16**, 1138.
22. P. Zatta *et al.*, *Coord. Chem. Rev.*, 2002, **228**, 271.
23. J.M.C. Gutteridge *et al.*, *Biochim. Biophys. Acta*, 1985, **835**, 441.
24. S.V. Verstraeten and P.I. Oteiza, *Arch. Biochem. Biophys.*, 2000, **375**, 340.
25. S. Kong *et al.*, *Free Radical Biol. Med.*, 1992, **13**, 79.
26. E. Mendez-Alvarez *et al.*, *Biochim. Biophys. Acta*, 2002, **1586**, 155.
27. R. Gerschman *et al.*, *Science*, 1956, **119**, 623.
28. J.M. McCord and I. Fridovich, *J. Biol. Chem.*, 1969, **244**, 6049.
29. J.M.C. Gutteridge and B. Halliwell, *Ann. NY Acad. Sci.*, 2000, **899**, 136.
30. J.M. Gebicki and B.H. Bielski, *J. Am. Chem. Soc.*, 1981, **103**, 7020.
31. E.J. Land and A.J. Swallow, *Biochim. Biophys. Acta*, 1971, **234**, 34.
32. S. Fukuzumi *et al.*, *Coord. Chem. Rev.*, 2002, **226**, 71.
33. C. Exley, *Free Radical Biol. Med.*, 2004, **36**, 380.
34. C. Walling, *Acc. Chem. Res.*, 1975, **8**, 125.
35. C.X. Xie and R.A. Yokel, *Arch. Biochem. Biophys.*, 1996, **327**, 222.
36. S.V. Verstraeten and P.I. Oteiza, *Arch. Biochem. Biophys.*, 1995, **322**, 284.
37. S.V. Verstraeten *et al.*, *Arch. Biochem. Biophys.*, 1997, **338**, 121.
38. S.V. Verstraeten and P.I. Oteiza, *Arch. Biochem. Biophys.*, 2000, **375**, 340.
39. P.I. Oteiza *et al.*, *Mol. Aspects Med.*, 2004, **25**, 103.
40. T.J. Shafer *et al.*, *Brain Res.*, 1993, **629**, 133.
41. T.J. Shafer and W.R. Mundy, *Gen. Pharmacol.*, 1995, **26**, 889.
42. A.C. Nostrandt *et al.*, *Toxicol. Appl. Pharmacol.*, 1996, **136**, 118.
43. S.V. Verstraeten and P.I. Oteiza, *Arch. Biochem. Biophys.*, 2002, **408**, 263.
44. R.P.H. Huijbregts *et al.*, *Traffic*, 2000, **1**, 195.
45. V.A. Sciorra *et al.*, *EMBO J.*, 1999, **18**, 5911.
46. A. Campbell *et al.*, *Brain Res.*, 2002, **933**, 60.
47. K. Lieb *et al.*, *J. Neurochem.*, 1996, **67**, 2039.
48. V.J. Johnson and R.P. Sharma, *Neurotoxicology*, 2003, **24**, 261.
49. Y. Christen, *Am. J. Clin. Nutr.*, 2000, **71**, 621S.
50. J. Savory *et al.*, *J. Inorg. Biochem.*, 2003, **97**, 151.
51. O. Ghribi *et al.*, *Brain Res. Mol. Brain Res.*, 2001, **96**, 30.
52. I. Klatzo *et al.*, *J. Neuropathol. Exp. Neurol.*, 1965, **24**, 187.
53. D.R. Crapper *et al.*, *Science*, 1973, **180**, 511.
54. D.G. Munoz, *Arch. Neurol.*, 1998, **55**, 737.
55. J.P. Landsberg *et al.*, *Nature*, 1992, **360**, 65.
56. P.F. Good and D.P. Perl, *Nature*, 1993, **362**, 418.
57. P. Zatta, *Trace Elem. Med.*, 1993, **10**, 120.
58. P.F. Good *et al.*, *Ann. Neurol.*, 1992, **31**, 286.
59. J.M. Candy *et al.*, *Lancet*, 1986, **1**, 354.

60. H. Murayama *et al.*, *Am. J. Pathol.*, 1999, **155**, 877.
61. J.R. McDermott, *Neurology*, 1979, **29**, 809.
62. J.H. Schuurmans Stekhoven *et al.*, *Neurosci. Lett.*, 1990, **119**, 71.
63. W.J. Lukiw *et al.*, *Neurobiol. Aging*, 1991, **13**, 115.
64. K.S. Jagannatha Rao, *Biochem. Int.*, 1992, **28**, 51.
65. V.B. Gupta *et al.*, *Cell. Mol. Life Sci.*, 2005, **62**, 143.
66. J. Savory and R.M. Garritto, *Nutrition*, 1998, **313**, 314.
67. P. Zatta *et al.*, *Brain Res. Bull.*, 2003, **62**, 15.

CHAPTER 7
Manganese in the Brain Functioning

7.1 Introduction

Among the redox-active metals that are essential trace elements, manganese is the most poorly understood. This is due to the low concentration of manganese in living tissues. In many human and animal tissues, manganese concentration is below 1 µg g^{-1} wet weight. Manganese is involved in the metabolism of proteins, lipids and carbohydrates, and serves as a co-factor for enzymes such as decarboxylases, hydrolases and kinases.[1] Because Mn^{2+} resembles Mg^{2+} in some physicochemical properties, a number of enzymes can substitute magnesium for manganese in their activation *in vitro*. Mitochondrial superoxide dismutase is a known manganese metalloprotein and exists ubiquitously.[2] Glutamine synthetase (GS) is a glia-specific manganese metalloprotein in the brain.[3] Mn dependent enzymes are essential for survival. Loss of expression of the Mn-binding superoxide dismutase, as shown by knockout mice,[4,5] leads to a severe deleterious phenotype and is lethal.

Specific transport proteins in the blood and specific uptake mechanisms for cells are unknown. Although Mn can be toxic at high doses and accumulation of excess amounts in the brain results in neurodegeneration, there is little evidence for a major role of the metal in any widespread neurodegenerative disease. Despite the apparent lack of specific regulatory mechanisms, Mn concentrations are kept low and cellular- and blood-Mn levels show little variation. Manganese is known to be essential for the development and functioning of the brain.[6] Manganese concentration in the human brain is higher in adults (approximately 0.25 µg g^{-1} wet weight) than in infants less than one year of age, suggesting that manganese is required for brain functions. Dietary manganese deficiency might affect manganese homeostasis in the brain, as evidenced by increased susceptibility of manganese-deprived rats to convulsions. Recent evidence points to a possible role in the prion diseases.[7,8] However, even here there is little evidence that elevated levels of Mn are anything other than a surrogate marker despite a possible role altering protein conformation by substitution for Cu.

The high-redox potential of Mn suggests that it can participate in the same kind of oxidative reactions as Cu or Fe. Indeed, this metal has 11 possible oxidation states and its outer shell can donate up to seven electrons. In animals, Mn exists in three possible forms Mn^{2+}, Mn^{3+} and Mn^{4+}.[9] When oxidized to the higher oxidation states, Mn readily precipitates in aqueous solutions. It is therefore likely that oxidized Mn only exists bound to specific ligands or chelators in biological systems. Therefore, it is likely that despite our lack of knowledge, there are likely to be specific proteins in cells and in the blood that are essential for the tight regulation of Mn activity in the body. It also implies, that despite our lack of knowledge of an active role of Mn in neurodegeneration, Mn has the potential to play an important role in neurodegenerative diseases. This could be due to simple, unchecked oxidative damage caused by Mn released from necrotic cells or due to altered availability of Mn and substitution of Mn for other metals as it has been shown for prion disease.[8] This chapter explores Mn metabolism and the potential of this metal to play a role in neurodegeneration.

7.2   Manganese Absorption

Manganese toxicity (manganism) is a rare CNS disorder that results from chronic high-dose exposure to Mn containing dust or fumes. Although manganese can be absorbed through digestion, the percentage of Mn in food that is absorbed is relatively low. Mn poisoning can occur through sources such as potassium permanganate. Unlike the metabolism of Cu and Fe, the biology of this redox-active metal is poorly understood. Disturbances that could alter Mn absorption through the gut are therefore unknown. The largest variability in manganese absorption is through the lungs and elevated Mn levels are most likely to arise through respiratory absorption. In general, Mn in the air comes from the combustion of fossil fuels and in particular from an additive to petrol known as methylcyclopentadienyl manganese tricarbonyl (MMT). However, the largest source of Mn pollution in the air results from industrial sources such as the ferromanganese industry. Mn can also enter the air from erosion of soil and rocks and is likely to be at high levels in dry areas prone to erosion. Even in urban areas, the levels are around 13 ng Mn L^{-1} in air.[10] Miners are often exposed to much higher levels. The lowest level known to cause manganism was 150 µg Mn L^{-1}. However, levels in mines can reach 1–5 mg Mn L^{-1}. Mn in soil is around 40–900 mg Mn kg^{-1}. Environmental sources of Mn in the air could potentially generate levels approaching this, but there is currently no evidence that such exposure could initiate manganism or another Mn-related disease. Levels of Mn in water are much higher than air (1–200 µg Mn L^{-1}). Absorption from water is less likely than absorption from food as there are higher concentrations in food than water. Vegetarian diets are higher in Mn than average with the daily intake being around 2–6 mg Mn and up to 11 mg Mn for vegetarians. The highest source of Mn in the diet are nuts, but consumption of large amounts of tea leads to increased Mn in dietary content. However, the amount of Mn absorbed from the diet is limited by the activity of the gastrointestinal tract and the liver.[11] The amount absorbed could be as little as 1% of that ingested.[12] However, increased dietary load will result in increased gastric absorption of Mn.[13] In some cases, toxicity can occur through high levels in drinking water.[14,15]

The other consideration that limits Mn absorption by any means is bioavailability. This can be due to either complexation of the metal or its valence state in particular salts. Some forms of Mn are not soluble and are therefore unlikely to be absorbed. In general, Mn absorption occurs for Mn in the divalent state. For dietary sources, this is not a consideration as the metal is already in an exchangeable state, but for air- or water-borne sources, this consideration is important. In this case, Mn that is the most weakly bound will be the most bioavailable.[16]

Mn deficiency is more likely than Mn toxicity or disease related to excess Mn absorption and retention by the body. The consequences of Mn deficiency are quite varied but can include, death in infants, reduced cognitive performance, loss of fertility and altered lipid and carbohydrate metabolism.[17,18] In animals, Mn deficiency is often observed in the form of changes in joints, deformed legs and bones. This is a result of altered proteoglycan biosynthesis and glycosylation that changes the form of cartilage.[19] Rats deprived of Mn have paroxysmal convulsion[20] indicating that Mn is essential for normal brain activity.[21]

Mn plays a vital role in both fertilization and gestation. During pregnancy, the levels of Mn absorbed rise dramatically. In some cases, hypermaganesemia can occur during pregnancy as a result of excessive absorption of Mn.[22] During pregnancy, there is also a higher level of Mn in the neonate. Levels of Mn drop after birth and stabilize at a lower level at 3–6 months of age in humans.[22,23] Increased Mn absorption can also occur under other conditions. In particular, Fe deficient individuals absorb increased levels of Mn.[24] This may be related to commonly utilized absorption pathways for both Fe and Mn.

7.3   Manganese Transport to the Brain

After absorption through digestion, Mn is transported to the liver *via* the portal vein. Possibly, divalent Mn is oxidized *via* interaction with ceruloplasmin. However, it is unclear whether this occurs in the plasma or in the liver.[25] The liver is the main depot for Mn but to what extent this is important is unclear. From here, Mn can be transported to the brain.[26] Absorption through the lungs also enters the blood supply, but from this source it can be transported to the brain directly.

The blood–brain barrier is important for the transport of essential trace elements into the brain from blood. Before the formation of the barrier in development, shortly after birth, manganese can easily enter the brain. Even after the formation of the barrier, it remains relatively leaky to manganese entry into the brain.[27] In adult life, brain permeability to Mn is much higher than for Fe and Zn.[28] Therefore, Mn can easily be concentrated in the brain and there is a potential for Mn to reach toxic concentrations due to less tight regulation of its transport. Unlike other metals, blood levels of Mn are not necessarily indicative of levels in the brain.

There is evidence from the use of radioactive Mn[54] that Mn applied into the nasal cavity is taken up through the olfactory epithelium and into the olfactory bulb.[29] From there, Mn can enter the brain. In the rat, the olfactory bulb is quite large. In humans, this area is much smaller and therefore presents only a small chance for direct passage of Mn into the brain. As well as entry *via* olfactory mucosa, significant amounts of Mn enter *via* absorption through the respiratory mucosa. This represents a very large surface area for the absorption of Mn. MMT is an organic manganese (Mn) compound added to unleaded petroleum. It has been suggested that the combustion products of MMT containing Mn, such as manganese phosphate, could cause neurological symptoms similar to Parkinson's disease in humans. When rats were chronically exposed to inhaled Mn phosphate, it was found that levels in the brain and the lung increased in relation to the level of exposure[30] with level in the lungs increasing up to 50 times at an exposure of 3000 mg m^{-3}. Such a significant change was not accompanied by any sign of neurodegeneration or altered behavior.

Mn in the blood exists as both the divalent and trivalent forms.[31] The relative affinity of these forms for ligands could result in specific transport across the blood–brain barrier. Oxidation of Mn to the trivalent form probably results from the activity of ceruloplasmin.[32] Transferrin is a plasma carrier of trivalent metals. It has been shown to bind Mn[33] and potentially this protein could bind Mn in the oxidized form and transport it to the brain.[34] Transferrin receptors are present on the surface of brain capillary endothelial cells.[35] This would suggest that Mn would enter the brain by a similar mechanism to Fe. Unfortunately, the method of entry of Fe is also not clearly defined. Two models have been put forward. One theory suggests that transferrin is stripped of metal in the endothelial cells and returned to the blood as apo-transferrin.[36] Then the metal is transported around the brain by an unknown mechanism. The second theory suggests that transferrin and the metal cross the endothelial cells together.[37]

Intravenous injection of Mn in the blood results in a proportion of this Mn entering the brain rapidly but remaining there only temporarily.[38] This is likely to be due to non-transferrin-mediated mechanisms that are currently unknown. Unlike trivalent Mn, the divalent form is largely free ($\sim$60%) and not associated with proteins or other ligands.[39] Studies with radioactive Mn showed much of the free Mn enters the brain within an hour after injection.[40] The divalent metal transporter (DMT-1) is also expressed on brain capillary endothelial cells and is currently the only candidate for a transporter of this free Mn into these cells.[41] However, this process of entry is poorly studied and so it cannot be concluded that this or any other protein is directly involved in Mn entry into the brain.

Studies with radioactive Mn[54] give some indication of the place of the highest levels of Mn accumulation within the brain. Six days after injection, the highest levels were found in the superior olivary complex, inferrior colliculi and red nuclei.[42,43] Mn also accumulates in the substantia nigra and striatum.[26,44]

Accumulation of Mn^{54} from transferrin bound Mn^{54} was much lower than when free or buffered Mn^{54} was used. Some amount of Mn enters the cerebral spinal fluid (CSF) and there is accumulation of Mn in the chorid plexus.[45] Transferrin in the CSF appears to sequester Mn and prevent its transport into brain parenchymal cells. The reason for this is unclear.

7.4 Manganese Uptake in the Brain

Understanding manganese metabolism is essential to understand its potential to damage the CNS. In particular, manganese transport into cells is poorly understood. Therefore, like the discussions above, most of the data concerning uptake of Mn is scant and potential mechanisms are largely speculative. Neurons have the potential to take up Mn bound to transferrin by virtue of the expression of the transferrin receptor.[46] Transferrin is secreted from oligodendrocytes providing a possible source of Mn transfer with in the brain.[47] Although transferrin might mediate uptake of Mn by neurons, a significant amount of Mn is taken up by astrocytes in the absence of transferrin.[48] Uptake of Mn by neurons or other cells could be mediated by DMT-1.

DMT-1 is an alternate name for NRAMP2, a member of a family of proteins that transport metals (Figure 1). NRAMP stands for natural resistance-associated macrophage protein and this family of proteins belongs to a large family of integral membrane proteins that are highly conserved and structurally similar.[49,50] In mammals there are two proteins, NRAMP1 and NRAMP2. NRAMP2 has two different splice variants. One of these possesses an iron response element (IRE) not present in the other splice variant.[51] In mammals, neither protein is selective for a particular metal.[52] However, given the higher concentration of Fe than other divalent metals, then the main metal that they transport is Fe. However, in the absence of other identified Mn transporters, NRAMP2 or DMT-1 remains the prime candidate for a Mn-transporting protein. A family of similar proteins has been identified in yeast termed Smf proteins.[53] These are also DMT-1s but they show greater selectivity than their mammalian homologues. Smf1p appears to be more selective for Fe uptake into the cell, while Smf2p is more specific for Mn. A third protein, Smf3p appears to play a role in metal transport within the cell.[54] NRAMP2 has been shown to transport Fe from the plasma membrane into the labile pool of Fe.[55] Other metals that DMT-1 can transport include Cd, Zn, Co, Ni, Cu and Pb.[40] The two isoforms of DMT-1 have different subcellular localizations.[56] In particular, the isoform lacking the IRE is found associated with the nucleus in neurons. DMT-1 with the IRE is mostly associated with vesicles within the cell and is found to colocalize with transferrin in recycling endosomes.[57] Therefore DMT-1 could represent the main way that Mn enters the cell and also the cytoplasm.

NRAMP1 appears to be expressed in the endoplasmic reticulum of neurons and extrudes metals into the cytoplasm.[58] Although it could also transport Mn, there is little direct evidence that it transports Mn into the cytoplasm.

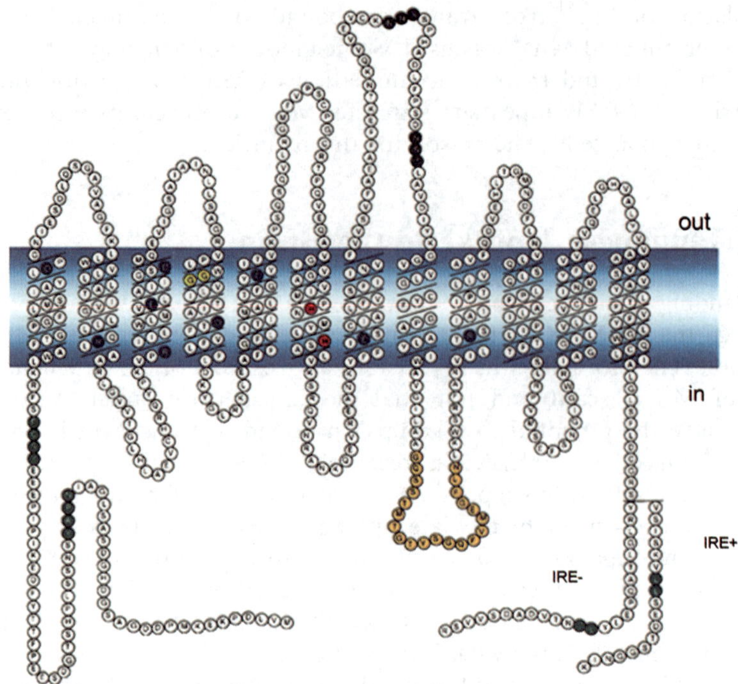

Figure 1   *Structure of NRAMP-2. A diagram of the way DMT-2 or NRAMP-2 is associated with the membrane of cells. There are two alternative splice variants of NRAMP-2. These differ in terms of the presence of a IRE. The two variants are shown as the one with the element (IRE+) and the one without (IR−)*

Mn transport into the mitochondria is mediated by the same mechanisms that transport Ca. Thus uptake into mitochondria is *via* the Ca^{2+} uniporter. During uptake, the Mn is believed to be in the divalent state. Mn transport out of mitochondria is *via* the slow Na-independent efflux mechanism, which is dependent on ATP. Within the mitochondria, Mn that remains in the divalent state can inhibit respiration by interfering with oxidative phosphorylation. This could potentially decrease cellular viability.[59]

Proteins that mobilise both Ca and Mn are also utilized in the other parts of the cell. Hailey–Hailey disease is an autosomal, dominant, blistering skin disease. The disease is caused by mutations in the gene *ATP2C1*.[60,61] The protein encoded by the gene is a Ca^{2+} transporting ATPase localized to the Golgi.[62] This protein also transports Mn. It is a necessary part of the secretary pathway and the regulation of its activity controls the activity of the Golgi in terms of protein sorting and glycosylation. A homologue of this protein, which exists in yeast (PRM1) can also transport Mn. The mammalian homologue can complement in the yeast system showing the same activity as PRM1.[63] Therefore, in terms of its role in regulating Ca movement and the correct activity of the secretory pathway, it is a "house keeping" gene. It is one of a family of proteins termed SPCAs (secretory pathway Ca transporters). As high-affinity

Mn transporters, the SPCAs would be the main way that Mn enters the Golgi,[64] but at a much slower rate than for Ca. SPCA can therefore transport both Ca and Mn with a similar affinity. A similar protein family, the SERCA pumps, allows entry into the endoplasmic reticulum but are far more selective for Ca.[62] It is unknown if SPCA-2 can transport Mn as all evidence relates to SPCA-1. It is currently unknown if SPCA-1 plays a role in detoxification of Mn. Potentially, this protein could aid in removal of excess Mn from cells by removing Mn from the cytoplasm into the secretory pathway. In yeast, PMR1 is believed to play this role. In PMR-1 mutants, Mn accumulates in the cytoplasm of yeast.[65] Surprisingly, these mutants that accumulate Mn show increased resistance to oxidative damage, but this is possibly because of increased incorporation of Mn into Mn-SOD.[66]

One of the essential requirements of cells for Mn is the enzyme galactosyl transferase. This enzyme requires Mn to glycosylate all N- and O-glycosylated glycoproteins.[67,68] Severe depletion of Mn from cells inhibits glycosylation.[69] The implication is that alterations in Mn levels could alter the glycosylation of proteins. Currently, there is no evidence of altered glycosylation playing a significant role in an Mn-associated disease.

7.5 Proteins that Utilize Manganese

The two main proteins in the body that utilize manganese are the mitochondrial specific superoxide dismutase (Mn-SOD or SOD-2) found ubiquitously in the body[70] and GS, a glial specific protein of the brain.[71] Another astrocytic enzyme that requires Mn is pyruvate carboxylase.[72] This implies that Mn is needed for cellular activities that include, energy metabolism, resistance oxidative stress and removal of potential neurotoxic molecules (Figure 2).

Analysis of Mn in mitochondria suggests that it mostly exists in the divalent form.[73] However, Mn-SOD, the major Mn-binding protein in the mitochondrion utilizes the redox cycling of the metal to catalyse superoxide dismutation. It is a very highly conserved enzyme.[74] In the absence of other SODs, this enzyme is able to compensate to provide significant protection. It is synthesized from the nuclear genome and transferred to the mitochondria from the ER and probably binds Mn in the ER directly following synthesis. The protein is tetrameric formed from four identical sub-units, which can all bind one atom of Mn. As mentioned above, knockout of Mn-SOD expression is lethal.[4,5] Homozygous mutant mice die between the first 10 days and 3 weeks of life with a dilated cardiomyopathy, accumulation of lipid in liver and skeletal muscle, and metabolic acidosis. Cytochemical analysis revealed a severe reduction in succinate dehydrogenase (complex II) and aconitase (a TCA cycle enzyme) activities in the heart and, to a lesser extent, in other organs. These findings indicate that Mn-SOD is required for normal biological function of tissues by maintaining the integrity of mitochondrial enzymes susceptible to direct inactivation by superoxide. When Mn-SOD knockout mice are treated with a Mn-SOD mimetic that is unable to cross the blood–brain barrier (5, 10, 15,

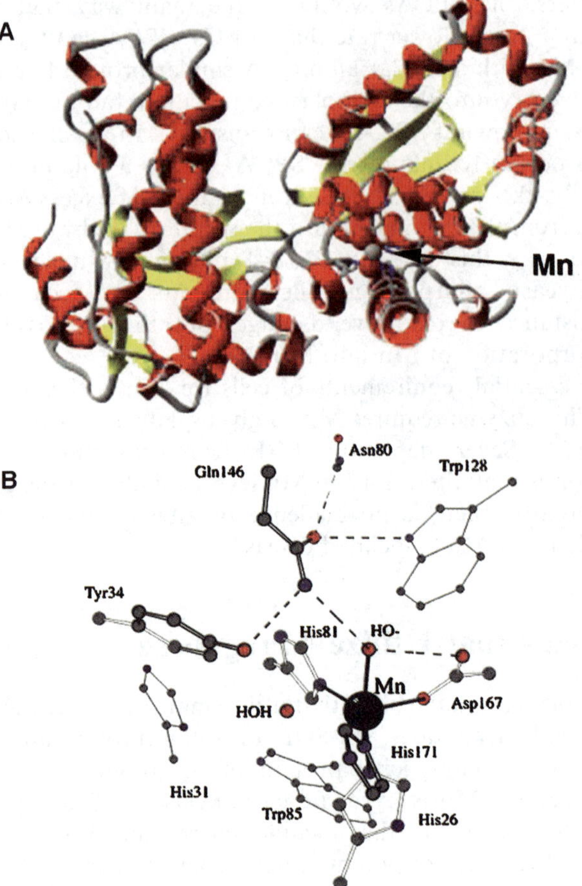

Figure 2 *Mn superoxide dismutase. (**A**) A ribbon diagram of a single subunit of Mn superoxide dismutase. The co-ordination site of the Mn atom is indicated by an arrow. (**B**) Details of the Mn co-ordination site*

20-tetrakis [4-benzoic acid] porphyrin), the mice are able to survive.[75] The mice instead develop a pronounced movement disorder progressing to total debilitation by three weeks of age. Neuropathologic evaluation reveals a striking spongiform degeneration of the cortex and specific brain stem nuclei associated with gliosis and intramyelinic vacuolization similar to that observed in cytotoxic edema and disorders associated with mitochondrial abnormalities, such as Leighs disease and Canavans disease. This shows the essential role of Mn-SOD to normal function of the nervous system.

The Mn- and Fe-co-factored SODs share extensive similarity in protein architecture,[76] and are structurally distinct from the Cu/Zn (and Ni) enzymes. Mn-SOD and Fe-SOD have an extremely broad phylogenetic distribution, being expressed in both prokaryotic and eukaryotic cells. The enzymes are typically isolated as dimers or tetramers (dimers of dimers) consisting of identical subunits. Each subunit is composed of two domains: a predominantly α-helical *N*-terminal domain and a mixed α/β *C*-terminal domain. The metal-binding site (Figure 3)[77] lies on the interface between these two domains in the interior of the protein, formed from ligands arising from both *N*-terminal (His-26, His-81) and *C*-terminal (Asp-167, His-171) regions.

The a₂ homodimer (represented by the *Escherichia coli* Mn-SOD structure) has 2-fold symmetry and is stabilized by an extended hydrophobic surface at the subunit interface. The intersubunit contacts include two residues (Glu-170, Tyr-174) that cross over in a "double bridge" motif to make structurally important hydrogen bonds with the complementary subunit.[78] In the tetrameric SODs (including the eukaryotic mitochondrial Mn-SOD and many thermophilic and hyperthermophilic Mn- and Fe-SODs), the intersubunit interactions are more complex, with additional contact regions between dimers.

In the active site of *E. coli* Mn-SOD, the metal ion is coordinated by four amino acid side chains and a buried solvent molecule (interpreted as H_2O or HO^-, depending on the oxidation state of the metal ion) that completes a nearly exact trigonal bipyramidal co-ordination complex. The organization of the inner co-ordination sphere is nearly identical between Mn- and Fe-SODs, and dominates the electronic properties of the active site. In addition to these inner-sphere interactions, the outer sphere of the complex plays important roles in its structure and catalytic function (Figure 4).[78]

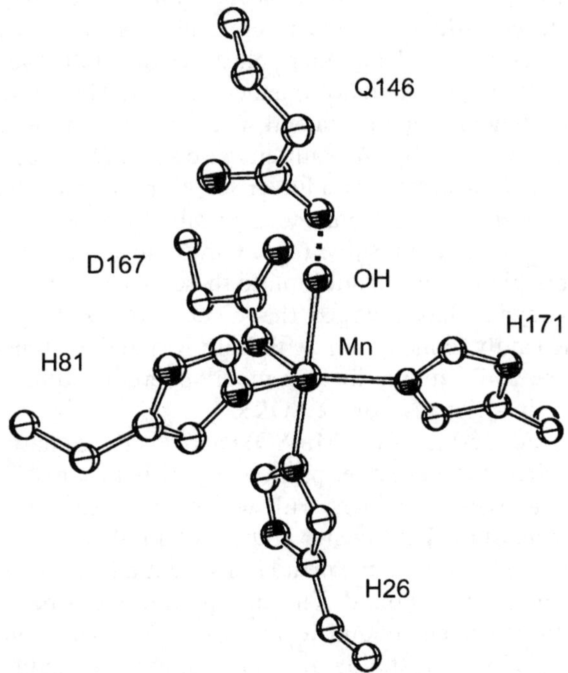

Figure 3 *Active site of Mn-SOD*
(Reprinted with permission from. ref. 77.)

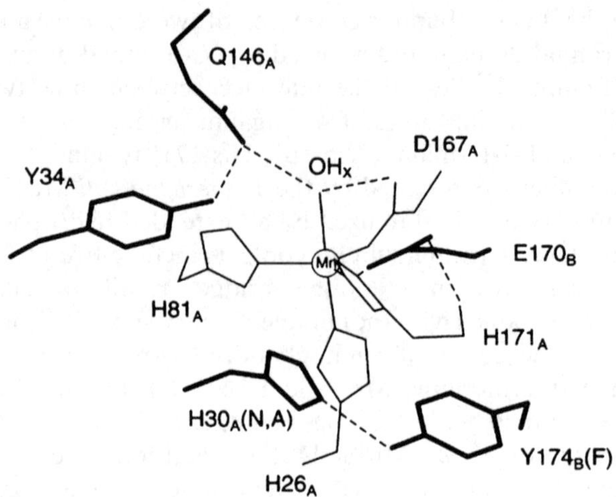

Figure 4 *Active-site structure of E. coli Mn-SOD. Outer-sphere and gateway residues are highlighted in bold*
(Reprinted with permission from ref. 78.)

A pair of "gateway" residues (His-30 and Tyr-34, in *E. coli* Mn-SOD sequence numbering) lie in front of the equatorial histidine ligands (His-81 and His-171) at the base of the substrate access funnel, restricting access to the metal center. The co-ordinated solvent molecule also makes an important hydrogen-bonding contact with an outer-sphere residue (Gln-146 in *E. coli* Mn-SOD) that modulates the reactivity of the metal center. This residue is the most important structural determinant of metal specificity within the (Mn, Fe) SOD family of enzymes, with the metal requirement correlating with the secondary structural region from which the residue arises. For enzymes that require Mn for activity, this residue arises from the *C*-terminal domain, whereas all Fe-SODs utilize an outer-sphere residue (*e.g.*, Gln-69 in *E. coli* Fe-SOD) arising within the *N*-terminal domain. In addition to these two canonical subfamilies, a third group of enzymes has emerged that is characterized by relatively low specificity for the catalytic metal ion, with both Mn and Fe functioning in the active site.[79] These SODs may utilize an outer-sphere histidine in place of the glutamine present in strict Mn- or Fe-SODs.

Substitution of Fe for Mn in the Mn-SOD active site dramatically decreases the catalytic activity of the enzyme, particularly at higher pH.[80] This unusual pH-sensitivity of activity correlates with a similar sensitivity of the optical absorption spectrum of the Fe^{3+} center in the protein, that has been interpreted in terms of hydroxide binding to the metal ion, essentially a "rusting" of the site when the wrong metal ion is bound. This interpretation has been confirmed by direct observation of solvent bound to the active-site metal ion in Fe-substituted Mn-SOD by X-ray crystallography, providing a structural explanation for the metal specificity of Mn-SOD.[81] This structural perturbation also affects the redox potential of the bound metal ion.[82]

The astrocytic protein GS plays a dual role in the brain as an enzyme that regulates the release of the neurotransmitter glutamate and also it is involved in the pathways that detoxifies ammonia. It also binds approximately 80% of all the Mn in the brain.[83] One of the most important roles of astrocytes in the brain is to protect neurons against cytotoxicity of various agents. Two such compounds are ammonia and glutamate. Glutamate released by neurons *via* normal neurotransmission can be captured by astrocytes through specific transporters. Within astrocytes, it is then converted back to glutamine.[84] Glutamine can then be recycled back to neurons where it is required for the synthesis of the neurotransmitter. As the reaction catalyzed by GS also consumes ammonia, then detoxification of ammonia is a by-product of glutamine synthesis. In situations where there is hyperammonemia, astrocytes show an increased activity of this enzyme.[85] Hyperammonemia is primarily responsible for hepatic encephalopathy.[86] Many other treatments cause an increase in the activity of GS, including hypoxia and ischemia[87,88] and is decreased in Alzheimer's disease and glucose deprivation.[89,90] Cellular responses to increased glutamate are mediated through a number of receptors. One of these is the *N*-methyl-D-aspartate (NMDA) receptor. It has the effect of inhibiting GS *via* nitric oxide signaling.[91] Therefore, it is possible that excess activation of NMDA receptors could induce ammonia toxicity. As glutamine synthase requires Mn for its activity, then Mn deficiency could also cause increased ammonia toxicity. However, there is currently no evidence that this occurs.

The enzyme GS is a key enzyme controlling the use of nitrogen inside cells. Glutamine, as well as being used to build proteins, delivers nitrogen atoms to enzymes that build nitrogen-rich molecules, such as DNA bases and amino acids. So, GS, the enzyme that builds glutamine, must be carefully controlled. When nitrogen is needed, it must be turned on so that the cell does not starve. But when the cell has enough nitrogen, it needs to be turned off to avoid a glut.

GS acts like a tiny molecular computer, monitoring the amounts of nitrogen-rich molecules. It watches levels of amino acids like glycine, alanine, histidine and tryptophan, and levels of nucleotides like AMP and CTP. If too much of one of these molecules is made, GS senses this and slows down production slightly. But as levels of all of these nucleotides and amino acids rise, together they slow GS more and more. Eventually, the enzyme grinds to a halt when the supply meets the demand.

The structure of GS has been determined by X-ray crystallography. The initial X-ray structural studies were on GS samples purified from a mutant *Salmonella typhimurium* strain, unable to adenylylate GS, to avoid heterogeneity from covalent modification. *S. typhimurium* GS has a molecular mass of 620 kDa. The large size and the complex regulation patterns of GS stem from its central role in cellular nitrogen metabolism. It brings nitrogen into metabolism by condensing ammonia with glutamate, with the aid of ATP, to yield glutamine.

Bacterial GS molecules are dodecamers formed from two face-to-face hexameric rings of subunits, with 12 active sites formed between monomers. Each active site can be described as a "bifunnel" in which ATP and glutamate bind at opposite ends (Figure 5).[92]

The ATP binding site is found at the top of the bifunnel, because it opens to the external 6-fold surface of GS. At the joint of the bifunnel are two divalent cation binding sites, n1 and n2, separated by 6 Å, where either magnesium or manganese bind for catalysis. The n2 ion is involved in phosphoryl transfer, while the n1 ion stabilizes an active GS and plays a role in binding glutamate. The affinity for metal ions at the n1 site is 50 times greater than at the n2 site. This is caused by greater negative charge toward the bottom half of the bifunnel in the vicinity of n1.

The GS molecule from bacteria is composed of 12 identical subunits, each of which has an active site for the production of glutamine. When performing its reaction, the active site binds to glutamate and ammonia, and also to an ATP molecule that powers the reaction. But, the active sites also bind weakly to other amino acids and nucleotides, partially blocking the action of enzyme. All of the many sites communicate with one another, and as the concentrations of competing molecules rise, more and more of the sites are blocked, eventually shutting down the whole enzyme. The cell also has a more direct approach

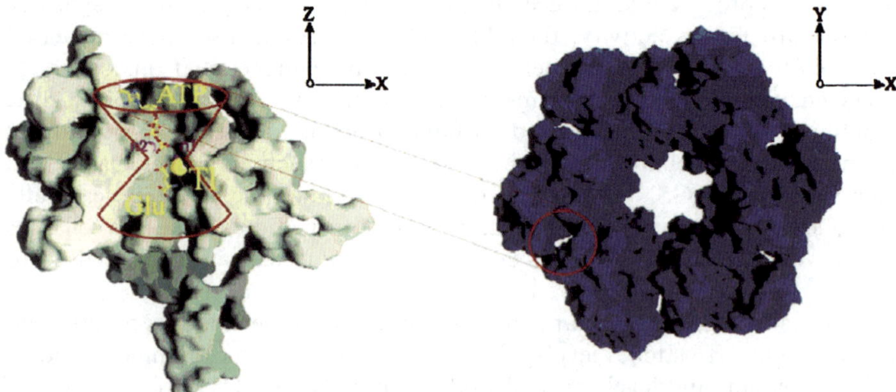

Figure 5   *Structure of bacterial GS. Bacterial glutamine synthetase is a dodecamer having 622 symmetry, with six 2-fold axes perpendicular to a six-fold axis. One of the two eclipsed hexameric rings is shown in blue. The dimensions of the dodecamer including the side chains are approximately 100 Å along the six-fold axis and 140 Å along a two-fold axis. The dodecamer has 12 active sites, one formed between every two neighboring subunits within a ring (shown by red circle). Each active site is a bifunnel, having entrances at the top and bottom for substrates ATP and glutamate, respectively. The green structure represents the C-terminal domain (residues 101–468) of one subunit. The molecular six-fold axis (Z) is shown to the right of the subunit. The bifunnel is about 30 Å wide at its opening, 15 Å wide at its middle and 45 Å deep. The two metal ions (n1 and n2) are 6 Å apart, located at the neck of the bifunnel. The location of the ammonium substrate has been determined from its Tl$^+$ analog in Tl$^+$–GS complexes. The distance between the n1 and Tl$^+$ sites is 4 Å, and between n2 and Tl$^+$ is 7 Å*
(Reprinted with permission from ref. 92.)

when it wants to shut down the enzyme. At a key tyrosine next to the active site, an ADP molecule can be attached to the protein, completely blocking its action.

GS catalyzes glutamine biosynthesis *via* the reaction:

$$\text{Glutamate} + \text{NH}_4^+ + \text{ATP} \xrightarrow{\text{M}^{2+}} \text{Glutamine} + \text{ADP} + \text{P}_i + \text{H}^+$$

where M can be magnesium or manganese. The reaction has been termed the "biosynthetic" reaction and is considered the most physiologically relevant reaction that GS catalyzes.

A two-step model for the mechanism of the biosynthetic reaction has emerged:

(i) formation of the activated intermediate γ-glutamyl phosphate. The n2 ion coordinates the γ-phosphate oxygens of ATP to allow phosphoryl transfer to the γ-carboxylate group of glutamate, yielding the intermediate; and

(ii) attack on the intermediate by ammonia, which releases free phosphate to yield glutamine.

On the basis of crystal structures of GS-effector complexes, a tentative enzymatic mechanism of Gln synthesis has been proposed:

(i) ATP and Glu bind to GS in order.
(ii) Glu attacks the gamma-phosphorus atom of ATP to produce γ-glutamyl phosphate and ADP.
(iii) ADP induces movement of Asp50$'$ enhancing binding of the third substrate, an ammonium ion. Deprotonation of the ammonium ion by Asp50$'$ permits ammonia to attack the γ-glutamyl phosphate forming a tetrahedral intermediate, which stabilize residues (324–328), covering the path of Glu entry.
(iv) Phosphate leaves and one proton from the γ-amino group of the tetrahedral intermediate is lost.
(v) The absence of interaction between Glu327 and Gln permits residues (324–328) to open, allowing Gln to leave the active site.

We make several versions of GS in our own cells. Most of our cells make a version similar to the bacterial one, but with 8 subunits instead of 12. Like the bacterial enzyme, it is controlled by the nitrogen-rich compounds down the synthetic pipeline. We also make a second GS in our brain. There, glutamate is used as a neurotransmitter, and GS is used when the glutamate is recycled after a nerve impulse is delivered. In the brain, GS is in constant action, so a highly regulated version is not appropriate. Instead, the alternate form is active all the time, continually performing its essential duty.

There are some new candidates for Mn-binding proteins. One of these has been identified as a homologue of the Mam3p. This protein is expressed in yeast and plays a significant role in the protection of yeast cells from Mn deficiency. The homologue in mammals is ancient conserved domain protein (ACDP-1). The function of this protein is unknown although mutations are linked to diseases such as Ochoa syndrome and urofacial syndrome. However, the clear link between Mam3p and Mn metabolism[93] implies that ACDP-1 could be potentially important in mammalian Mn metabolism.

Earlier it was mentioned that individuals with manganism exhibit symptoms similar to Parkinson's disease. Therefore, it is not surprising that a protein associated with Parkinson's disease, Parkin, has recently been suggested to play a role in defense against Mn toxicity.[94] Inherited forms of Parkinson's disease are linked to point mutations in the *Parkin* gene. Expression of Parkin in the dopaminergic cell line SH-SY5Y was found to protect those cells from Mn toxicity. It is currently unknown how this occurs, but it is possible that Parkin plays some role in Mn metabolism.

7.6 Manganese Neurotoxicity

Manganese has the potential to be more toxic than any other divalent metal. The brain is more permeable to Mn than to Fe[28] and Mn can easily be concentrated in certain brain regions, especially the basal ganglia.[95] Cu is tightly regulated with many specific mechanisms to control its level and is not found free in the body. Similarly, Fe has specific transport system and only a small proportion is free in cells. In comparison, there are virtually no specific Mn-transporting proteins and a large percentage of Mn is apparently free in biological systems. The implication is that when Mn is elevated, it can have catastrophic consequences. Mn is a pro-oxidant and direct toxic effects of Mn on dopaminergic neurons have been observed.[96,97] Within the *substantia nigra*, high Mn causes a decrease in antioxidant enzymes that clear hydrogen peroxide.[98] As well as generating radical by Fenton chemistry, Mn, like Fe, can cause the breakdown of dopamine to 6-hydroxydopamine.[99] Such metabolites are highly toxic and might explain the specific destruction of dopaminergic neurons in manganism. However, studies in which dopamine levels were repressed did not show a significant reduction in their sensitivity to Mn toxicity.[100] The implication is that there is currently no direct evidence that dopaminergic neurons are more sensitive to Mn toxicity than other neurons. Therefore, it is possible that Mn toxicity only relates to the types of cells that accumulate high levels rather than a cell-specific effect.

Mn toxicity to neurons may have specific characteristics. In particular, it has been noted that Mn causes reduction in the levels of hydrogen peroxide degrading enzymes, such as catalase and thiol containing compounds such as glutathione.[9,101] Mn-induced cell death can be inhibited by antioxidants such as *N*-acetyl cysteine.[100] Although, high levels of toxicity can be observed in many models of Mn-induced cell death, the exact method of cell execution, as with many aspects of Mn biology, remains elusive.

7.7 Effects on Brain Function

Manganese, like a number of other trace elements including Cu, is released into the synaptic cleft during neurotransmission. The amount released is low but likely to be higher than that found in CSF (0.83–1.50 mg L^{-1}).[102] There is evidence that Mn released in this way could modulate neurotransmission. Altered responses of a variety of neurotransmitter systems have been shown. These include the neuromuscular synapse[103] and synapse releasing GABA, glutamate or aspartate in the CNS.[104–107]

Mn is present at high levels in the striatum and *substantian nigra*. Mn is axonally transported within the GABAnergic striato-nigral and dopaminergic nigro-striatal pathways.[26] It is not clear if Mn is released at the synapses connecting these regions or not. However, Mn is present at synapses and is often released with neurotransmitters.[108] Mn is released from the mitral cell termini in the piriform cortex and the amygdala.[109] Mn can block voltage-dependent calcium channels.[110] High Mn can also alter other cellular systems. Mn causes significant down-regulation of the glutamate uptake system in astrocytes by inhibiting transcription of GLAST transporter.[111] This change has the potential of enhancing excitotoxcity.

Manganese deficiency does not occur often in animals or humans. Deficiency can cause problems in growth and skeletal development and reduce fertility.[10,17] Dietary Mn deficiency seems to alter Mn homeostasis in the brain. Offspring of Mn deficient rats are often ataxic. Mn-deficient rats are more susceptible to seizures induced by electroshock.[112] Blood Mn levels in epileptic patients are lower than normal.[113,114] Blood levels of Mn remain tightly regulated and low blood Mn is highly predictive of decreased brain Mn levels. This possibly implies that alterations in Mn homeostasis in the brain, is associated with susceptibility to epileptic seizures.

7.8 Neurological Disease Related to Manganese

The main way in which Mn affects the activity of the brain is through toxicity or deficiency. In addition to this is the emerging potential of altered Mn metabolism to play a role in well-known neurodegenerative diseases. The two diseases that are directly relevant to this discussion are prion diseases (transmissible spongiform encephalopathies) and Parkinson's disease. Both of these diseases and the role of metals in their etiology are dealt with elsewhere in this book. The most common of these two diseases is Parkinson's disease. In particular, Mn-related brain damage often results in what is termed "Parkinson's-like" symptoms. This is probably because both diseases affect similar regions of the brain. In other words, the neuronal damage associated with manganism or Parkinson's disease is similar. In contrast, the neuronal death resulting from prion disease is not associated with the neurotoxic effects of Mn. In prion disease, Mn binds to the prion protein altering its conformation. Chelation of Mn in scrapie-infected mice results in a significant reduction in the level of abnormal prion protein occurring in the disease (M. Brazier, personal communication). It is unclear, if this interaction between Mn and the prion protein initiates the disease. However, Mn has been found to be accumulated in the brains of scrapie-infected mice well before the onset of clinical signs of disease.[8] The brains of patients with Creutzfeldt–Jakob disease show high levels of Mn.[115] Whether or not this deposition of Mn is causative or a symptom of prion disease remains unknown. However, the potential of Mn to exacerbate accumulation of the abnormal, disease-associated isoform of the prion protein implies that elevated Mn levels are a clear risk factor for the prion disease. Another condition that has been related to Mn is chronic hepatic encephalopathy.[116]

The main Mn toxicity syndrome in humans is manganism, which has similarities to Parkinson's disease largely because a similar part of the brain is affected. The syndrome is observed in those who have high exposure to airborne Mn. These include miners, ferroalloy and battery manufacturers, automotive mechanics, fungicide manufacturers and welders.[117–121] It can also occur in patients that have had nutritional supplements applied by parentral administration.[122] Manganism causes dystonia, a neurological sign associated with damage to the globus pallidus. Patients have reduced movement or bradykinesia and widespread rigidity in the body. Some patients have a tremor

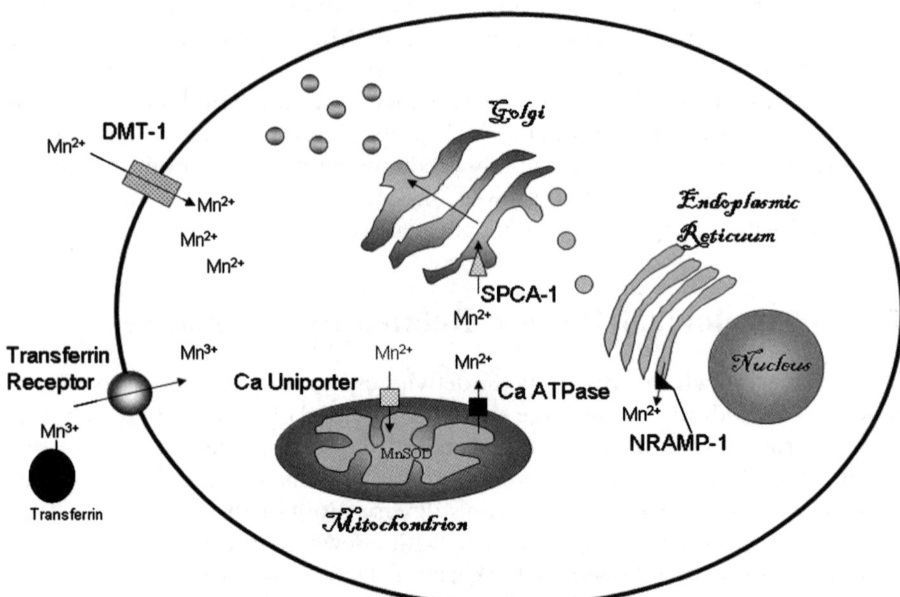

Figure 6 *Mn metabolism in cells. Mn enters cells either through DMT-1 or via tranferrin binding to the transferrin receptor. Mn with cells can exist either as the Mn²⁺ or Mn³⁺, but the majority is believed to exist as Mn²⁺. Mn can enter the mitochondria by the Ca uniporter and exit via the Ca ATPase. Excess Mn can possibly be released through the golgi by the Mn transporter SPCA-1. It is unclear what the function of NRAMP-1 is but one possibility is that it is involved in release of Mn from the endoplasmic reticulum*

when at rest and often fall backwards. Treatments with dopamine analogues as used for the treatment of Parkinson's disease do not result in any improvement. Treatment with Mn chelators at an early stage is usually able to reverse the condition. Whereas. Parkinson's disease results from target destruction of dopaminergic neurons, manganism results from the deposition of toxic Mn in particular brain regions. It is possibly just co-incidental that the basal ganaglia is the main site of neurodegeneration in both diseases. Despite the frequent references and discussion, there is no evidence that Mn plays any role in Parkinson's disease.

7.9 Conclusions

Despite the potential of Mn to cause neurological damage understanding the potential role of Mn in neurodegeneration is limited by the poor understanding of Mn metabolism and the dearth of studies on Mn in activities of neurons (Figure 6). Mn has been linked to a small number of neurodegenerative diseases but even here, evidence for a role remains thin. Manganism is the only true neurodegenerative disorder that is clearly linked to Mn. However, this disease is easily recognized, low level and can be treated if diagnosed early. There have been scares about potential increase in manganism because of the rise of the use of Mn additives in petroleum but there is little evidence for any danger. Currently, in the age where much research funding drives researchers to make their work "relevant" to human health, Mn is not a major issue. The only exception to this rule is prion disease where the possible link to Mn needs to be further explored.

References

1. F.C. Wedler, Biological significance of manganese in mammalian systems, in *Progress in Medicinal Chemistry*, G.P. Ellis and D.K. Luscombe (eds), vol 30, Elsevier, Amsterdam, 1993, 89–133.
2. R.A. Weisiger and I. Fridovich, Mitochondrial superoxide dismutase. Site of synthesis and intramitochondrial localization, *J. Biol. Chem.*, 1973, **248**, 4793–4796.
3. M.D. Norenberg, The distribution of glutamine synthetase in the central nervous system, *J. Histochem. Cytochem.*, 1979, **27**, 469–475.
4. Y. Li *et al.*, *Nature Genet.*, 1995, **11**, 376.
5. R.M. Lebovitz *et al.*, *Proc. Natl. Acad. Sci. USA*, 1996, **93**, 9782.
6. J.R. Prohaska, Function of trace elements in brain metabolism., *Physiol. Rev.*, 1987, **67**, 858–901.
7. D.R. Brown *et al.*, *EMBO J.*, 2000, **19**, 1180.
8. A.M. Thackray *et al.*, *Biochem. J.*, 2002, **362**, 253.
9. F.S. Archibald and C. Tyree, *Arch. Biochem. Biophys.*, 1987, **256**, 638.
10. D.G. Barceloux, *Clin. Toxicol.*, 1999, **37**, 293.

11. M.E. Anderson *et al.*, *Neurotoxicology*, 1999, **20**, 161.
12. J.E. Furchner *et al.*, *Health Phy.*, 1966, **12**, 1415.
13. E.M. Abrams *et al.*, *J. Anim. Sci.*, 1976, **42**, 630.
14. R. Kawamura *et al.*, *Kitsato Atch. Exp. Med.*, 1941, **18**, 145.
15. X.G. Kondakis *et al.*, *Arch. Environ. Health.*, 1989, **44**, 175.
16. K.V. Ragnarsdottir and L. Charlet, in *Prion Disease and Copper Metabolism*, D.R. Brown (ed), Horwood Publishing, Chichester, 2002, 156.
17. L.S. Hurley, *Philos. Trans. R. Soc. Lond. B*, 1981, **294**, 145.
18. E.J. Underwood, *Trace Elements in Human and Animal Nutrition*, Academic Press, New York, 1977, 13.
19. J.L. Greger, *Neurotoxiciology*, 1999, **20**, 205.
20. L.S. Hurley *et al.*, *Am. J. Physiol.*, 1963, **204**, 493.
21. A. Takeda, *Brain Res. Rev.*, 2003, **41**, 79.
22. A. Spencer, *Nutrition*, 1999, **15**, 731–734.
23. S. Hatano *et al.*, *J. Pediatr. Gastroenterol. Nutr.*, 1985, **4**, 87–92.
24. A.B.R. Thomson, D. Oltunbosun and L.S. Valberg, *J. Lab. Clin. Med.*, 1971, **78**, 642.
25. R.A. Gibbons *et al.*, *Biochim. Biophys. Acta*, 1976, **444**, 1.
26. A. Takeda *et al.*, *Brain Res. Bull.*, 1998, **45**, 149.
27. J. Daubing, *Prog. Brain Res.*, 1968, **29**, 417.
28. N. Sotogaku, N. Oku and A. Takeda, *J. Neurosci. Res.*, 2000, **61**, 350.
29. G. Gianutsos, G.R. Morrow and J.B. Morris, *Fundam. Appl. Toxicol.*, 1997, **37**, 102.
30. L. Normandin *et al.*, *Toxicol. Appl. Pharmacol.*, 2002, **183**, 135.
31. A.M. Scheuhammer and M.G. Cherian, *Biochem. Pharmacol.*, 1985, **34**, 3405.
32. M. Aschner and J.L. Aschner, *Brain Res. Bull.*, 1990, **24**, 857.
33. L. Davidsson *et al.*, *J. Nutr.*, 1989, **119**, 1461.
34. P. Aisen, P. Aasa and A.G. Redfield, *J. Biol. Chem.*, 1969, **244**, 4628.
35. W.A. Jefferies *et al.*, *Nature*, 1984, **312**, 162.
36. R.L. Roberts, R.E. Fine and A. Sandra, *J. Cell Sci.*, 1993, **104**, 521.
37. L. Descamps *et al.*, *Am. J. Physiol.*, 1996, **270**, H1149.
38. V.A. Murphy *et al.*, *J. Neurochem.*, 1991, **57**, 115.
39. P.M. May, P.W. Linder and D.R. Williams, *Experimenta*, 1974, **32**, 1442–1444.
40. O. Rabin *et al.*, *J. Neurochem.*, 1993, **61**, 50.
41. H. Gunshin *et al.*, *Nature*, 1997, **388**, 482.
42. A. Takeda *et al.*, *Brain Res.*, 1994, **640**, 341.
43. A. Takeda, S. Ishiwatari and S. Okada, *J. Neurosci. Res.*, 2000, **59**, 542.
44. W.N. Sloot and J.B.P. Gramsbergen, *Brain Res.*, 1994, **657**, 124.
45. A. Takeda, J. Sawashita and S. Okada, *Brain Res.*, 1994, **658**, 252.
46. J.R. Connor, *J. Neurosci. Res.*, 1990, **27**, 595.
47. J.R. Connor, *Dev. Neurosci.*, 1994, **16**, 233.
48. A. Takeda, A. Devenyi and J.R. Connor, *J. Neurosci Res.*, 1998, **51**, 454.
49. M. Cellier *et al.*, *Proc. Natl. Acad. Sci. USA*, 1995, **92**, 10089.
50. M. Tabuchi *et al.*, *Biochem. J.*, 1999, **344**, 211.

51. S. Gruenheid *et al.*, *Genomics*, 1995, **25**, 514.
52. J.R. Forbes and P. Gos, *Trends Microbiol.*, 2001, **9**, 397.
53. M.E. Portnoy, X.F. Liu and V.C. Culotta, *Mol. Cell Biol.*, 2000, **20**, 7893.
54. A. Cohen, N. Nelson and N. Nelson, *J. Biol. Chem.*, 2000, **275**, 33388.
55. V. Picard *et al.*, *J. Biol. Chem.*, 2000, **275**, 35738.
56. J.A. Roth *et al.*, *J. Neurosci.*, 2000, **20**, 7595.
57. S. Gruenheid *et al.*, *J. Exp. Med.*, 1999, **189**, 831.
58. C.A. Evans *et al.*, *Neurogenetics*, 2001, **3**, 69.
59. C.E. Gavin, K.K. Gunter and T.E. Gunter, *Neurotoxicology*, 1999, **20**, 445.
60. Z. Hu *et al.*, *Nature Genet.*, 2000, **24**, 61.
61. R. Sudbrak *et al.*, *Hum. Mol. Genet.*, 2000, **9**, 1131.
62. K. Van Baelen *et al.*, *Biochim. Biophys. Acta*, 2004, **1742**, 103.
63. V.-K. Ton *et al.*, *J. Biol. Chem.*, 2002, **277**, 6422.
64. M. Chiesi and G. Inesi, *Biochemistry*, 1980, **19**, 2912.
65. P.J. Lapinskas *et al.*, *Mol. Cell Biol.*, 1995, **15**, 1382.
66. E.E. Luk and V.C. Culotta, *J. Biol. Chem.*, 2001, **276**, 47556.
67. A. Varki, *Trends Cell Biol.*, 1998, **8**, 34–40.
68. J.T. Powell and K. Brew, *J. Biol. Chem.*, 1976, **251**, 3645.
69. R.J. Kaufmann, M. Swaroop and P. Murtha-Riel, *Biochemistry*, 1994, **33**, 9813.
70. R.A. Weisiger and I. Fridovich, *J. Biol. Chem.*, 1973, **248**, 4793.
71. M.D. Norenberg, *J. Histochem. Cytochem.*, 1979, **27**, 756–762.
72. M.C. Scrutton and A.S. Mildvan, *Biochemistry*, 1968, **7**, 1490.
73. T.E. Gunter *et al.*, *J. Neurochem.*, 2004, **88**, 266.
74. J.W. Whittaker, *Biochem. Soc. Trans.*, 2003, **31**, 1318.
75. S. Melov *et al.*, *Nature Genet.*, 1998, **18**, 159.
76. M.S. Lah *et al.*, *Biochemistry*, 1995, **34**, 1646.
77. J.W. Whittaker, *Biochem. Soc Trans.*, 2003, **31**, 1318.
78. R.A. Edwards *et al.*, *Biochemistry*, 2001, **40**, 4622.
79. S. Sugio *et al.*, *Eur. J. Biochem.*, 2000, **267**, 3487.
80. M.M. Whittaker and J.W. Whittaker, *Biochemistry*, 1997, **36**, 8923.
81. R.A. Edwards *et al.*, *J. Am. Chem. Soc.*, 1998, **120**, 9684.
82. C.K. Vance and A.-F. Miller, *J. Am. Chem. Soc.*, 1998, **120**, 461.
83. F.C. Wedler and R.B. Denman, *Curr. Top. Cell. Regul.*, 1984, **24**, 153.
84. L. Herz *et al.*, *Neurochem Res.*, 1978, **3**, 1.
85. A.J. Zamora, J.B. Cavanagh and M.H. Kyu, *J. Neurol. Sci.*, 1973, **18**, 25.
86. R.A. Hawkins and J. Jessy, *Biochem. J.*, 1991, **277**, 697.
87. C.K. Petito *et al.*, *Brain Res.*, 1992, **569**, 275.
88. J.A. Kelleher *et al.*, *Neurochem.*, 1994, **19**, 209.
89. N. Seiler, *Neurochem. Res.*, 1993, **18**, 2.
90. F. Rosier *et al.*, *Biochem. J.*, 1996, **315**, 607.
91. E. Kosenko *et al.*, *Neurochem. Int.*, 2003, **43**, 493–499.
92. D. Eisenberg *et al.*, *Biochim. Biophys. Acta*, 2000, **1477**, 122.
93. M. Yang *et al.*, *Biochem. J.*, 2005, **386**, 479.
94. Y. Higashi *et al.*, *J. Neurochem.*, 2004, **89**, 1490.

95. H. Kabata *et al.*, *Nippon Eiseigaku Zasshi*, 1989, **44**, 667.

96. V. Anantharam *et al.*, *J. Neurosci.*, 2002, **22**, 1738.

97. M. Parenti *et al.*, *Brain Res.*, 1987, **473**, 236.

98. G. Cohen and R.E. Heikkila, *J. Biol. Chem.*, 1974, **249**, 2447.

99. D.G. Graham *et al.*, *Mol. Pharmacol.*, 1978, **14**, 644.

100. D.L. Stredrick *et al.*, *Neurotoxicology*, 2004, **25**, 543–553.

101. J.J. Liccione and M.D. Maines, *J. Parm. Exp. Ther.*, 1988, **247**, 156.

102. G.C. Cotzias and P.S. Papavasiliou, *Nature*, 1962, **195**, 823.

103. I. Meiri and R. Rahamimoff, *Science*, 1972, **176**, 308.

104. K. Kostial *et al.*, *Br. J. Pharmacol.*, 1974, **51**, 231.

105. H. Kita, K. Narita and W. Van der Kloot, *Brain Res.*, 1981, **205**, 121.

106. S.P. Butcher, P.J. Roberts and J.F. Collins, *J. Neurochem.*, 1984, **43**, 1039.

107. D. Centonze *et al.*, *Exp. Neurol.*, 2001, **172**, 469.

108. A. Takeda, N. Sotogaku and N. Oku, *Neuroscience*, 2002, **114**, 669.

109. A. Takeda, S. Ishiwaari and So. Okada, *Brain Res.*, 1998, **810**, 147.

110. S. Hagiwara and S. Nakajima, *J. Gen. Physiol.*, 1966, **49**, 793.

111. K. Erikson and M. Aschner, *Neurotoxicology*, 2002, **23**, 595–602.

112. L.S. Hurley *et al.*, *Am. J. Physiol.*, 1963, **204**, 493.

113. A.L. Dupont and E.R. Harpur, *Clin. Biochem.*, 1977, **10**, 11.

114. P.S. Papavasiliou *et al.*, *Neurology*, 1979, **29**, 1466.

115. B.-S. Wong, *J. Neurochem.*, 2001, **78**, 1400.

116. G. Pomier-Layrargues, L. Spahr and R.F. Butterworth, *Lancet*, 1995, **345**, 735.

117. H.B. Ferraz *et al.*, *Neurology*, 1988, **38**, 550.

118. H. Roels *et al.*, *Am. J. Intern. Med.*, 1987, **11**, 307.

119. P. Sierra *et al.*, *Am. Ind. Hyg. Assoc. J.*, 1995, **56**, 713.

120. D.X. Wang *et al.*, *Toxicol. Sci.*, 1998, **42**, 24.

121. L. Lu *et al.*, *Neurotoxicology*, 2005, **26**, 257.

122. J.M.E. Fell *et al.*, *Lancet*, 1996, **347**, 1996.

CHAPTER 8
Alzheimer's Disease: Which Metal Now?

The disease that has become known as Alzheimer's disease (AD) was first described by the German neurologist Alois Alzheimer in 1902. It is the best known and most common of the dementing neurodegenerative conditions. The exact cause of the disease remains elusive despite the extraordinary volume of research on this disease. The neurological and behavioral changes that occur are progressive and irreversible, but often take years to develop and progress to the fatal stage. Symptoms include memory loss, anxiety, aggression, delusion, depression, disorientation and loss of intellectual facilities, including cognition and eventually loss of physiological functions due to the advancing dementia. It affects one in ten people over the age of 65 and one out of every two people over the age of 80. The AD comes in two main forms, the inherited or early onset disease that affects people in their 40s and 50s and the more common sporadic form that has a later onset. However, although the inherited forms are directly related to mutations in certain genes, the late onset forms of the disease are also influenced by "risk factors", some of which are genetically linked. The genes associated with early onset Alzheimer's do not account for all cases observed suggesting that other genes are involved. However, the three most important are the Alzheimer precursor protein (APP) and the two presenilin proteins implicated in the metabolic breakdown of APP to form the β-amyloid fragment.[1–4] As a result of these findings, there has been a considerable amount of research into the potential of altered metabolism or activity of these proteins suggesting they may be the cause of AD. In addition to these findings, other results have suggested that altered expression of other proteins can be risk factors for late onset AD. As these proteins are risk factors, their true role in the disease process remains speculative. Of these, the most discussed is change in the expression of isoforms of apolipoprotein E in which a switch to the ε-4 isoform is associated with the earlier onset of the disease.[5,6] Other proteins considered to influence the onset of AD α-2 macroglobulin,[7] insulin-degrading enzyme,[8] urokinase plasminogen activator,[9] angiotensin converting enzyme[10] and α-T catenin.[11] The other protein that is associated with AD is a polymerized form of the microtubule-binding protein tau. Phosphorylated and deposited in paired-helical filaments (PHF), neurofibrillary tangles are one of the common markers for AD. Some observers have suggested that PHFs are the cause of AD as opposed to Aβ. Recently, there has been renewed interest in a possible role for metal imbalance as a cause of AD. For some time it was suggested that aluminum deposits played some role but without any real supportive evidence those theories have been largely dismissed. Now copper has been proposed to possibly play a role given the ability of both APP and Aβ to bind the metal. In addition, altered iron metabolism has been found to occur in AD. Thus, the metal metabolism in AD continues to be a potential influencing factor.

8.1 Amyloid Precursor Protein

The major hypothesis as to the cause of AD relates to the deposition of a small peptide in the brain termed β-amyloid. This peptide is either a 40 or 42 amino acid digestion product of a large protein termed the amyloid precursor protein (APP). In particular, the longer peptide Aβ1-42 is thought to be the form that causes the disease. It readily aggregates and it is the major component of the amyloid plaques of AD (Figure 1).

The implication of this finding has resulted in intense study of the APP protein. Once again, a neurodegenerative disease has become associated with a protein that generated an aggregating fragment, is potentially a metal-binding protein and the function which people do not agree upon. However, despite the unknowns, there is a considerable body of literature relating to the production, metabolism and biological activity of APP and its two homologues, the amyloid precursor – like proteins (APLP-1 and APLP-2).

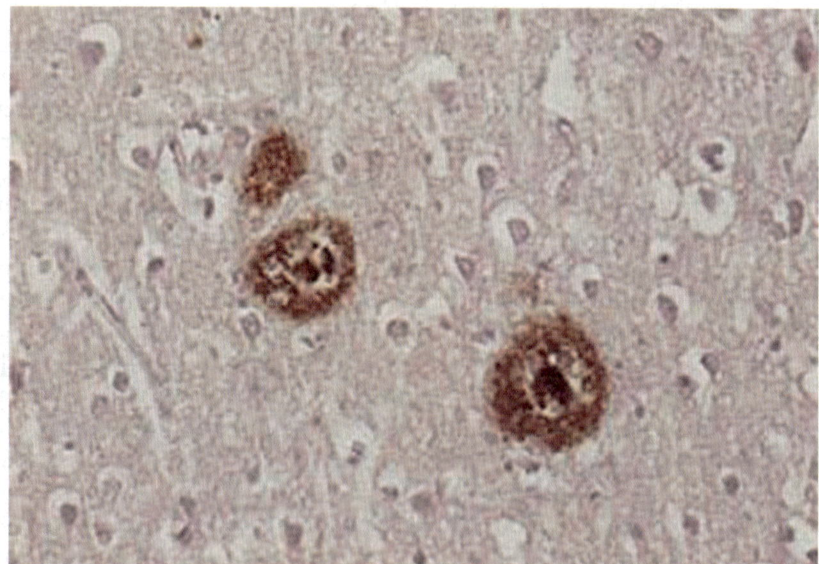

Figure 1   *Amyloid senile plaques. This figure shows a transverse section of human cortex from an AD patient. Plaques-containing Aβ can be observed due to immuno-detection with a specific antibody*

The gene for APP is localized on human chromosome 21q21.2 and contains 18 exons spanning more than 300 kb of DNA.[12–14] APP is a type-I integral membrane-associated glycoprotein. As such, it has a transmembrane domain and a large extracellular *N*-terminal domain. The region of the *C*-terminus inside the cell is relatively short. The region of APP that forms Aβ covers the most *C*-terminal extracellular domain (28 amino acid residues) and part of the transmembrane domain (14 amino acid residues). The *N*-terminus of APP contains a conserved type-II copper-binding domain (CuBD) associated with a cysteine-rich region and next to a growth factor like domain.[15,16] There is also a suggestion that APP contains a zinc-binding site, believed to have a role maintaining the structure of the protein.[17]

A major consideration for the activity of this protein is that APP has a number of splice variants dependent upon the exons incorporated into the protein.[18–20] The different splice variants are designated by the length of the protein which varies between 695 and 770 amino acid residues (Figure 2). The important members of the splice variation family are APP695, APP714, APP751 and APP770. The most abundant of these in the brain is APP695. APP770 is found most commonly in peripheral tissues.[21,22] Both APP770 and APP751 have an additional 55 amino acid domain that has homology to the Kunitz family of protease inhibitors.[23,24] This region is often termed the Kunitz domain. APP770 also contains an additional domain that has homology to another protein of unknown function known as the MRC OX-2 antigen.[25,26]

APP shares high homology to the two APLP proteins.[27] The proteins have been identified in other mammals and orthologues of the proteins have been identified in lower vertebrates, insects and worms.[28–30] Homology between APP and APLP2 is higher than between APP and APLP1.[31] APP and APLP2 are generally co-expressed during development while APLP1 is localized to the nervous system.[32,33] The proteins seem to be cleaved in similar ways.[34–36] However, cleavage of APLPs does not result in the formation of Aβ. In contrast, all three proteins have potential CuBDs in the *N*-terminus.[37,38] This suggests all three proteins have a role in copper metabolism.

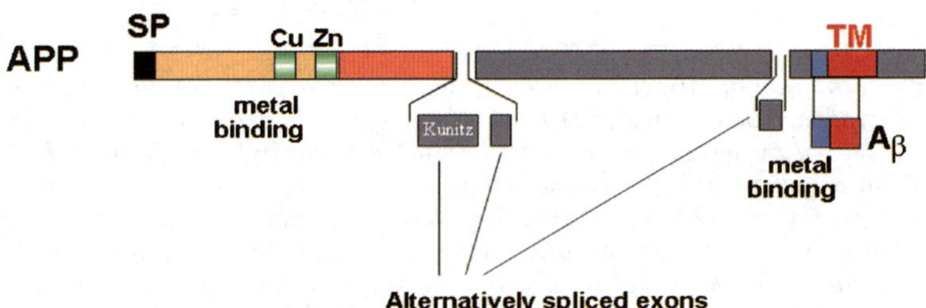

Figure 2 *Amyloid precursor protein. The protein generated from the APP gene can be different sizes due to alternative splicing. Shown are the binding sites for Cu and Zn as well as the signal peptide (SP) and the transmembrane domain (TM). The region of the protein that forms Aβ is also indicated*

8.2 Cleavage of APP and Formation of β-Amyloid

The cleavage of APP plays a pivotal role in the formation of Aβ and especially determining where the short or long version of the fragment is generated. The cleavage pattern that generates Aβ also results in the release of a soluble but truncated version of APP. The three enzymes that result in first steps in metabolic break down of the APP protein are termed secretases and are likely to function extracellularly. The three secretases are termed α, β and γ. The first step in break down follows either by the action of the α- or β-secretases. Digestion by β-secretase cleaves APP at the *N*-terminal end of the Aβ sequence liberating a soluble *N*-terminus of APP and leaving a membrane bound *C*-terminal fragment termed C99. This fragment then serves as the substrate for the γ-secretase which will cleave the protein at a number of sites resulting in the release of the Aβ fragment with sizes varying between 39 and 43 amino acid residues and the APP intracellular domain (AICD).[39,40] This pathway is termed the amyloidogenic pathway (Figure 3). The alternative non-amyloidogenic pathway results when

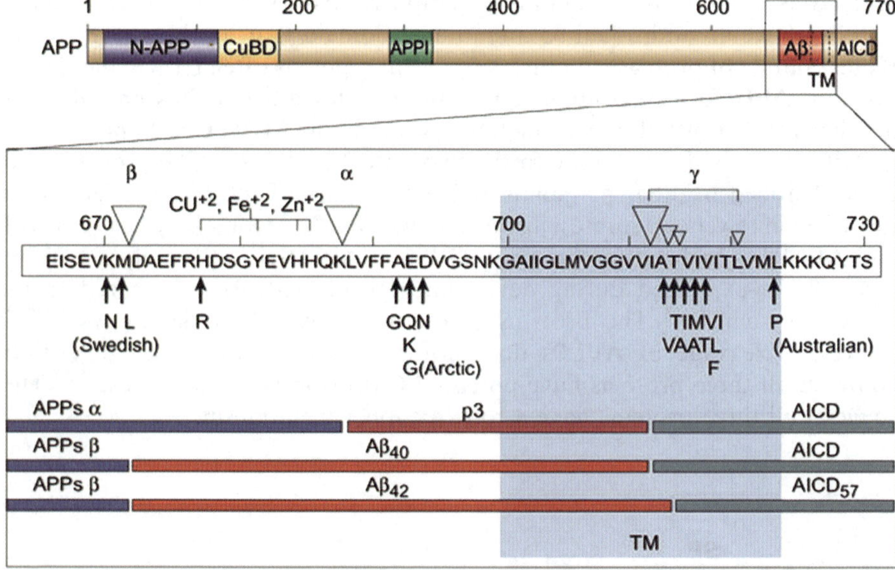

Figure 3 *Formation of Aβ. Aβ is formed by cleavage of APP by three secretases. The expanded box from the APP sequences shows the sequence of the protein that is cleaved by secretases (indicated by large triangles). Single triangles show the sites that α- and β-secretases cut. Multiple triangles show the potential sites of γ-secretase. The mutations that causes inherited AD are shown below the protein sequence as indicated by arrows. Below this are shown the three main cleavage patterns that result from the action of the secretases. The action of α- and γ-secretase results in sAPP, P3 and AICD fragments. The action of β- and γ-secretases results in sAPP plus Aβ and AICD fragment as well as Aβ. The length of Aβ depends on the site of γ-secretase digestion. The main two products being either Aβ1-40 or Aβ1-42. Mutations around the γ-secretases site influence the site of cleavage*

APP is cleaved by the α-secretase after amino acid residue 16 of the Aβ sequence, thus preventing Aβ formation but releasing soluble APP and a *C*-terminal fragment C83. C83 is then cleaved by the γ-secretase to release AICD and a small peptide termed p3.[41,42]

The identity of the proteins that form the secretases remains unresolved. In particular, the identity of the α-secretases is not known although there are a number of candidates. Two proteins from the ADAM family (a disintegrin and metalloprotease) have been proposed. These are TACE (tumor necrosis factor-α converting enzyme) otherwise termed ADAM17 and ADAM10.[43,44] Formal proof that either of these enzymes is α-secretases has not yet been produced. In contrast, genetic screen and enzyme purification has yielded the identity of the β-secretase.[45–48] The enzyme is now commonly referred to as BACE (β-site APP cleaving enzyme). BACE is co-expressed in many of the same regions as APP.

Of far greater importance to the understanding of Aβ formation is the identification of γ-secretase. It is the variation in the activity of this enzyme complex that results in the formation of the long form of Aβ associated with AD. Although this complex is not yet fully identified, key components have been discovered. These proteins are the presenilins, nicastrin (Nct) and (Aph-1/Pen-2). The presenilins (PS-1 and PS-2) provide the active core of the complex and are clearly the most important component. Inherited mutations in the genes for PS-1 and PS-2 are tightly linked to inherited forms of AD. Cleavage of the PS protein is necessary to produce the active enzyme in which the catalytic site contains aspartyl residues from either peptide, typical of an aspartyl protease.[49,50] In order to form the complex, the two PS fragments must be stabilized in the cell membrane. Nct is one of the components that does this.[51]

The protein Aph-2 is invertebrate homologue of Nct (also termed Pen-1). It was identified as a presenilin in the worm, caenorhabditis elegans along with Aph-1 (Pen-2).[52] Aph-1, like presenilin, is a protein with multiple membrane spanning domains and has similar activity to Nct.[53]

Despite the range of studies of APP cleavage and the identification of the amyloidogenic pathway, the formation of Aβ may not be aberrant. The short version, Aβ1-40, is secreted by cultured cells under normal conditions.[54] Nevertheless, increased production of Aβ results from some of the inherited mutations in the APP protein. In particular, the so-called Swedish mutation increases the percentage of protein that is cleaved by the β-secretase.[55,56] The site of the mutation is very close to the site at which β-secretase cleaves. Other mutations around the α-secretase cleavage site reduce the frequency at which the α-secretase is able to cleave APP and will also increase the amount of Aβ that is formed. Although formation of Aβ1-40 is considered to be unlikely to increase the risk of acquiring AD, anything that increases cleavage at the β-site is likely to increase the amount of Aβ1-42 produced. It is estimated that 5–10% of Aβ produced is of the longer form.[57] Mutations such as the Swedish mutation which increase Aβ production six fold will thus cause a significant increase in the amount of Aβ1-42 generated.[56] Although these changes are significant, factors that alter the site of cleavage of the γ-secretase are more likely to play a role in generation of the form of Aβ that is more likely to aggregate and initiate the pathological changes associated with AD. The mutations of APP termed London, British and Indiana alter the site at which γ-secretase acts.[58] In this case, there is no increase in the amount of Aβ but an increase in the amount of the long form that is generated. More significant than this is that mutations in the presenilin genes result in increased production of the longer Aβ peptide.[59] These mutations are causal to a percentage of familial forms of AD. From this evidence, it is logical to suggest that strategies to inhibit the formation of, or clear Aβ1-42 could have benefits in treating AD.

It has been suggested that the two forms of Aβ are generated at different sites in the cell.[60] Although non-neuronal cells only produce the peptide at the cell surface, neurons generate Aβ1-40 in the *trans*-Golgi network, but Aβ1-42 is generated in the endoplasmic reticulum. There is an emerging view that intracellular accumulation of Aβ1-42 is the first step in the pathway to developing AD.[61] Due to multiple copies of the chromosome with the APP gene, Down's syndrome patients usually develop AD. In the brains of sufferers who died young, intracellular accumulation of Aβ are very common.[62] The possibility that Aβ forms inside cells and not at the cell surface opens the possibility that interactions with Aβ and other cellular components could play a role in disease development. This may be important especially with the emerging role of metals in disease development.

8.3   Neurofibrillary Tangles

Neurofibrillary tangles are one of the diagnostic criteria for AD. For some time there has been an alternative hypothesis that the formation of PHFs is the true cause of AD. The existence of tangles in the disease correlates better with the presence of dementia than Aβ plaques.[63] Diseases with PHFs have been termed tauopathies. Unlike AD, PHFs and mutations in the tau gene are causal to the frontal-temporal dementias including Pick's disease.[64,65] The PHFs that constitute them are composed of tau, which has six isoforms in the central nervous system,[66] all of these, in a hyperposphorylated form, are found in tangles.[67] The fibrillarization pathway is thought to consist of two key steps. First, the microtubule-binding function must be neutralized so that free tau can accumulate with the cell. Second, the tau molecules must self-associate through the microtubule-binding repeat regions. It has been suggested that the formation of PHFs of the neurofibrillary tangles develop as a result of the interaction between Aβ or APP and tau. Tau will bind to plaques of Aβ in brain sections and it was also found that tau showed specific binding to the *C*-terminal domain of APP.[68] Also treatment of hippocampal neuronal cultures with Aβ resulted in the generation of tau with high levels of phosphorylation.[69] As abnormal phosphorylation of tau is a marker of PHFs, then this suggests that interactions between the two proteins may play some role in the formation of PHFs in the Alzheimer brain. Despite the interest in tau, recent studies have shown that Aβ is deposited earlier than PHFs in the AD brain.[70]

8.4   Copper, APP and Aβ

The CuBD is found in the *N*-terminus of APP within amino acid residues 135–175.[15] The CuBD contains the typical *His*-X-*His* motif of a type-II site similar to that seen for superoxide dismutase and lysyl oxidase copper homeostasis protein groups. APP reduces the bound copper to Cu^+ suggesting the protein could have a copper reductase activity.[71] This reduction leads to oxidation of cys-144 and cys-158 and resulting in the formation of a new disulfide bridge.[72] The dissociation constant for Cu^{2+} binding to APP has been suggested to be approximately 10 nM.[17]

The potential for APP to be internalized with Cu bound from an extracellular source suggests that APP could play a role in internalization of Cu. Studies from CHO cells suggest that copper stimulated the activity of α-secretase resulting in increased secreted APP and reduced Aβ formation.[73] The protein released could potentially take Cu away from the cell implying a role for APP in Cu efflux. The cerebral cortex and livers of double knockout mice (APP and APLP2) show increased levels of copper.[74] This further supports the notion that APP is involved in Cu efflux.

APP can reduce Cu^{2+} to Cu^+ in a cell-free system potentially leading to increased oxidative stress in neurone.[71] The domain that contributes to such activities is the CuBD[15] residing between residues 135 and 158 of APP, a region that shows strong homology to APLP2 but not to APLP1. Potentially, APP-Cu^+ complexes are involved that reduce hydrogen peroxide to form an APP Cu(II)-hydroxyl radical intermediates.[72,75] APP residues 135 to 158 consisting of cysteine and Cu-coordinating histidine residues can modulate copper-mediated lipid peroxidation and neurotoxicity in culture of APP knockout (APP%) and wild-type neurone.[74] Wild-type neurons were found to be more susceptible than APP neurons to physiological concentrations of copper but not other metals.

APP knockout mice have significantly increased copper, but not zinc or iron levels, in the cerebral cortex and liver as compared to age and genetically matched WT mice. APLP2 knockout mice also revealed increases in copper in cerebral cortex and liver. These findings suggest that the APP family can modulate copper homeostasis and that APP/APLP2 expression may be involved in copper efflux from liver and cerebral cortex.[71,72] Most importantly, copper was found to influence APP processing in a cell-culture model system when copper was observed to greatly reduce the levels of amyloid Aβ peptide and copper also caused an increase in the secretion of the APP ecto-domain.[73,76] An increase in intracellular APP levels which paralleled the decrease in Aβ generation suggested that additional copper was acting on two distinct regulating mechanisms, one on Aβ production and the other on APP synthesis.[73] Taken together, APP and APLP2 are most likely involved in copper homeostasis.

The expression of APP and APLP2 in the brain suggests they could have an important direct or indirect role in neuronal metal homeostasis. In APP and APLP2 knockout mice copper levels were significantly elevated in both APP knockout and APLP2 knockout mouse cerebral cortex (40% and 16%, respectively) and liver (80% and 36%, respectively) compared with matched wild-type

(WT) mice.[74] These findings indicate APP and APLP2 expression specifically modulates copper homeostasis in the liver and cerebral cortex, the latter being a region of the brain particularly involved in AD. Perturbations to APP metabolism and in particular, its secretion or release from neurons, may alter copper homeostasis and explain a disturbed metal-ion homeostasis observed in AD.

Zinc up to concentrations of 50 μM or the presence of 1,10-phenanthroline specifically increases the level of secreted APP in APP-transfected CHO-K1 cells.[76] By contrast, the level of secreted APP in copper-resistant CHO-CUR3 cells remained unaffected. APP holoprotein increases dramatically in CHO-CUR3 cells compared with CHO-K1 cells. The large decrease of Aβ release seen in both cell lines at elevated extracellular zinc levels was due to specific inhibition of secretion. These results indicate that a disturbed zinc-homeostasis may be an important factor influencing APP production, transport and processing.

Adding copper to APP-transfected CHO cells greatly reduces the levels of Aβ peptide in both parental CHO-K1 and in copper resistant CHO-CUR3 cells which have lower intracellular copper levels.[72] Copper also causes an increase in the secretion of the APP ecto-domain indicating that the large decrease in Aβ release is not due to a general inhibition in protein secretion. There was an increase in intracellular full-length APP levels, which paralleled the decrease in Aβ generation, suggesting the existence of two distinct regulating mechanisms, one acting on Aβ production and the other one on APP synthesis. Thus, our findings suggest that copper or copper agonists might be useful tools to discover novel targets for anti-Alzheimer drugs since copper promoted the non-amyloidogenic pathway of APP.[72]

The deposition of Aβ is found predominantly in the hippocampus and temporal lobe cortex and is probably closely related to the primary pathogenesis of AD, with consequent neuronal death and increase in oxidative stress.[77–79] Many studies have confirmed that Aβ is neurotoxic in cell culture. There is an obvious similarity between AD and prion diseases, in that both are characterized by the deposition of a disease-causing form of a normal cellular protein. Unlike in prion diseases, the length of the Aβ species is considered to be the important factor in AD pathogenesis as Aβ is a proteolytically cleaved version of the normal APP.

The metallochemistry of Aβ has been investigated in some detail.[80] Aβ can be rapidly precipitated by Zn^{2+} ions at low physiological concentrations and it was recently reported that other metal ions like Cu^{2+}, Fe^{3+}, unlike Zn^{2+}, produced a greater aggregation of Aβ under weakly acidic conditions (pH 6.8–7.9).[81] Such a mildl acidic environment probably resembles conditions occurring in the brain. The significance of these *in vitro* studies with Aβ and metal ions is emphasized by other data showing that the homeostasis of Zn, Cu and Fe, are significantly altered in AD brain.[77] Experiments utilizing micro-particle-induced X-ray emission (PIXE) analysis of cortical and accessory basal nuclei of the amygdala showed that these metals accumulated in the neuropile of AD brain, and that their concentrations were increased 3–5 fold compared with age-matched controls. The metals were found to be particularly high in the Aβ deposits.[82] Zn^{2+} in Aβ amyloid deposits was recently detected by histological fluorescent techniques in human brain.[83] It was also noted that the apolipoprotein E4 (apoE4) allele, which commonly appears in late onset AD cases, is associated with increased serum levels of Cu^{2+} and Zn^{2+} in AD. This suggests that the underlying perturbations in metal homeostasis associated with AD are systemic and not just confined to the brain.[84]

Zn/Cu-selective chelators reportedly enhanced the resolubilization of pathological Aβ deposits from post-mortem AD brain samples, suggesting that Cu and Zn ions may play a role in assembling these deposits.[85] The metals could also play a more significant role other than purely facilitating fibril formation. *In vitro* work from Ashley Bush's lab reported that Aβ is a redox active protein, which reduces Cu^{2+} or Fe^{3+} and then produces H_2O_2 by electron transfer to O_2.[86,87] Aβ cytotoxicity was shown to be mostly mediated by H_2O_2, produced directly by the Aβ variant, as the toxicity of the peptide was augmented by Cu^{2+} correlating with the degree of metal reduction by the same peptide. Aβ is very vulnerable to Cu^{2+}-mediated auto-oxidation, which leads to oxidative effects, such as carbonyl-adduct formation, histidine loss and dityrosine cross-linking. Such modifications have been located on the Aβ deposits extracted from AD amyloid.[88]

The metal-mediated redox activity of Aβ may well play a significant role in the pathogenesis of AD *in vivo*, though this still remains to be shown. The affinities of the Zn^{2+}-binding sites on Aβ1-40 were measured as 100 nM and 5 µM, indicating that they may be occupied under physiological conditions.[89] The highest affinity Cu^{2+}-binding site on Aβ1-42 has a measured association constant (Ka) of 10–15 atoM.[90] With such strong affinity for Cu^{2+}, Aβ species like Aβ1-42 are likely to bind Cu^{2+} *in vivo*. The increase in Cu^{2+} affinity of Aβ1-42 over the normal APP is related to APP proteolysis. APP is a membrane spanning protein and the Cu^{2+}-binding site is probably hidden within the protein, becoming exposed in the proteolyzed Aβ fragment. Also the peptide Aβ1-42 has higher β-sheet content, and these structures are frequently found in the tertiary structure of Cu^{2+} catalytic sites.

8.5 Metals and Alzheimer's Disease

Links between metals and AD date back many years. The first metal experimentally linked to this disease was aluminium. Injection of Al into rodents resulted in the formation of neurofibrillary tangles.[91] Since that time, it has become clear that the formation of tangles is not specific to AD. Neurodegeneration is also known to be associated with Al poisoning.[92] Other diseases have been associated with Al, including dialysis-induced dementia,[93] osteomalacia[94] and microcytic anaemia.[95] There actually seems to be no reliable evidence for any link between AD and Al. Those researchers who still support the notion that Al could be a risk factor usually quote a Canadian publication linking high levels of Al in local drinking water to an increase in relative risk of developing AD from 1.999 to 2.14.[96,97] However, there is a general consensus that Al is not a major risk factor if a risk factor at all.[98]

Since then, several other studies have implicated imbalances of other elements including, Si,[99] Pb,[100] Hg,[101] Br,[102] Zn,[103] Cu[104] and Fe[105] in AD. A disruption in the homeostasis of the latter two redox-active metals is particularly significant in light of the increases in oxidative stress parameters, such as lipid peroxidation, and the oxidation of PHFs, plaques[106–108] and nucleic acids.[109] Micro particle-induced X-ray emission, recently found that Zn^{2+}, Fe^{3+} and Cu^{2+} are significantly increased in AD brain neuropil and that these metals are more concentrated within the core and periphery of amyloid plaques.[82] The interest in Zn and Cu is clear from their interactions with APP and Aβ. Therefore, these three metals are clearly of the most interest to Alzheimer research.

8.6   A Balance Between Copper and Zinc

A study in which Alzheimer patients were treated with the antioxidant vitamin E showed that this treatment could delay functional decline due to the disease.[110] This finding highlights the connection between AD and oxidative stress. As indicated above, oxidation adducts are present in the brains of patients. Therefore, it is logical to assume that if Aβ binds Cu and that Cu is a redox active metal that the accumulation of Cu by Aβ could cause the oxidative damage seen in the disease. Amyloid plaque load has been reported not to correlate with dementia.[111] The soluble form of Aβ1-42 extracted from brains has a higher toxicity than that from plaques.[112] This would imply that although plaques are a dramatic sign of AD, they may not play as active role as non-plaque Aβ. As indicated above, Zn is significantly increased in the neocortex of Alzheimer's patients[82] and has been shown to be present in the plaque cores from the brains of patients and transgenic mice.[83,113] Aβ possesses a high affinity Zn^{2+}-binding site that is selective for the metal, but also a second low affinity site that can bind other metals. This lower affinity site binds metals in a way that can potentiate protease resistance of the protein and its aggregation and precipitation.[114] Zn^{2+}-mediated assembly of Aβ is reversible with chelation.[115] Metal chelators are also able to extract Aβ from post-mortem tissue, possibly because of the removal of Zn.[85] Although a divalent metal, Zn^{2+} is redox inactive and could therefore have antioxidant properties when substituted for other metals.[116] Therefore, it is possible that if Zn^{2+} substitutes for Cu^{2+} or is co-localized it could reduce the redox activity of Aβ. Cu^{2+} bound to Aβ has been shown to cause the generation of hydrogen peroxide.[86] Evidence has been provided that addition of Zn to Cu loaded Aβ will quench its ability to generate hydrogen peroxide and abrogate any toxic effects it has.[117] It is possible then that amyloid plaques, high in Zn form as a result of interaction with Zn and represent an inert form of Aβ that has little consequence for the cause of AD but is only a "tombstone" marking the diseased brain.

The sequence of neurochemical events leading to metal-mediated plaque formation is still uncertain. One model contemplated is that the affinity of Aβ for Cu^{2+} and Zn^{2+} could be increased by the generation of a rogue, oxidized, Aβ form. The oxidized form of Aβ could be generated by the peptide's vulnerability to Cu-mediated oxidative damage.[118] Abnormalities of brain metal homeostasis in AD, or as a consequence of aging, could also contribute to Aβ deposition. Recently, it was found that Cu and Fe levels rise markedly and invariably with aging in the Tg2576 model for AD, and its background strain.[119] The age-dependent rise in these metals may be set up chemical conditions where the Aβ toxicity is promoted.

On the basis of this growing pool of information about Zn^{2+} and Cu^{2+} interactions with Aβ in AD, the group of Ashley Bush recently embarked on a trial of copper/zinc-chelators to attempt to inhibit Aβ accumulation in APP2576 transgenic mice. Oral treatment with CQ (clioquinol), a retired USP antibiotic and orally bioavailable Cu/Zn chelator, induced a 49% decrease in brain Aβ deposition in a blinded study of APP2576 transgenic mice treated orally for 9 weeks.[120] There was no evidence of neurotoxicity or increased non-amyloid pathology. General health and body weight parameters were significantly more stable in the treated animals, which were conspicuously improved after only 16 days of treatment. The drug may work by a combined action that facilitates metal-mediated disaggregation of the Aβ collections, while also inhibiting H_2O_2 production even by soluble forms of Aβ. The drug's rapid onset of benefit in the transgenic mice[120] may have been due to a rapid fall in cerebral H_2O_2 concentrations caused by the inhibition of its catalytic production by Aβ. The 16-day interval might be too soon to expect any marked difference in the amount of amyloid accumulation itself between the CQ-treated and control mice. Importantly, CQ treatment did not induce a loss in metal levels systemically, probably because it is a relatively weak chelator and the metals are redistributed rather than excreted. Therefore, the benefits of the drug appear to be due to its ability to bind selectively to the Aβ–metal complex, and are not due to metal depletion of brain tissue. The affinity of CQ for Cu^{2+} and Zn^{2+} is not as strong as standard chelators, like EDTA or TETA. Nevertheless, CQ was as effective as high-affinity chelators in blocking the production of H_2O_2 by Aβ *in vitro*, in preventing precipitation of synthetic Aβ by Zn^{2+} and Cu^{2+} ions, and in extracting Aβ from post-mortem AD brain specimens.[120]

CQ treatment of non-transgenic mice significantly decreases brain levels of Cu, Zn and Fe.[121] However, treatment of 21-month-old APP2576 mice with CQ for 9 weeks paradoxically elevated brain Cu by 19% and Zn by 13% while markedly inhibiting Aβ deposition. This paradoxical result is explained by our recent findings that brain Cu and Zn levels are relatively decreased by APP transgene expression in APP2576 mice, despite Aβ levels accumulating several hundred-fold from 2.8 to 18 months. The decrease in Cu and Zn levels that were observed were, therefore, not a consequence of plaque pathology, and were relative to a marked increase ($\approx 50\%$) in Cu and Fe levels that occurred after 6 months of age.[119] This relative decrease must either be due to secreted APP and/or Aβ promoting the efflux of the metal ions, or APP/Aβ preventing their uptake. Supporting this latter possibility is evidence that Aβ scavenges extracellular Cu^{2+}, possibly to prevent oxidation.[122] In light of our recent findings, the paradoxical increase in Cu and Zn in CQ-treated APP2576 mice may be explained by CQ preventing Cu^{2+} and Zn^{2+} from complexing with extracellular Aβ, so securing metal for uptake into metal-deficient brain tissue instead of being sequestered into amyloid. Therefore, despite being a chelator, CQ treatment may be able to restore homeostatic defects of normal brain metal metabolism which may occur in AD. The consequent lowering of extracellular metal concentrations inhibited the formation, or facilitated the dissolution, of amyloid deposits.

In contrast, TETA (triene), a high-affinity copper/zinc chelator that does not penetrate the BBB, failed to inhibit amyloid deposition in the transgenic mice.[120] TETA, like penicillamine, is used for the treatment of the copper overload disorder, Wilson's disease. Wilson's disease commences as a hepatic disorder. Advanced complications include cataracts and neurological complications. There is no known association with AD, which is not surprising because copper accumulation is intracellular in Wilson's disease, whereas it is extracellular in AD. The ineffectiveness of a traditional copper chelator (TETA) in inhibiting amyloid depostion in transgenic mice,[120] indicates that systemic metal depletion (e.g. "chelation therapy"), is not likely to be a useful therapeutic strategy for AD. Indeed, the transgenic mice successfully treated with CQ exhibited a small but significant increase in brain zinc and copper suggesting that these metals had been trapped by the amyloid aggregates, and prevented from constitutive entry into the tissue.

Despite the success in the use of CQ in transgenic mice, it should be noted that transgenic mice that have amyloid deposits do not develop AD. Use of these drugs in human trials has been less promising although some benefit was observed.[123] In addition, other studies have suggested that Cu supplementation increases the survival of transgenic mice that have amyloid plaques in their brains[124] and reduces the level of plaques formed in their brains. This therefore, represents an opposing view. On one hand Cu chelators appear to reduce plaque load, but in addition Cu supplementation can also reduce plaque load in similar transgenic mice. In the case of the Cu diet the suggested mechanism of action of Cu is that it increases the activity of antioxidants such as Cu/Zn SOD and increased the amount of Cu binding to APP. This decreases the formation of homodimers of APP.[125] Decreased formation of homodimers then decreases the extent to which the protein is cleaved by the β-secretase (Figure 4). The problem then is how does CQ act? Some chelators do not necessarily remove metals from organs but might deliver them.

Recent results using a yeast model system may provide some insight as to what CQ actually does. Treiber *et al.*[126] have shown that adding CQ to the yeast culture medium drastically increased the intracellular Cu concentration but there was no significant effect observed on zinc levels. This finding suggests that CQ can act therapeutically by changing the distribution of copper or facilitating Cu uptake rather than by decreasing Cu levels. The overexpression of the human APP or APLP2 extracellular domains decreases intracellular copper levels. The expression of a mutant APP deficiency for Cu binding increases the intracellular copper levels several-fold. These data support the suggested function of APP and APLP2 in copper efflux and provide a new conceptual framework for the formerly diverging theories of Cu supplementation and chelation in the treatment of AD. In mammals CQ-copper complexes could form in the intestinal tract and cross the blood-brain barrier to enter the brain and thus explain why soluble copper and zinc levels were increased by

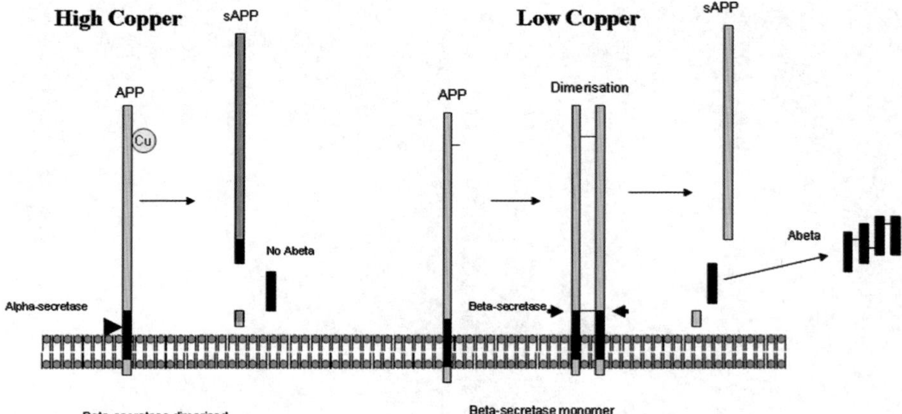

Figure 4 *Copper binding and APP cleavage. Copper binding might influence the rate at which Aβ is formed. Copper binding to APP could block potential dimerization of APP. Dimerization of APP has been suggested to cause a preferential switch to cleavage of the protein by β-secretase. This would increase levels of Aβ. Therefore decreased-Cu levels might favor Aβ formation. In addition, binding of Cu to the β-secretases, BACE-1 could inhibit its activity*

15% upon CQ treatment in mice brain.[120] Therefore, CQ-copper complexes could selectively and markedly elevate copper levels in the brain of individuals with AD and counterbalance the changes in copper levels observed in AD, most probably mediated through the APP export function. On the background that aged APP transgenic mice treated with copper lived longer than untreated mice,[124] these issues may be important in understanding the integration of copper homeostasis in AD and with other physiological processes such as aging.

8.7 The Rise of Iron

Free iron has been implicated in neurodegenerative diseases through it redox transitions *in vivo*. Abnormal high levels of Fe have been found in a variety of diseases which include Parkinson's disease, multiple system atrophy, progressive supranuclear palsy and AD[108,127] (Figure 5). However, an increased level of Fe does not imply increased Fe-related oxidative damage if the increase in Fe is accompanied by an increase in the appropriate storage proteins, as this keeps Fe in a redox inert state.

One of the main storage proteins for Fe is ferritin. In the brain most ferritin is found in microglia. Other proteins associated with iron management are the ferritin receptor, lactotransferrin receptor, melanotransferrin and ceruloplasmin[128] and also the divalent metal transporter (DMT-1). In addition, the iron regulatory proteins (IRP-1 and IRP-2) play the most prominent role in regulating iron metabolism. In particular, it has been found that these proteins are altered in AD.[129] IRPs are cytoplasmic mRNA-binding proteins involved in intracellular regulation of iron homeostasis. IRPs regulate expression of ferritin

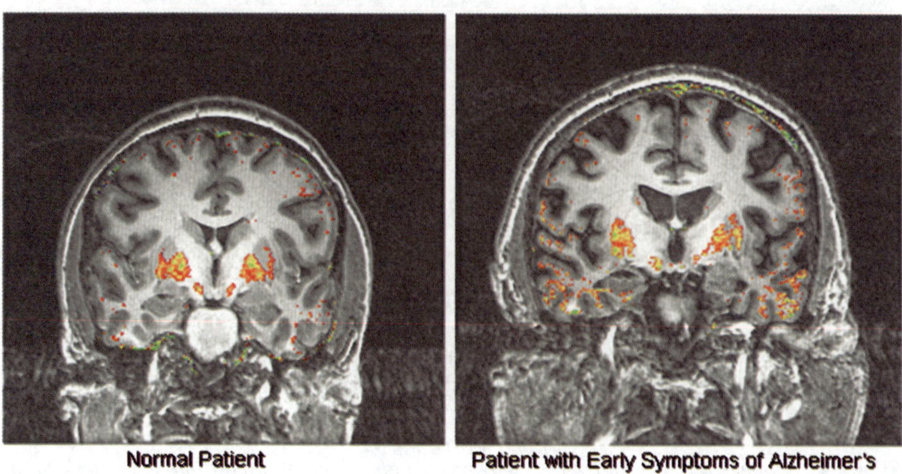

Figure 5 *Iron is increased in AD. MRI analysis of the brains of patients with and without AD. MRI analysis shows higher levels of Fe in the brain of the Alzheimer patient*

and transferrin receptor at the mRNA level by interacting with a conserved RNA structure termed the iron-responsive element (IRE). This concordant regulation of transferrin receptors and ferritin is designed so a cell can obtain iron when it is needed, and sequester iron when it is in excess. Iron accumulates in the brain in AD without a concomitant increase in ferritin.[105] Another protein, transferrin is essential for cellular uptake of iron. In AD, this protein's expression is also unaltered.[130] Although there is no change in expression levels, the cellular expression of both proteins is greatly altered. An increase in iron without proper sequestration can increase the vulnerability of cells to oxidative stress. It has been demonstrated that in normal human brain extracts, the IRP is detected as a double IRE/IRP complex by RNA band shift assay, but in two of six Alzheimer's brain extracts examined a single IRE/IRP complex was obtained.[131] Furthermore, the mobility of the single IRE/IRP complex in Alzheimer's brain extracts is decreased relative to the double IRE/IRP complex. Western blot and RNA band super shift assay demonstrate that IRP1 is involved in the formation of the single IRE/IRP complex. *In vitro* analyses suggest that the stability of the doublet complex and single AD complex are different. The single complex from the AD brain is more stable. A more stable IRE/IRP complex in the AD brain could increase stability of the transferrin receptor mRNA and inhibit ferritin synthesis. At the cellular level, the outcome of this alteration in the molecular regulatory mechanism would be increased iron accumulation without an increase in ferritin; identical to the observation reported in AD brains. The appearance of the single IRE/IRP complex in Alzheimer's brain extracts is associated with relatively high endogenous ribonuclease activity. It is possible that elevated RNase activity is one mechanism by which the iron regulatory system becomes dysfunctional.

The accumulation of iron in plaques may have a protective role. As indicate above Aβ can bind iron and Fe will cause the aggregation of the protein. A number of studies have suggested that Aβ can decrease the toxicity of Fe.[132,133] However, even though the toxicity of iron is decreased by the Fe-Aβ complex, the toxicity of Aβ is increased.[134,135] This may at first seem paradoxical but as Fe is far more toxic than Aβ even with Fe bound, then the simple answer is that Aβ can decrease the toxicity of Fe. This implies an emerging picture of metal Aβ interactions suggesting a balance between toxic "free" Aβ enhanced by metal interactions (Cu and Fe) and plaque sequestration of metals (Cu, Zn, Fe) that are protective and inhibit cell death. However, this does not explain how Fe could cause cell death in AD.

One of the organelles in AD that has repeatedly been shown to be damaged is the mitochondria. These organelles are very sensitive to oxidative damage and in AD this sensitivity is increased.[136] Mitochondria are the site of heme synthesis and therefore a major site for the handling of Fe in the cell. There is also evidence that the synthesis of a number of iron-binding proteins associated with mitochondria, including heme, are altered and that this could potential result in free Fe in the cell.[137] It has been suggested that changes in mitochondrial handling of Fe are part of the normal aging process.[138] In AD, the majority of oxidative damage is present in the cytoplasm rather than in the mitochondria. However, altered mitochondrial generation of radicals could be a trigger to the changes that are occurring.

Mitochondria are a potential major source of oxidative radicals and oxidative precursors, in the form of O_2^- and H_2O_2, respectively, since their production is linked to metabolism. In early studies, increases in mitochondrial DNA in the cell soma of AD susceptible neurons were found,[139] which in itself might cause increased oxidative potential. Perhaps more importantly, *in situ* hybridization studies with a chimeric cDNA probe to a common mitochondrial mutation (5 kb common deletion) have shown at least a three-fold increase in AD cases compared to controls. Ultrastructural localization of mtDNA with colloidal gold shows that deleted mtDNA is mainly found in abnormal mitochondria (*i.e.*, those lacking cristae, swollen and in many cases fused with lipofuscin). These findings suggest that the mtDNA *in situ* hybridization detected mtDNA proliferation, deletion and duplication in abnormal mitochondria, many of which have been fused with lysosomes, indicating that they are being turned over. Quantitative analysis of the co-localization of the mtDNA deletion and 8OHG (8-hydroxyguanosine) in AD cases demonstrate a strong positive correlation.[139] However, mitochondrial DNA, even that containing the 5 kb deletion, is relatively spared from oxidative damage, like the formation of 8OHG, in comparison to cytoplasmic nucleic acid (*i.e.*, RNA). We therefore suspect that mitochondrial abnormalities correlate with, but do not directly cause, reactive oxygen species (ROS). This may be due to the fact that hydroxide radicals which are responsible for the formation of 8OHG, have a sphere of diffusion of only 2 nm and are fairly short lived. Therefore, since damage is topographically distinct, it is likely that OH• radical formation occurs in the cytoplasm rather than the mitochondria and that they are unable to diffuse through the mitochondrial membrane to affect mtDNA. However, abnormal mitochondria may produce excess H_2O_2 through the conversion of

O_2^- by mitochondrial SOD. Such H_2O_2 is readily diffusible and relatively stable, that is, until interacting with redox-active transition metals where the Fenton reaction produces hydroxyl radicals.

Although these findings suggest that mitochondria could be the source of the abnormalities in cellular Fe and H_2O_2 which would combine to generate oxidative damage, it remains unclear how mitochondria are more dysfunctional in AD than in the other members of the aging population. Very recent work has suggested that either Aβ or mutant tau[140,141] could alter mitochondrial membrane depolarization. Such interesting ideas require further investigation.

8.8 Any More Metals in Alzheimer's Disease?

The true situation regarding metals and AD remains unresolved. However, it is quite clear that redox active metals play some role. Possibly, the true picture involves Cu and Fe in different roles. However, currently metal chelation treatment is not proving as successful as first thought. This does not imply that the metals do not play an important role. Indeed, as both APP and Aβ bind Cu it is very likely that Cu is somehow involved. The question remains, is there any other metal involved? Given the extent of recent studies it does not seem likely. The only other metal that could play a role is Mn but there is no evidence that this is the case.[142] However, perhaps the true picture will not come from working out which metal plays a role but how cellular regulation of redox active metals influences disease progression. Perhaps another protein, involved in binding a metal is involved. A recent paper has suggested that BACE1, the β-secretase, also binds Cu^{2+}.[143] Perhaps this is another way altered Cu metabolism could result in changes in activity of key proteins involved in APP processing. Clearly, metals and AD will remain one of the expanding horizons of research into neurodegenerative disease.

8.9 Coordination Chemistry of Metal Ions Interacting with APP

As mentioned above there are at least two domains of APP which may bind metal ions effectively and this bindings are potentially biologically relevant: the Aβ peptide region and the cysteine-rich region. The CuBD is located in the latter region.[15,16] This domain comprises residues 124–189 but the major binding sites are located close to Cys-144 – Cys-158 region, where three His residues are inserted in the –His-Leu-His-Trp-His- sequence. The presence of three imidazoles close to each other makes this site very effective in the Cu(II)-ion binding.[144,145] Around pH 5.5–6 the complex with three-bound imidazoles predominates, while within physiological pH range the major complex is the species with a very unusual coordination mode involving {$3N_{imid}, 2N^-_{amide}$} donor set (Figure 6).[145] The 3xHis motif is a much more effective binding site than the 2xHis site found, *e.g.*, in SOD and could be critical for the specific

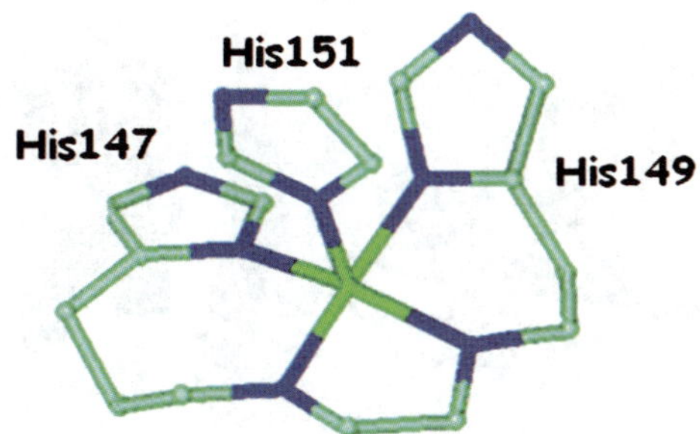

Figure 6 *Structural model of Cu(II)-APP complex. In the coordination sphere of the metal-peptide complex the imidazole nitrogens of the His-149 and the amide nitrogen of the Leu-148 and His-149 are in the plane. The His-151 residue is bound in apical position*
(Reproduced from ref. 145.)

binding of Cu^{2+} by cysteine-rich domain. This is a characteristic-binding feature for the short unstructured peptide fragments.

The NMR studies of the whole CuBD (APP124–189) have indicated that the peptide in this region may be well organized (Figure 7).[146] The structure consists of α-helix (APP147–159) and triple-stranded β-sheet comprising residues 133–139, 162–167 and 181–186. Three disulfide bonds between Cys-133 and Cys-187 linking two β-strands, Cys-158 and Cys-186 linking α-helix to β-strand as well as Cys-144 and Cys-174 connecting loops at the other ends of the molecule stabilize the CuBD structure (Figure 8).[146] This structure, due to steric reasons, could favor the involvement of only two His side chains in copper-ion coordination. The tetrahedral geometry proposed for such coordination may favor Cu^+-ion binding (Figure 9).[146] Although, coordination of two imidazoles to a copper ion in CuBD is very likely the binding of Tyr-168 and Met-170 side chains is not well documented. The copper site in CuBD is redox active and the favoring of Cu^+ could be critical for the biological consequences of the copper binding to this region of APP.

Much more attention was given to Cu^{2+} interactions with the Aβ peptide, which contributes to formation of amyloid fibrils and plaques critical for AD. The metal-ion binding to Aβ and oxidation reaction products were detected in the intact amyloid plaques from AD brains.[147] The Raman spectra obtained for senile plaque cores isolated from different AD brains were essentially the same within and among different brains. The major component is the Aβ peptide having β-sheet structure with Cu and Zn ions bound to His residues.[147] Thus, in the intact senile plaques Aβ is a metallopeptide. The role of metal-peptide interactions in the polymerization process has been often suggested (see, *e.g.*, ref. 81). It seems that even trace metal contamination may induce the auto-aggregation and oligomerization of Aβ.[148]

Aβ could be considered as a low-molecular-weight protein consisting of 39–43 amino acid residues. The most abundant fragments are Aβ1-40 and Aβ1-42 peptides. The sequence of human Aβ1-42 is as follows: DAEFRHDSGYEVHH QKLVFFAEDVGS-NKGAIIGLMVGGVVIA with first 28 residues occupying

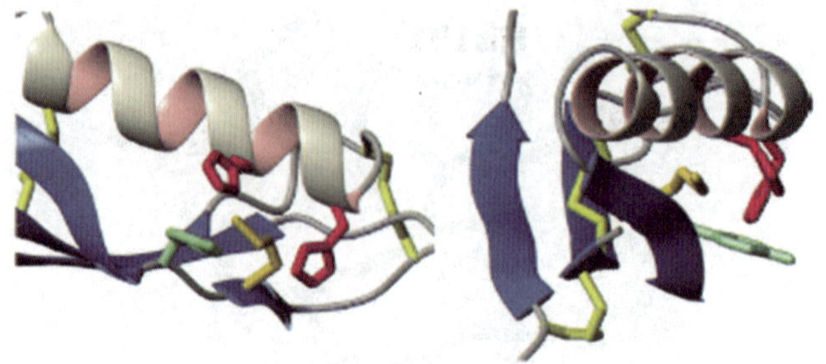

Figure 7 *The metal-binding site in APP. Metal-binding site consisting of the residue His-147, His-151, Tyr-168 and Met-170*
(Reproduced with permission from ref. 146.)

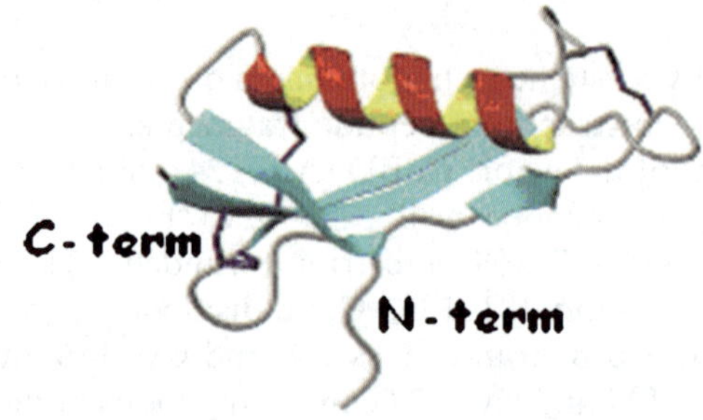

Figure 8 *The structure of CuBD determined by NMR spectroscopy. The structure consists of an α-helix (residue 147–159) packed against a triple-stranded β-sheet (residues 133–139, 162–167 and 181–188). Disulfide bonds between Cys residues (133 and 187, 153 and 186, 144 and 176) stabilize the CuBD structure*
(Reproduced with permission from ref. 146.)

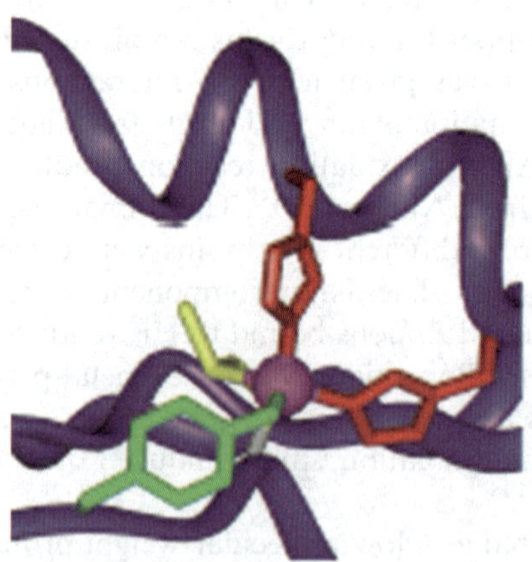

Figure 9 *Model of Cu(I)-binding to CuBD. Model of Cu(I) coordinated in tetrahedral geometry to His-147, His-151, Tyr-168 and Met-170*
(Reproduced with permission from ref. 146.)

the extra-cellular region and residues 29–42 inserted in the transmembrane region. The hAβ peptide contains four potential Cu^{2+} anchoring sites at His-6, His-13 and His-14 side-chain imidazoles and the *N*-terminal amino nitrogen.

The human Aβ peptide fragments 1–6, 1–9 and 1–10 containing only one His residue (His-6) have very similar coordination ability including binding modes and the complex stabilities.[149] The *N*-terminal amino nitrogen of Asp is a very effective anchor due to possibility of the chelate formation by {NH_2, β-COO^-} donor set.[150] Around neutral pH range Cu^{2+} ions are coordinated by three donors the amino, imidazole and amide nitrogens and additionally β-carboxylate of Asp-2.[149] It is noteworthy that mouse peptide 1–6, 1–9 and 1–10 fragments which differ at position 5 (Arg→Gly) and 10 (Tyr→Phe) behave similarly although the stability of the mouse peptide complexes are distinctly higher.[149] The phenolate donor of Tyr-10 is not involved in the Cu^{2+} binding by Aβ1-10.

In the 11–28 fragment of human β-amyloid peptide the two His residues are located at positions 13 and 14. The studies on N-protected fragments 11–16,[151] 11–20 and 11–28[152] have shown that, in the species dominant around neutral pH, both His are involved in the Cu^{2+}-ion coordination. Additionally one or two amide nitrogens may complete the metal-ion coordination sphere. The mouse peptide contains only His-14 and Arg instead of His-13 and behaves very differently. As only one imidazole nitrogen may be involved in the Cu^{2+}-ion binding the formed complexes are weaker and (around pH 7) the ability of the human peptide to bind Cu^{2+} is much higher than that of mouse variant (Figure 10).[151] Both ligands become similar to each other in Cu^{2+} binding above pH 9 when coordination mode in dominant complexes is the same {$N_{imid},3N^-$}.[151]

The human peptides Aβ1-16 and 1–28 contain three His residues and additionally *N*-terminal amino group as possible anchoring and binding sites. In the N-protected peptides the major coordination mode in the pH range 6–7.5 involves three imidazoles of His-6, His-13 and His-14 residues[153] and at higher pH sequential amide nitrogens are deprotonated and coordinated to Cu(II). The deprotection of the *N*-terminal group results in the involvement of amino nitrogen in the metal-ion binding as well (Figure 11).[153,154]

The comparison to mouse variant[153] and various analogues of Aβ1-28 with His substituted by Ala[154] has suggested that the *N*-terminal amino group and His imidazoles are basic for metal-ion coordination around neutral pH. The most important His residue was suggested to be His-13.[154] However, its specific ability to interact with Cu^{2+} ions could derive from the vicinity of His-14 residue. The human peptide is much more effective in Cu^{2+}-ion binding than mouse or rat variant due to simultaneous involvement of His-13 and His-14 residues. The earlier suggestions[155,156] that Tyr-10 may be involved in the metal-ion coordination were not confirmed.[153,154] The possibility of the bridging by an imidazole moiety[155] to form dimeric species has not been supported.[153,154] The coordination of Cu^{2+} to Aβ induces some ordering of the Aβ peptide, however, according to CD spectra, it does not form a typical β-sheet conformation.[154] According to EPR studies the Cu^{2+}-binding mode to soluble and fibril Aβ is the

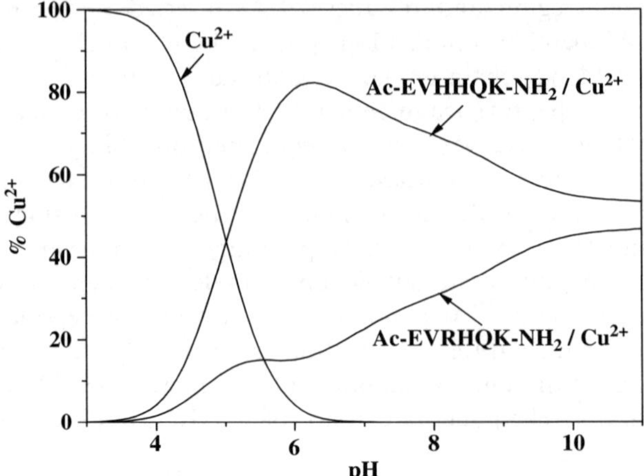

Figure 10 *Competition diagram between copper complexes. The distribution of Cu(II) between human (Ac-EVHHQK-NH$_2$) and mouse (Ac-EVRHQK-NH$_2$) blocked fragments of β-amyloid peptide in aqueous solution for molar ratio 1:1:1*

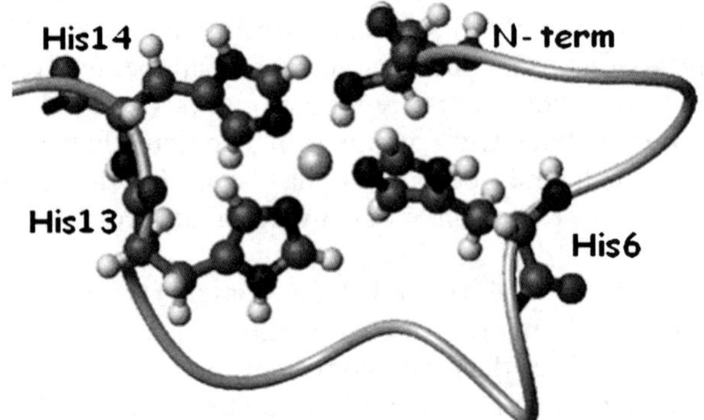

Figure 11 *Model of Aβ coordinating Cu(II) ion at physiological pH. Copper ion is coordinated by N-terminal amino group and three histidine residues (His-6, His-13 and His-14) in a square-planar geometry*
(Reproduced with permission from ref. 154.)

same.[157] The metal-binding site is monomeric and the donor set does not change when soluble peptide aggregates to fibrils in the presence of Cu^{2+}.

Zn^{2+} ions are usually much less effective in the interactions with peptides than Cu^{2+} ions. However, similarly to Cu^{2+} the histydyl imidazoles as well as the *N*-terminal amino group are quite attractive as the binding sites for Zn^{2+}. The NMR work has suggested that human Aβ peptide involves in the coordination of Zn^{2+} ion three His residues[158] and possibly water of hydroxyl ligand.[155] The interaction of extracellular fragment hAβ1-28 with Zn^{2+} is quite specific and it is resulting in the peptide aggregation. It is likely that Zn^{2+} ions may also form intermolecular cross-links between Aβ peptide molecules facilitating aggregation process.[159] The substitution of H-13 for Arg (rat or mouse peptide) changes distinctly the binding ability and the zinc-induced aggregation is lower than in the case of hAβ.[158]

Human Aβ(1-28) in DMSO has well defined *C*-terminal domain(16–28) including helical region spanning between Ala21 and Val24, and a flexible *N*-terminal region.[160] The structure of the rat peptide rAβ(1–28) is much better organized[161] with the helical region occurring between Glu16 and Val24. The binding of Zn^{2+} ions to rAβ(1–28) increases the peptide organization especially on its *N*-terminus. The Zn^{2+}-coordinated peptide has five distinct structural regions: the *N*-terminus (residues 1–3), helix 1 (residues 4–8), extended chain (residues 9–15), helix 2 (residues 16–22) with the turn-like structure between Arg13 and Gln15, and the *C*-terminus (residues 23–28).[161] The binding sites were proposed to be His-6 and His-14 imidazoles and Arg13 guanidinyl sidechain. The more ordered conformation of rat peptide may contribute to better stabilization of the helical peptide conformation protecting it against transition into pathological β-strand observed for human Aβ.[161]

References

1. A. Goate *et al.*, *Nature*, 1991, **349**, 704.
2. R. Sherrington *et al.*, *Nature*, 1995, **375**, 754.
3. E. Levy-Lahad *et al.*, *Science*, 1995, **269**, 973.
4. E.I. Rogaev *et al.*, *Nature*, 1995, **376**, 775.
5. E.H. Corder *et al.*, *Science*, 1993, **261**, 921.
6. W.J. Strittmatter *et al.*, *Proc. Natl. Acad. Sci. USA*, 1993, **90**, 1977.
7. D. Blaker *et al.*, *Nat. Genet.*, 1998, **19**, 357.
8. L. Bertram *et al.*, *Science*, 2000, **290**, 2302.
9. N. Ertekin-Taner *et al.*, *Science*, 2000, **290**, 2303.
10. J. Hu *et al.*, *Neurosci. Lett.*, 1999, **277**, 65.
11. A. Myers *et al.*, *Science*, 2000, **290**, 2304.
12. H.G. Lemaire *et al.*, *Nucleic Acids Res.*, 1989, **17**, 517.
13. M. Hattori *et al.*, *Nucleic Acids Res.*, 1997, **25**, 1802.
14. S. Yoshikai *et al.*, *Gene*, 1990, **87**, 257.
15. L. Hesse *et al.*, *FEBS Lett.*, 1994, **349**, 109.
16. J. Rossjohn *et al.*, *Nat. Struct. Biol.*, 1999, **6**, 327.
17. A.I. Bush *et al.*, *J. Biol. Chem.*, 1993, **268**, 16109.
18. D. Goldgaber *et al.*, *Science*, 1987, **235**, 877.
19. J. Kang *et al.*, *Nature*, 1987, **325**, 733.
20. N.K. Robakis *et al.*, *Lancet*, 1987, **1**, 384.
21. R. Sandbrink *et al.*, *J. Biol. Chem.*, 1994, **269**, 1510.
22. J. Kang *et al.*, *Biochem. Biophys. Res. Commun.*, 1990, **166**, 1192.
23. P. Ponte *et al.*, *Nature*, 1988, **331**, 525.
24. T. Oltersdorf *et al.*, *Nature*, 1989, **341**, 144.
25. G. Konig *et al.*, *J. Biol. Chem.*, 1992, **267**, 10804.
26. S.J. Richards *et al.*, *Brain Res. Bull.*, 1995, **38**, 305.
27. W. Wasco *et al.*, *Proc. Natl. Acad. Sci. USA*, 1992, **89**, 10758.

28. D.R. Rosen *et al.*, *Proc. Natl. Acad. Sci. USA*, 1989, **86**, 2478.
29. H. Okado *et al.*, *Biochem. Biophys. Res. Commun.*, 1992, **189**, 1561.
30. I. Daigle *et al.*, *Proc. Natl. Acad. Sci. USA*, 1993, **90**, 12045.
31. W. Wasco *et al.*, *Nat. Genet.*, 1993, **5**, 95.
32. K. Lorent *et al.*, *Neuroscience*, 1995, **65**, 1009.
33. S. Heber *et al.*, *J. Neurosci.*, 2000, **20**, 7951.
34. S. Naruse *et al.*, *Neuron*, 1998, **21**, 1213.
35. M.H. Scheinfeld *et al.*, *J. Biol. Chem.*, 2002, **277**, 44195.
36. D.M. Walsh *et al.*, *Biochemistry*, 2003, **42**, 6664.
37. A. Simons *et al.*, *Biochemistry*, 2002, **41**, 9310.
38. A.R. White *et al.*, *J. Neurosci.*, 2002, **22**, 365.
39. P. Cupers *et al.*, *J. Neurochem.*, 2001, **78**, 1168.
40. Y. Gao *et al.*, *Proc. Natl. Acad. Sci. USA*, 2001, **98**, 14979.
41. F.S. Esch *et al.*, *Science*, 1990, **248**, 1122.
42. S.S. Sisodia *et al.*, *Science*, 1990, **248**, 492.
43. J.D. Buxbaum *et al.*, *J. Biol. Chem.*, 1998, **273**, 27765.
44. S. Lammich *et al.*, *Proc. Natl. Acad. Sci. USA*, 1999, **96**, 3922.
45. I. Hussain *et al.*, *Mol. Cell. Neurosci.*, 1999, **14**, 419.
46. S. Sinha *et al.*, *Nature*, 1999, **402**, 537.
47. R. Vassar *et al.*, *Science*, 1999, **286**, 735.
48. L. Lin *et al.*, *Proc. Natl. Acad. Sci. USA*, 1999, **96**, 12108.
49. G. Thinakaran *et al.*, *J. Biol. Chem.*, 1996, **271**, 9390.
50. M.S. Wolfe *et al.*, *Biochemistry*, 1999, **38**, 4720.
51. D. Edbauer *et al.*, *Proc. Natl. Acad. Sci. USA*, 2002, **99**, 8666.
52. C. Goutte *et al.*, *Proc. Natl. Acad. Sci. USA*, 2002, **99**, 775.
53. N. Takasugi *et al.*, *Nature*, 2003, **422**, 438.
54. C. Haass *et al.*, *Nature*, 1992, **359**, 322.
55. M. Mullan *et al.*, *Nat. Genet.*, 1992, **1**, 345.
56. M. Citron *et al.*, *Nature*, 1992, **360**, 672.
57. H.F. Dovey *et al.*, *Neuroreport*, 1993, **4**, 1039.
58. N. Suzuki *et al.*, *Science*, 1994, **264**, 1336.
59. M. Citron *et al.*, *Nat. Med.*, 1997, **3**, 67.
60. T. Hartmann *et al.*, *Nat. Med.*, 1997, **3**, 1016.
61. O. Wirths *et al.*, *J. Neurochem.*, 2004, **91**, 513.
62. K.A. Gyure *et al.*, *Arch. Pathol. Lab. Med.*, 2001, **125**, 489.
63. P.V. Arriagada *et al.*, *Neruology*, 1992, **42**, 631–639.
64. M.G. Spillantini and M. Goedert, *Trends Neurosci.*, 1998, **21**, 428.
65. J. Kuret *et al.*, *Biochim. Biophys. Acta*, 2005, **1739**, 167.
66. M. Goedert *et al.*, *Neuron*, 1989, **3**, 519.
67. I. Grundke-Iqbal *et al.*, *Proc. Natl. Acad. Sci. USA*, 1986, **86**, 2853.
68. M.A. Smith *et al.*, *Nat. Med.*, 1995, **1**, 365.
69. J. Busciglio *et al.*, *Neuron*, 1995, **19**, 823.
70. P. Fernandez-Vizarra *et al.*, *Histol. Histopathol.*, 2004, **15**, 823.
71. G. Multhaup *et al.*, *Science*, 1996, **271**, 1406.
72. G. Multhaup *et al.*, *Biochemistry*, 1998, **37**, 7224.
73. T. Borchardt *et al.*, *Biochem. J.*, 1999, **344**, 461.

74. A.R. White *et al.*, *Brain Res.*, 1999, **842**, 439.
75. G. Multhaup *et al.*, *Met. Ions. Biol. Syst.*, 1999, **36**, 365.
76. T. Borchardt *et al.*, *Cell. Mol. Biol.*, 2000, **46**, 785.
77. C.S. Atwood *et al.*, *Met. Ions Biol. Syst.*, 1999, **36**, 309.
78. M.A. Smith *et al.*, *J. Neurochem.*, 1998, **70**, 2212.
79. M.E. Calhoun *et al.*, *Nature*, 1998, **395**, 755–756.
80. A.I. Bush, *Curr. Opin. Chem. Biol.*, 2000, **4**, 184.
81. C.S. Atwood *et al.*, *J. Biol. Chem.*, 1998, **273**, 12817.
82. M.A. Lovell *et al.*, *J. Neurol. Sci.*, 1998, **158**, 47.
83. S.W. Suh *et al.*, *Brain. Res.*, 2000, **852**, 274.
84. C. Gonzalez *et al.*, *Eur. J. Clin. Invest.*, 1999, **29**, 637.
85. R.A. Cherny *et al.*, *J. Biol. Chem.*, 1999, **274**, 23223.
86. X. Huang *et al.*, *Biochemistry*, 1999, **38**, 7609.
87. X. Huang *et al.*, *J. Biol. Chem.*, 1999, **274**, 37111.
88. C.S. Atwood *et al.*, *Soc. Neurosci. Abstr.*, 1997, **23**, 1883.
89. A.I. Bush *et al.*, *J. Biol. Chem.*, 1994, **269**, 12152.
90. C.S. Atwood *et al.*, *J. Neurochem.*, 2000, **75**, 1219–1233.
91. I. Klatzo *et al.*, *J. Neuropathol. Exp. Neurol.*, 1965, **24**, 187.
92. A.I.G. McLaughlin *et al.*, *Brit. J. Indust. Med.*, 1962, **19**, 253.
93. A.C. Alfrey *et al.*, *N. Engl. J. Med.*, 1976, **294**, 184.
94. D.A. Bushinsky *et al.*, *J. Bone Miner. Res.*, 1995, **10**, 1988.
95. M. Touam *et al.*, *Clin. Nephrol.*, 1983, **19**, 295.
96. D.R. McLaughlan *et al.*, *Neurology*, 1996, **46**, 401.
97. A. Campbell, *Dephrol. Dial. Transplant.*, 2002, **17**, 17–20.
98. D.G. Munoz, *Arch. Neurol.*, 1998, **55**, 737.
99. F.M. Corrigan, G.P. Reynolds and N.I. Ward, *Biometals*, 1993, **6**, 149.
100. M. Prince, *Epidemiology*, 1998, **9**, 618.
101. C. Hock *et al.*, *J. Neural. Transm.*, 1998, **105**, 59.
102. D. Wenstrup *et al.*, *Brain Res.*, 1990, **533**, 125.
103. C.M. Thompson *et al.*, *Neurotoxicology*, 1988, **9**, 1.
104. E. Andrasi *et al.*, *Arch. Gerontol. Geriatr.*, 1995, **21**, 89.
105. J.R. Connor *et al.*, *J. Neurosci. Res.*, 1992, **31**, 75.
106. M.A. Smith *et al.*, *Proc. Natl. Acad. Sci. USA*, 1994, **91**, 5710.
107. M.A. Smith *et al.*, *J. Neurosci.*, 1997, **17**, 2653.
108. L.M. Sayre *et al.*, *J. Neurochem.*, 2000, **74**, 270–279.
109. A. Nunomura *et al.*, *J. Neurosci.*, 1999, **19**, 1959.
110. M. Sano *et al.*, *N. Engl. J. Med.*, 1997, **336**, 1216–1222.
111. B.T. Hyman *et al.*, *J. Neuropathol. Exp. Neurol.*, 1993, **52**, 594–600.
112. A.E. Roher *et al.*, *J. Biol. Chem.*, 1996, **271**, 20631.
113. J.-Y. Lee *et al.*, *J. Neurosci.*, 1999, **19**, 1.
114. A.I. Bush *et al.*, *Science*, 1994, **265**, 1464.
115. X. Huang *et al.*, *J. Biol. Chem.*, 1997, **272**, 26464.
116. T.M. Bray and W.J. Bettger, *Free Radical Biol. Med.*, 1990, **8**, 281.
117. M.P. Cuajungco *et al.*, *J. Biol. Chem.*, 2000, **275**, 19439.
118. C.S. Atwood *et al.*, *Cell. Mol. Biol.*, 2000, **46**, 777.
119. C.J. Maynard *et al.*, *J. Biol. Chem.*, 2002, **277**, 44670.

120. R.A. Cherny *et al.*, *Neuron*, 2001, **30**, 665.
121. D. Vigo-Pelfrey *et al.*, *J. Neurochem.*, 1993, **61**, 1965.
122. A. Kontush *et al.*, *Free Radical Biol. Med.*, 2001, **30**, 119.
123. C.W. Ritchie *et al.*, *Arch. Neurol.*, 2003, **60**, 1685.
124. T.A. Bayer *et al.*, *Proc. Natl. Acad. Sci. USA*, 2003, **100**, 14187–14192.
125. A. Schmechel *et al.*, *J. Biol. Chem.*, 2003, **278**, 35317.
126. C. Treiber *et al.*, *J. Biol. Chem.*, 2004, **279**, 51958.
127. M.A. Smith *et al.*, *Proc. Natl. Acad. Sci. USA*, 1997, **94**, 9866.
128. R.J. Castellani *et al.*, *Free Radical Biol. Med.*, 1999, **26**, 1508.
129. M.A. Smith *et al.*, *Brain Res.*, 1998, **788**, 232.
130. D.A. Loeffler *et al.*, *J. Neurochem.*, 1995, **65**, 710.
131. D.J. Pinero, J. Hu and J.R. Connor, *Cell. Mol. Biol.*, 2000, **46**, 761.
132. G.M. Bishop and S.R. Robinson, *J. Neurosci. Res.*, 2003, **73**, 316.
133. G.M. Bishop and S.R. Robinson, *Brain Pathol.*, 2004, **14**, 448.
134. D. Schubert and M. Chevion, *Biochem. Biophys. Res. Commun.*, 1995, **216**, 702–707.
135. C.A. Rottkamp *et al.*, *Free Radical Biol. Med.*, 2001, **30**, 447–450.
136. U. Kumar, D.M. Dunlop and J.S. Richardson, *Life Sci.*, 1994, **54**, 1855.
137. H. Atamna, *Ageing Res. Rev.*, 2004, **3**, 303.
138. H.M. Schipper, *Ageing Res. Rev.*, 2004, **3**, 265.
139. K. Hirai *et al.*, *J. Neurosci.*, 2001, **21**, 3017.
140. H. Qiao *et al.*, *Neurobiol. Aging*, 2005, **26**, 849.
141. D.C. David *et al.*, *J. Biol. Chem.*, 2005, **280**, 23802–23814.
142. W.R. Markesbery *et al.*, *Neurotoxicology*, 1984, **5**, 49.
143. B. Angeletti *et al.*, *J. Biol. Chem.*, 2005, **280**, 17930.
144. M. Luczkowski *et al.*, *J. Chem. Soc. Dalton Trans.*, 2002, 2266.
145. D. Valensin *et al.*, *Dalton Trans.*, 2004, **16**.
146. K.J. Barnham *et al.*, *J. Biol. Chem.*, 2003, **278**, 17401.
147. J. Dong *et al.*, *Biochemistry*, 2003, **42**, 2768.
148. X. Huang *et al.*, *J. Biol. Inorg. Chem.*, 2004, **9**, 954.
149. T. Kowalik-Jankowska *et al.*, *J. Inorg. Biochem.*, 2001, **86**, 535.
150. J.F. Galey *et al.*, *J. Chem. Soc. Dalton Trans.*, 1991, **2281**.
151. T. Kowalik-Jankowska *et al.*, *J. Chem. Soc. Dalton Trans.*, 2000, **4511**.
152. T. Kowalik-Jankowska *et al.*, *J. Inorg. Biochem.*, 2002, **92**, 1.
153. T. Kowalik-Jankowska *et al.*, *J. Inorg. Biochem.*, 2003, **95**, 270.
154. C.D. Symel *et al.*, *J. Biol. Chem.*, 2004, **279**, 18169.
155. C.C. Curtain *et al.*, *J. Biol. Chem.*, 2001, **23**, 20466.
156. T. Miura *et al.*, *Biochemistry*, 2000, **39**, 7024.
157. J.W. Karr *et al.*, *J. Am. Chem. Soc.*, 2005, **126**, 13534.
158. S. Liu *et al.*, *Biochemistry*, 1999, **38**, 9373.
159. C.D. Syme and J.H. Viles, *Biochim. Biophys. Acta*, 2006, **1764**, 246–256.
160. K. Sorimachi and D.J. Craik, *Eur. J. Biochem.*, 1994, **291**, 237.
161. J. Huang *et al.*, *J. Biol. Inorg. Chem.*, 2004, **9**, 627.

Prion Diseases and Redox Active Metals

9.1 Introduction

In 1982, Stanley Prusiner and colleagues purified an abnormal protein from the brains of mice experimentally infected with a rare sheep disease called scrapie.[1] This protein was called the prion protein. Earlier work had suggested that this disease and others, loosely collected together as transmissible spongiform encephalopathies (TSEs), were not transmitted by conventional infectious agents. Prusiner suggested that this new protein was the infectious agent in all these diseases.[2] Such a contentious suggestion has lead to ferocious debates. Many researchers still maintained that there was no such thing as an infectious protein. Despite this, by 1990 most people accepted that the cause of the TSEs was the abnormal isoform of the prion protein Pruisner's research group had identified. The most convincing evidence for this had come from the work of Charles Weissmann, whose prion protein (PrP)-knockout mice could not be infected because they lacked expression of the protein that was now forever linked to these diseases.[3,4] Since then, it has become more widely accepted for these diseases to be termed prion diseases. In 1997, Stanley Prusiner won the Nobel Prize for his work on prion diseases.[5] Even then, there was still an element of resistance in the scientific community. It was considered that, in order for the transmissible agent to truly be a protein only, the protein would have to be generated from a recombinant source.

In 2004 that evidence emerged. Recombinant protein injected into mice led to a prion disease that could then be transmitted to other mice.[6] Naturally, scepticism still continues about this novel theory. Those who work in the prion field know that this is simply part of the game. Intense scepticism of any findings on any aspect of research in prion diseases makes progress in the field very slow. Part of the problem is probably due to a fundamental misunderstanding of the nature of prion diseases. Although the disease can be transmitted experimentally, prion diseases are not contagious diseases as are bacterial or viral diseases.

Some forms of prion diseases are inherited, such as Gerstmann–Sträussler–Scheinker syndrome (GSS).[7,8] GSS is linked to point mutations in the prion protein gene (*prnp*). Despite inheriting what amounts to a dominant lethal genetic mutation, GSS patients usually reach their fifth decade of life before symptoms of the disease emerge. This clearly links onset of these diseases to something inherent in the normal aging process. Other forms of transmission have occurred because of human intervention. Iatrogenic CJD results from use of tissues or hormone derivatives of tissue from people who had the most conventional disease, sporadic CJD (sCJD) (Figure 1).

Kuru, a disease of the native of New Guinea occurred because of the ritual practice of eating brains from older relatives.[9] The new disease, variant CJD (vCJD) had been linked to the eating of food contaminated with the bovine spongiform encephalopathy (BSE) agent. BSE and vCJD share many similarities and the two diseases clearly have a similar origin.[10,11] It is widely accepted that BSE caused vCJD, but there is also considerable doubt that vCJD arose from the eating of BSE-contaminated meat. BSE was also mostly the result of human intervention. The feeding of rendered animal remains back to dairy

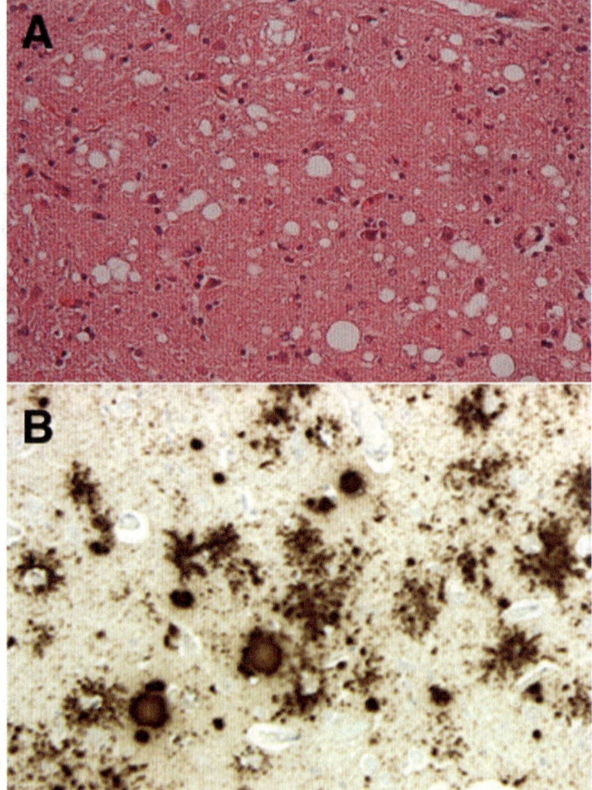

Figure 1 *Pathology in CJD. (A) A section through the brain of a CJD patient stained with eosin and haematoxilin. The characteristic vacuoles can be seen indicative of spongiform degeneration along with an absent of neuronal nuclear profiles. (B) An equivalent section stained with an antibody against the prion protein. Both plaques and diffuse aggregates of the abnormal prion protein can be seen*

cattle resulted in tens of thousands of cases of BSE carrying cows. Yet, now that this practice is banned and BSE numbers have dropped dramatically, there still remain a significant number of BSE cases. The cause of these cases of BSE remains unknown. This is similar to the majority of cases of human prion disease. The major form of human prion diseases is sCJD. This disease cannot be linked to any form of infection. Similarly, the disease of sheep called scrapie can also not be linked to any specific infection event. Scrapie is the first described prion disease with reports dating back to the 15th century.

As panic over the BSE epidemic subsides and the predicted exponential increase in vCJD cases has not happened, more rational thought has entered into the arena to assess the possible cause of the major forms of prion diseases. The two logical explanations that have been put forward are the following: The sporadic forms of prion diseases could arise through a freak event in the normal aging process. As mentioned above, GSS does not manifest until late in life. This implies that the kinetics of prion formation are very slow and take upwards of 30 years to result in abnormal prion protein forming in the brain. Alternatively, some changes that occur as we grow older may be needed to trigger protein conversion to make the normal cellular form of the prion protein to flip conformation and generate PrP^{Sc}. Once a significant amount of this protein is formed, it is able to catalyze its own conversion and spontaneous deposition of large amounts of PrP^{Sc} results in the pathological changes leading to CJD.[12] The possible change in the aging brain is the gradual decay in the balance between oxidative damage and antioxidant defense. The second hypothesis about the cause of sporadic prion diseases is that exposure to an agent in the environment may trigger protein conversion. Evidence for this comes from the existence of localized hot spots for different prion diseases.[13] The disease of deer, chronic wasting disease, is very heavily localized to small areas within USA, such as Colorado. In Iceland, some farms have recurrent scrapie problems while others remain consistently scrapie free. What in the environment could have such an effect remains to be verified. The prime candidate has been manganese.[14] Manganese accumulates in the brains of patients and animals with prion diseases.[15,16] However, it remains to be shown whether manganese is causative, a factor that enhances the incidence of ever present disease or is simply co-incidental.

The implication of all the forgoing is that prion diseases are fundamentally a result of a normal brain protein becoming conformationally altered (Figure 2). An event or a series of similar events result in the stabilization in a conformational switch in the isoform of the protein generated by the brain.[12] The trigger of this can be one of three possibilities. The first is the introduction into the cell of preformed PrP^{Sc} aggregates. This is then able to catalyze conformation alteration of prion protein generated by that cell. The second possibility is that the normal cellular isoform of the prion protein (PrP^{C}) encounters a different agent which then catalyzes conversion. This could be interaction with a metal that does not normally bind to the protein such as manganese.[17] Lastly, conversion to the abnormal isoform occurs naturally but with a low probability. This implies that the kinetic equilibrium does not favor PrP^{Sc} formation,

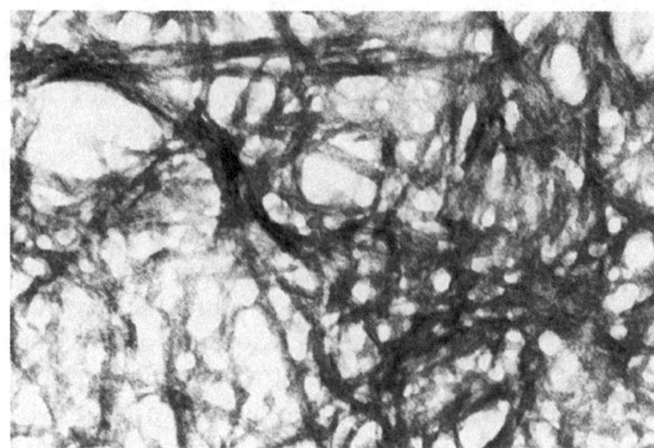

Figure 2 *Prion Fibrils. Electron microscopy and shadowing were used to analyse purified PrP[Sc]. Within aggregates of the abnormal prion protein, fribrils can be observed*

but that in time a small amount will form that is the sufficient to catalyze further conversion by the first mechanism. An alternative version of this last hypothesis is that PrP[Sc] is formed in the brain all the time, but mechanisms are in place which rapidly clear it away before it can auto-catalyze further PrP[Sc] formation. Disease develops when this corrective process falters, possibly as a result of aging.

Understanding the common threads in all these theories and the clear link between disease progress and expression of the prion protein and the conformation it assumes, makes discussion of any "contagion" causing these diseases seem absurd. Panic among both the lay and the scientific communities about the inherent infectiousness of prion disease is purely a hysterical response to misinformation or wanton ignorance of the fundamental truth of the cause of this disease. A normal brain glycoprotein becomes converted to a protease resistant isoform, by whatever mechanism, and initiates a series of pathological changes in the brain resulting in death. This implies that to understand these diseases we must know what causes the patient's own prion protein to change conformation and understand how production of this abnormal conformation relates to the pathological changes that cause the death of the patient.

Until recently, study of the prion protein has focused on PrP[Sc] and the normal form of the protein has been overlooked. However, the gene for PrP was first described[18] at the same time that BSE first emerged. The protein is a glycoprotein anchored to the outside of the cell by a glycosylphophatidylino-sitol (GPI) anchor.[19] This means that the protein is attached to the membrane by a sugar group. Although there has been some speculation about there being a transmembrane form of the protein, this has largely been dismissed.[20–22] Similar reports of dimeric forms of the protein and subcellular localization of the protein to the cytosol or nucleus are isolated and unconfirmed.[23–25] The protein is highly expressed by neurons and is concentrated in the synapse.[26] The expression of the protein is not specific to neurons and low level expression can be detected in many cell types. The age of the cellular prion protein began in

1995 with the first suggestion that a fragment of the protein could bind Cu.[27] This largely was ignored until 1997 when the first accepted evidence that the protein binds Cu *in vivo* was produced.[28] Since then there has been overwhelming support for the idea that PrP^c is a metalloprotein.

The function of the protein has been the subject of a number of investigations. Despite numerous different approaches the emerging consensus is that lack of expression of PrP^c causes cells to respond poorly to stress.[29] These changes can range from altered electro-physiological parameters,[30] altered sleep patterns,[31] modified cell adhesion characteristics[32] and disturbed cell signaling pathways.[33] More substantial evidence points to PrP^C being some form of antioxidant. Other research has suggested that PrP^C is a molecule with the ability to clear away superoxide radicals that would otherwise damage cell components.[34] This would make PrP^C a superoxide dismutase (SOD). Alternative research has shown that it can alter Cu uptake into cells[35] and that binding of Cu to PrP^C is important to the mechanism by which it is internalized from the cell surface.[36] These theories are not contradictory, as sequestering Cu is, in itself, an antioxidant effect. Cu has the potential to generate molecules that can cause oxidative stress. Therefore, the leading theory as to the function of PrP^c is that it is an antioxidant.

Conversion of PrP^C to PrP^{Sc} results in the loss of function of the protein without the loss of its expression. In PrP-knockout mice, lack of expression of PrP^C could be compensated by rapid changes in expression of other proteins that could perform similar functions. As a potential antioxidant, such rapid compensation is very likely considering the wide range of antioxidants that the body can mobilize and the high inducibility of many cellular antioxidants. One implication is that loss of PrP^C function could expose neurons to assaults that cause to initiate cell death. Therefore understanding the function of this protein and how to compensate for it could be one possible way to counteract the cell death that occurs in prion disease.

Prion diseases are neurodegenerative conditions. The result is a very rapid loss of neurons in specific areas of the brain. This neuronal loss occurs late in the disease and corresponds to onset of neurological and behavioral symptoms. In experimentally induced prion disease, there is a long incubation time between challenge with the prion disease agent and the onset of symptoms. In the case of BSE this can be years. Following onset of symptoms death from complications follows very rapidly. In humans with CJD this can be a matter of months. If cell death could be stopped then possibly, the CJD patient could recover. Clearly, knowing what causes this cell death is central to understanding these diseases. Surprisingly, until recent years, there has been little research on the mechanism of cell death in prion disease. Reviews on the subject of "neurodegeneration and prion disease" often failed to mention mechanisms of cell death in any detail. These models showed that PrP^{Sc} or a peptide derivative could kill neurons by an apoptotic mechanism.

The first models of the mechanism of neurodegeneration emerged from cell culture studies in the early 1990s. These models showed that PrP^{Sc} or a peptide derivative could kill neurons by an apoptotic mechanism.[37–39] The first finding that was of any significance was that neuronal cell death requires the expression of PrP^C by the target cell.[39] Cell culture studies were the first to show that neurons from PrP-knockout mice were resistant to toxic prions. This was later confirmed in animal models.[40,41]

Advancement in the field of neurodegeneration and prion diseases has occurred in recent years. Many individual and complementary approaches have been taken, providing a wealth of information that has the potential to one day provide us with a possible way forward in finding preventative treatments to halt the advance of neurodegeneration. Prion diseases are rare but so are reliable models of most human neurodegenerative diseases. In this regard, prion diseases are the exception as experiment infection of mice provides us with an accurate and essential tool for research. The implication of this is the study of prion disease, which might provide insights into neurodegeneration that are relevant to other diseases like Alzheimer's disease (AD) where animal models do not exist.

The main model used by researchers to study prion diseases is the mouse model using mouse passaged scrapie. Some researchers also use hamsters but the availability of transgenic mice makes mice a more attractive choice. Studies with such mice have lead to a whole range of interesting research. The sheep disease scrapie can be divided into a series of strains or different forms. These strains retain a range of characteristics when used to infect mice of a similar genetic background. These characteristics include the length of time the animal takes to fall ill (incubation time), localization within the brain of pathology and extent and localization PrP^{Sc} deposition as well as the ratio between the amounts of the three glycol-forms (di-, mono-, or non-gycosylated) of PrP^{Sc} detected.[42] Challenge with the scrapie agent is usually performed either by force feeding mice scrapie agent laced food (oral challenge) or direct injection of the agent into the brain. Oral challenge is a less successful route of infection but it has provided insight into the mechanism of oral transmission of prion disease.

In terms of the study of neurodegeneration, the mouse model has proved a difficult one to provide a mechanism of action. This is because it is difficult to separate the cause of neuronal death from the necessity to introduce PrP^{Sc} into the brain from an external source. Apoptotic cell death occurs in the brain and this is preceded by the activation of microglia and occurs in parallel with increased astrogliosis.[43–46] Very early changes to neurons can be detected such as loss of dendritic spines.[47] Use of conditional PrP-knockout mice has shown that stopping expression of PrP^C during the disease progress, after considerable PrP^{Sc} has been formed, results in cessation of cell death and recovery of the animal.[41] This really just confirms what was first identified in 1994 using cell culture models.[39] Namely, that PrP^C expression is necessary by the target cell and without it toxic prions cannot kill neurons.

As can be seen by the foregoing discussion research into prion diseases crosses a vast territory. The wealth of research in recent years means that many aspects have been studied in depth. Metal binding to the prion protein and its consequential changes in function and structure are one of these expanding areas. Of all the diseases related to metals, prion diseases have been the most dramatically altered by the single finding of the metal binding to PrP^c. This chapter will review these findings.

9.2   Cu-Binding to PrP^c

Cu binds to PrP^C *via* the repeat region in the *N*-terminal unstructured domain of the protein (Figure 3, *vide infra*).

This region is either an octameric repeat in mammals or a hexameric repeat in other species. Depending on species, the number of repeats varied from 4 to 6 units, each with a single histidine. Although the octarepeat motifs within PrP^C have no sequence homology to classical Cu-binding proteins,[48] evidence suggests that these motifs bind Cu(II) with a remarkable degree of selectivity; as such they may comprise the prototype of a new class of Cu-binding motif. Recombinant PrP^C ($rPrP^C$) can be expressed and purified in large amounts from inclusion bodies in an *Escherichia coli* expression system.[49] The protein is denatured in 8 M urea and subsequently refolded by gradual removal of urea. $rPrP^C$ can be refolded in the presence of Cu(II), which binds specifically in the octameric repeat region and also was demonstrated to have increased solubility.[34] Cu-binding to the *N*-terminal octarepeat region of human recombinant PrP23–98 has now been demonstrated using equilibrium dialysis.[28] It has also been demonstrated for synthetic peptides using mass spectrometry,[27,50] fluorescence spectroscopy,[50] Raman Spectroscopy,[51] circular dichroism; proton nuclear magnetic resonance (NMR), spectroscopy,[52] electron paramagnetic resonance (EPR) and electron spin-echo envelope modulation spectroscopy.[53]

Cu^{2+}-binding has also been reported for nearly full-length forms of Syrian hamster (residues 29–231)[54] and human PrP (residues 91–231)[55] and full-length mouse PrP (residues 23–231).[56] Affinity chromatography using immobilized Cu^{2+} ions has been used to purify mature, glycosylated PrP^C isolated from hamster brain.[57] With regard to stoichiometry, the number of Cu(II)-binding sites in the *N*-terminal region of PrP^C has been variously reported as between 2 and 5.6 and pH dependent.[28,50,52–54,58] At neutral pH, Cu^{2+}-binding to the *N*-terminal domain occurred in the micromolar range with positive co-operativity. There was a remarkably close correlation of Hill co-efficients calculated by different laboratories: 3.4 (PrP23–98),[28] 3.3 (PrP58–91)[52] and 3.6 (PrP23–98).[56] More recently, the suggestion was made that there are two, independent high affinity Cu^{2+}-binding sites of 10^{-14} and 4×10^{-14} M, deduced from the analysis of PrP58–98 and PrP91–231, respectively.[55] However, the copper-binding affinity to PrP^C is still a controversial issue with K_d varying from micromolar to nanomolar range.[50,56]

Studies with deletion mutants confirm that up to 4 Cu^{2+} ions bind to the octameric repeat. They found that four Cu^{2+} ions were missing from each molecule of the mutant protein compared to full-length $rPrP^C$. This corresponded well with evidence showing that the histidines in the octameric repeat region coordinate with the four Cu^{2+} ions.[17,34] Other mutants lacking one, two or three octarepeats were also studied and showed that the amount of

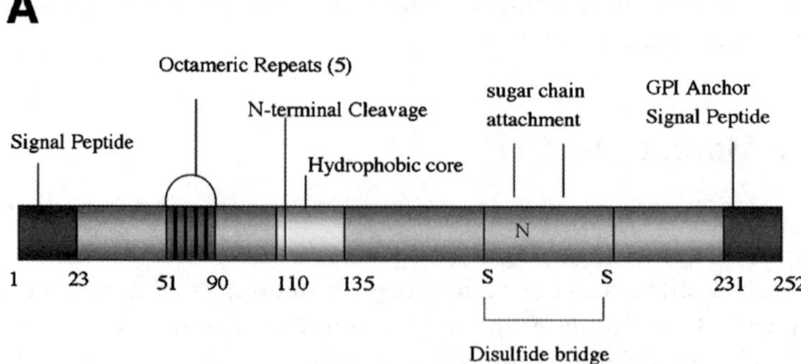

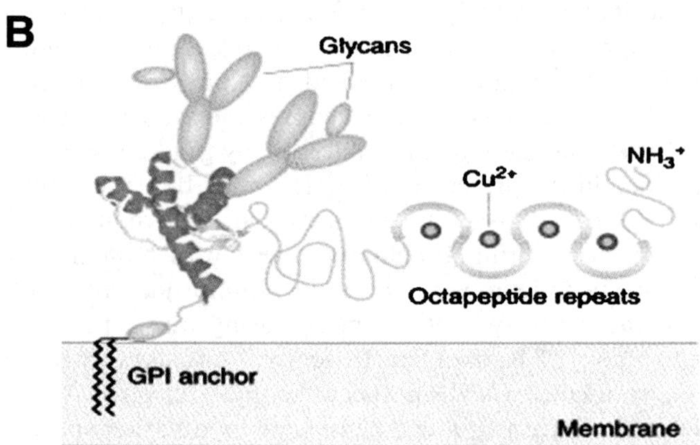

Figure 3 *Schematic representation of PrP. (A) Linear representation of the PrP sequence. Numbers are based on the mouse sequence. (B) Graphic representation showing the secondary structure of the globular domain and the approximate location of the sugars chains (glycans). The relation of the protein to the membrane is also shown. This protein is anchored to the cell membrane by a GPI anchor. The signal peptide for entry into the endoplasmic reticulum and the GPI signal peptide are cleaved off before the protein reaches the cell surface. Glycosylation can occur at one, two or none of the asparagine residues indicated. A hydrophobic region envelopes a cleavage point where the protein is cleaved during normal metabolic breakdown. A disulphide bond links two regions of the protein which form separate alpha-helices in the three dimensional structure of the protein. The complete octarepeats can bind up to four Cu atoms. Most mammals also have an incomplete repeat prior to this*

Cu^{2+}-binding is directly proportional to the number of repeats, although affinity values were not calculated.[17] Several other authors also suggested that Cu^{2+} can bind to PrP^C along the more structured *C*-terminal domain of the protein.[59,60] Continuous wave EPR studies demonstrated that on binding, Cu^{2+} first fills the *C*-terminal-binding sites before occupying the octarepeats at the *N*-terminal.[59] However, these findings were not confirmed by studies of mutant proteins lacking the *N*-terminal. These mutants do not show any Cu^{2+}-binding to the *C*-terminal of the protein.

9.3   Details of Cu^{2+} Co-Ordination to Mammalian PrP^C and its Fragments

9.3.1   Binding of Cu^{2+} Ion by a Single Octapeptide Repeat Pro-His-Gly-Gly-Gly-Trp-Gly-Gln

The known structures of several mammalian PrP^C show two characteristic protein domains, a flexible unstructured *N*-terminal tail consisting of about 100 amino acid residues and the globular three-dimensional domain of the *C*-terminal comprising in human protein residues 125 to 228[61–65] (Figure 4). A globular domain of mammalian PrP^C consists of three helices and two short anti-parallel sheets. The structures of the *C*-terminal globular domain of human, bovine, murine and Syrian hamster prion proteins are very similar to each other with human being closest to bovine PrP^C.[64] The rigidity and some structure organization of the *N*-terminal unstructured domain may strongly depend on pH.[66] The X-ray structures of the globular domain[67,68] agree well with those evaluated by NMR.

The impact of *C*-terminal domain on Cu^{2+}-binding to prion protein seems to be of minor importance (*vide infra*).[69]

The major Cu^{2+}-binding domain involves 4–6 His residues being located in the *N*-terminal unstructured domain, including octarepeat region comprising amino acid residues 60–91 (4 His residues) and peptide fragment PrP 91–126 (with 2 His residues, in human protein they are His-96 and His-111).

Single octarepeat peptide sequence (PHGGGWGQ) contains only one effective anchoring site for Cu^{2+} ion: imidazole nitrogen of the His side-chain[70] which seems to play a fundamental role in Cu(II) ion coordination to PrP^C. In the major species formed around neutral pH, Cu(II) binds the octapeptide *via* imidazole nitrogen and two amide nitrogen donors.[71] Earlier study[51–53] proposed the involvement of His side-chains and one[52] or two[51,53] amide nitrogens. Imidazole was suggested to act as the bridging unit in 2-, 3- and 4-octarepeat fragments.[52] The detailed EPR and ESEEM (electron spin-echo envelope

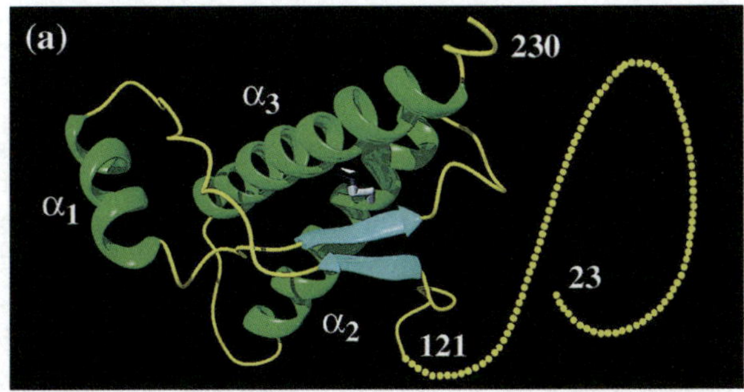

Figure 4   *Three dimensional structure of bovine prion protein (23–230)* (Reproduced with permission from Ref. 64)

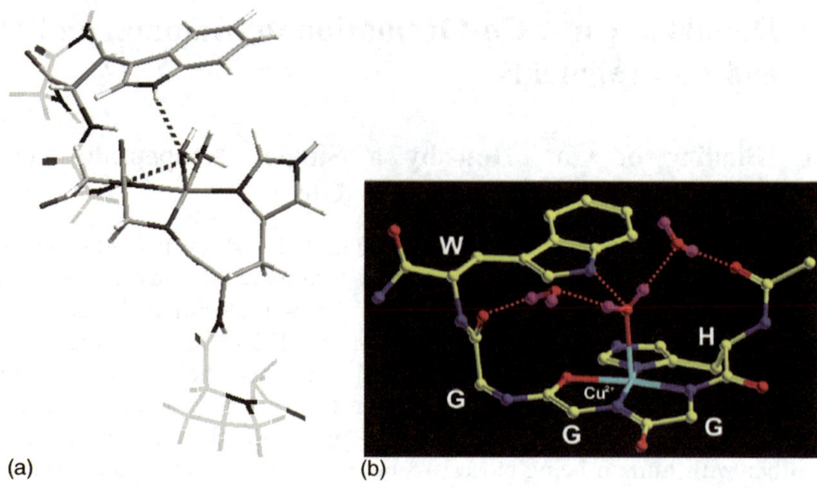

(a)   (b)

Figure 5   *(A) Molecular structure for the CuH$_{-2}$L complex in solution for the Ac-PHGGGWG-NH$_2$ peptide. (B) Crystal structure of the HGGGW fragment in complex with Cu^{2+} ion*
(Reproduced with permission from Ref. 73)

modulation) studies[53] have shown that around pH 7.4 imidazole does not act as the bridging unit and Cu^{2+} ion is coordinated by imidazole and two amide nitrogens of vicinal Gly residues as suggested earlier by Raman study.[51] Coordination of Cu^{2+} ion with imidazole nitrogen and three amide nitrogen donors was also suggested.[72] It is rather surprising that Cu^{2+} ion interactions with the simple octapeptide could lead to so many different models of coordination. The metal ion coordination to a single octapeptide was finally supported by careful potentiometric, NMR[71] and X-ray studies.[73] The coordination mode involved imidazole nitrogen and two amide nitrogens of two Gly[71,73] residues as well as carbonyl oxygen of the Gly residue.[73] The NMR[71] and X-ray[73] studies have shown also the new very important feature, the close contact between Cu^{2+} ion and Trp side-chain resulting from the hydrogen bond formation *via* metal ion bound water (Figure 5). The latter finding allows to understand the ability of the octarepeat region to reduce Cu^{2+} ion to Cu^{+} with a basic role of Trp (*vide infra*).[74,75]

9.3.2   pH-Dependence of Cu^{2+} Binding to Octapeptide Fragment

The structure of the major species discussed above (CuH$_{-2}$L) dominates only within relatively narrow pH range. The Cu^{2+} binding to protected PHGGGWGQ octapeptide is strongly pH dependent and the potentiometric titrations have indicated that lowering pH from 7.4 to below 6.5 releases free Cu^{2+} ions from the peptide complex (Figure 6).[71] In the milimolar concentrations at pH 6 about 40% of Cu^{2+} is released from the peptide. The high pH sensitivity of Cu^{2+} binding to the octarepeat unit may have a basic influence on

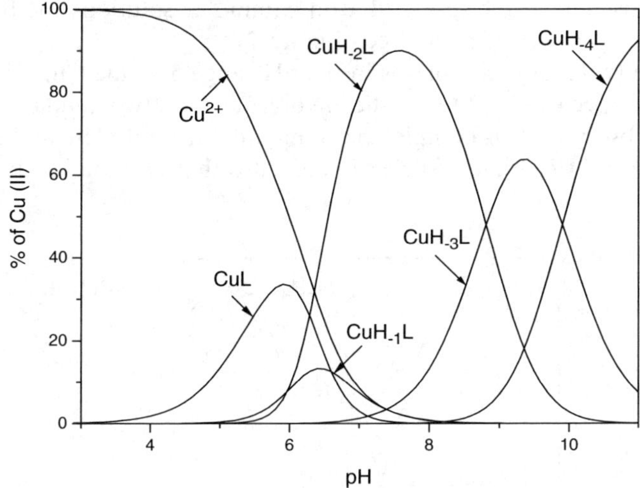

Figure 6   *Species distribution profile for Cu^{2+} complexes of Ac-PHGGGWGQ-NH$_2$; Cu^{2+} to peptide molar ratio 1:1, [Cu^{2+}] = 0.001 M*

the biological function of PrP^C, especially on its transport properties. As pH in the cell environment could vary from 6 to 8, these changes in the Cu^{2+}-binding ability might be biologically relevant, *e.g.* allowing PrP^C to release metal ion inside the cell as hypothesized by Burns *et al.*[73]

The interesting impact on the stability of the major complex of Cu^{2+} with single octarepeat around pH 7.4 have three Gly residues located on the *C*-terminal side of His residue. The substitution of (Gly)$_3$ motif by other residues *e.g.* (Ala)$_3$ decreases distinctly the ability to bind Cu^{2+} ion by octapeptide.[76] This property could be critical for the pH-dependent transport ability of PrP^C toward Cu^{2+} ions.[71,73]

One of the major difficulties when discussing the biological relevance of Cu^{2+} interactions with single octapeptide repeat is the relatively low stability of the complex formed around neural pH. The coordination mode N$_{im}$, 2N$^-$, CO is usually found in the Cu^{2+} complexes with very simple peptide containing one His residue.[70] Human PrP^C contains, however, four such octarepeats and the binding properties of 32 amino acid peptide fragment with four His residues differ considerably from that of a single octapeptide unit.

9.3.3   Binding of Cu^{2+} Ions to Dimeric and Tetrameric Octapeptide Fragments

Protected dimeric octarepeat peptide Ac-(PHGGGWGQ)$_2$-NH$_2$, (2-Oct), contains two potential anchoring sites for a Cu^{2+} ion: two His side-chain imidazoles. Each of these imidazoles is able to coordinate one metal ion, however it is also likely that both of them will coordinate simultaneously to one Cu^{2+} ion when metal is present in lower amount. The calculations based on the potentiometric data have shown that in equimolar solutions the formation of the mono-copper species is favored (Figure 7).[77]

There are two major complexes in the pH range 6–8, the CuL and $CuH_{-2}L$. In the CuL species (L=2-Oct), the involvement of two imidazoles in Cu^{2+} binding is observed. This complex predominates around pH 6 and its structure evaluated by NMR (Figure 8) clearly indicates that two imidazoles are bound

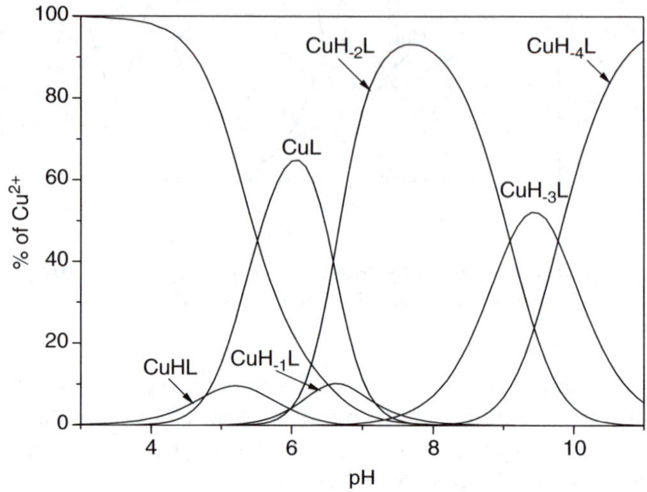

Figure 7   *Species distribution profile for Cu^{2+} complexes of Ac-$(PHGGGWGQ)_2$–NH_2 at 25°C and I = 0.1 M KNO_3. $[Cu^{2+}] = 1 * 10^{-3}$ M, Metal to ligand ratio 1:1*

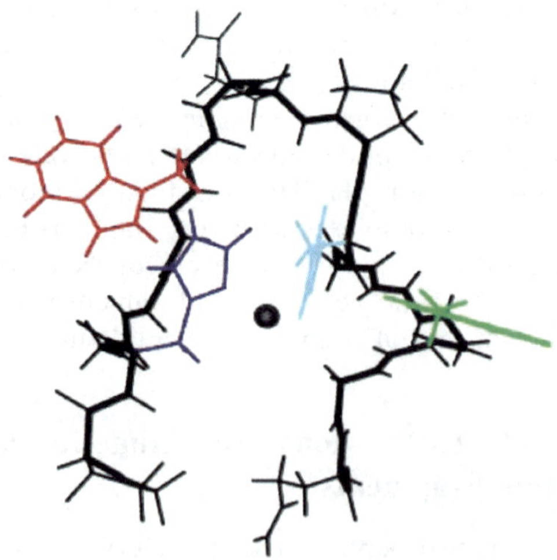

Figure 8   *Structural details of structure obtained for the CuL complex of prion dimeric octapeptide repeat (Ac-$(PHGGGWGQ)_2$–NH_2). Structure calculation was performed by restrained molecular dynamics with simulated annealing in the torsional angle space. The Figure was created with MOLMOL 2K.1.0*

to one Cu^{2+} ion by two different nitrogen donors, N_π and N_τ, respectively.[77] This coordination pattern is often called "inter-repeat" binding.[78] Around pH 7.4 the major species, ($CuH_{-2}L$), is the same as in the case of a monomeric protected octapeptide 1-Oct, involving imidazole(s) and two amide nitrogens of Gly residues ("intra-repeat" binding). However, the stability of the complex with 2-Oct is distinctly higher when compared to that of 1-Oct.[77]

For the higher metal amount, *e.g.* 2 to 1 metal to ligand molar ratio, around pH 7 2-Oct coordinates two copper ions, one on each His-binding site plus 2 or 3 amide nitrogen donors.[77]

Extension of the peptide sequence to tetrameric unit Ac-(PHGGGWGQ)$_4$-NH$_2$ (4-Oct), domain existing in mammalian PrP^C, changes coordination ability of the peptide ligand quite considerably when compared to 1-Oct or even 2-Oct. In equimolar conditions Cu^{2+} is able to bind simultaneously four imidazoles forming the CuL species dominating at pH range 5.5–7.5 (Figure 9). This binding mode leads to a very stable complex, which could be biologically more relevant than those formed for the simple mono-His containing systems.[70,77] Its stability is several orders of magnitude higher than the respective complexes with 1-Oct or 2-Oct. The comparison of the concentration of uncoordinated Cu^{2+} for *e.g.* 1-Oct and 4-Oct shows that at pH 6 about 40% of Cu^{2+} is ligand-free, while in 4-Oct only several % of total metal is unbound.

It is interesting to note that in CuL, Cu^{2+} ion is surrounded by four imidazoles (Figure 10) resembling closely the copper-catalytic center of SOD1.

The other species being in equilibrium with CuL at pH 7.4 is the complex, which, besides imidazoles, involves also one amide nitrogen ($CuH_{-1}L$).[77]

Extensive EPR studies have indicated that at pH 7.4 poli-imidazole binding occurs indeed involving three *N*-terminal His side-chain (His-60, His-68 and

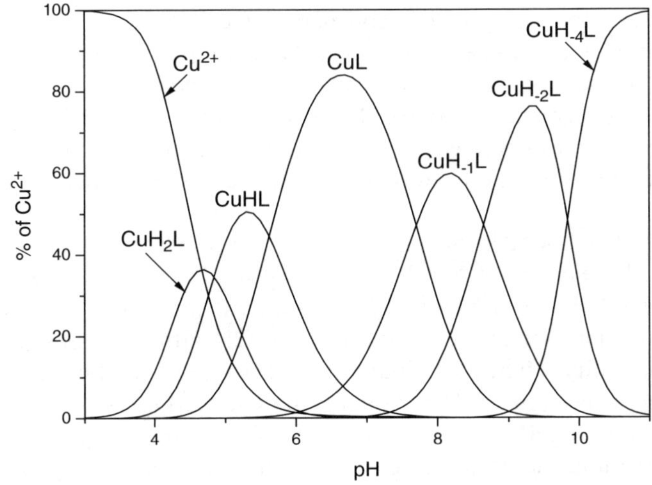

Figure 9 *Species distribution profile for Cu^{2+} complexes of Ac-(PHGGGWGQ)$_4$–NH$_2$ at 25°C and I = 0.1 M KNO$_3$. [Cu^{2+}] = 1 * 10^{-3} M, Metal to ligand ratio 1:1*

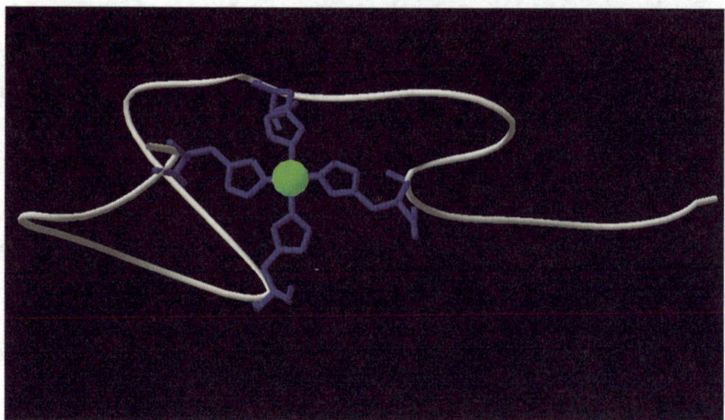

Figure 10   *Schematic representation of structure of Cu^{2+} complex of octarepeat tetramer*

His-76) or all four imidazoles.[79] This species dominates around pH 6.5 as it was found also by the potentiometric studies.[77]

At higher concentrations of Cu^{2+} the formation of poly-Cu^{2+} complexes is observed. 4-Oct is able to coordinate one, two, three and four Cu^{2+} ions per peptide unit as it is seen in the case of PrP^C. The coordination pattern depends on the amount of coordinating Cu^{2+} ions. For 4 Cu^{2+} ions around neutral pH the binding mode follows that seen for the Cu^{2+}_1-Oct system and around pH 7 in the major species each metal ion is bound by imidazole and two (or three) amide nitrogens.[77] However, the binding ability of 4-Oct seems to be much higher than 1-Oct or 2-Oct. The higher ability of 4-Oct to bind Cu^{2+} ions may derive from the fact that the binding of the first metal ion causes pre-organization of the peptide structure (multi-imidazole or inter-repeat binding) which could induce easier coordination of the successive Cu^{2+} ions. This could be one of the reasons for the positive co-operative effect proposed earlier in the coordination of Cu^{2+} ion to PrP^C.[28,52,56] The number of bound Cu^{2+} ions to octarepeat domains has a critical impact on the structure of the *N*-terminal domain and its chemical reactivity. The coordination of one Cu^{2+} ion to 4-Oct sequence creates the effective species able to reduce Cu^{2+} to Cu^+. This could be critical reactivity necessary for Cu-PrP^C to act as a copper reductase (*vide infra*).

For a fully Cu^{2+} loaded protein, assuming four bound metal ions according to established structures[51,53,71,73,77,79] implies that the metal–ion binding sites in the *N*-terminal PrP^C domain are like beads on a string, where each bead is a Cu-HGGGW segment separated by intervening Gly-Gln-Pro links. Interestingly, glycine and proline often participate in β-turns and thus the intervening links may provide a mechanism for allowing the Cu-binding segments to fold and perhaps collapse together. Coordination dominated by a single histidine would make a PrP^C molecule bound to Cu^{2+} pH sensitive, supporting the idea that it may function *in vivo* as a Cu transporter, taking Cu into the acidic environment of lysosomes and then releasing it.

The coordination observed in the fully loaded tetrameric repeat is, however, not very effective for Cu^{2+}/Cu^+ redox site. The tetragonal coordination -including imidazole and two or three amide nitrogens could stabilize very effectively Cu^{2+} oxidation state not allowing its easy reduction to Cu^+.

9.4 The Fifth and Sixth Binding Sites Located in *N*-Terminal Domain

Although there is now little doubt that copper binds to the octameric repeat region, the "fifth Cu-binding sites"[28,55] are still a matter of controversy. Some authors propose that Cu binds in the form of Cu^{2+} in the *C*-terminal domain of PrP^C. A mutation in amino acid residue 198 abolishes this binding, suggesting that the histidine involved in binding might be His-187. Other authors suggest that the site of Cu^{2+} binding is in the region of the toxic peptide PrP106–126. If the latter is true, the histidine within the peptide would be the one involved in binding, with additional coordination from the appropriate nitrogen atoms in the vicinity.

Recent work by Qin *et al.*[80] used mass spectroscopic-based foot-printing techniques to attempt to position the His-dependent metal coordination sites in Cu bound to PrP^C. This technique allowed for the total number of histidines involved in metal coordination to be determined by measuring mass differences between apo- and metal coordinated proteins or peptides. They confirmed the Cu^{2+} coordination sites at the four octarepeat histidines PHGGG/SWGQ (residues 60, 68, 76 and 84 of mouse PrP^C) and also suggested a second type of site involving His-95, in the related sequence GGGTHNQ, possibly in conjunction with His-110 (equivalents to His-96 and His-111 in human PrP).[55,81] These sites have binding affinities estimated in a range from 2.2×10^{-6} to 10^{-14} M for the octarepeats and 5×10^{-6} to 4×10^{-14} M for the site involving His-95.

Further analysis of this site suggested that the two histidines (His-96, His-111) are necessary for Cu^{2+} binding outside the octameric repeat.[82] A competition study between two peptides one consisting of the 5th site and the other of the octameric repeat regions suggested that the 5th site would compete the octameric repeat region for Cu^{2+} binding. Studies with Ni^{2+} instead of Cu^{2+} suggested that each of these two histidines could bind one metal ion, independently of the other.[83] The study also suggested that metal binding at this 5th site could initiate structural transitions in the protein that would favor its conversion to the abnormal form. However, studies on fragments and peptides should be taken with a note of caution. Binding of metals are influenced by the side groups adjacent to the key ligands involved in coordination of the metals and the flexibility of the protein is also influential in metal binding. Studies from our group suggest that Cu^{2+} binding to the 5th site in the full-length protein is much lower than that of the octameric repeat region and is much more sensitive to pH changes being absent as an effective metal binding site at pH 6.6. Truncation of the protein at the beta cleavage site (around amino acid residue 89) exposes this 5th site. In such truncated molecules, this site could play a role in metal binding. However, in the full-length protein, Cu^{2+} binds to the octameric repeat region first,[84] most like due to formation of the tetra-imidazole binding to Cu^{2+} ion.[77]

9.5 Binding of Cu^{2+} and Other Metals to PrP91–126 Region Cu(II) Coordination to PrP106–126 (KTNMKHMAGAAAAGAVVGGLG)

As mentioned above PrP106–126 has been shown to be highly fibrillogenic, resistant to proteinase K and toxic to neurons *in vitro*[38,85] and *in vivo*.[86] To be toxic, however, it requires expression of PrP^{C}[39] as it has been described for the toxicity of PrP^{Sc}.[87] PrP106–126 interacts with PrP^{C} in the same region as PrP^{Sc} (hydrophobic region PrP112–119) and that is why PrP106–126 is very interesting model peptide for neurotoxic activity of PrP^{Sc}.[88,89] Copper may be involved in the peptide interactions and aggregation,[39,90,91] thus detailed chemistry of the metal binding to the nerotoxic PrP peptide is very much needed. The neurotoxic peptide has two potential anchoring sites for Cu^{2+} ions, the *N*-terminal amino group and imidazole of His-111. There are also other functions able to interact with Cu^{2+} in like Lys side-chain amino groups and Met thioether sulfur. Neither of these groups is an effective binder but they may complete the metal ion coordination sphere or interact with metal ion *via* a metal-bound water molecule involving hydrogen bond system.

Jobling *et al.*[91] studying the interactions of Cu^{2+} and Zn^{2+} ions with PrP106–126 have shown that both metals induce peptide aggregation, which could be abolished by addition of chelating agent reducing levels of metal ions. Although neurotoxic effects are the same for N-protected and unprotected peptide the Cu^{2+} ion interactions with neurotoxic peptide could differ due to involvement of the *N*-terminal amino group in metal–ion binding in the case of unprotected peptide. On the basis of the study of Cu^{2+} coordination to unprotected PrP106–126 and a set of its derivatives the latter work suggested the involvement of *N*-terminal amino nitrogen, His-111 imidazole nitrogen, carbonyl oxygen of peptide bond and thioether sulfur of Met-112.[91] According to this work the Cu-S bond participates in the aggregation process. Mutagenesis of His-111, Met-109 or Met-112 abolished neurotoxicity of PrP106–126 and its ability to form fibrils. The coordination of sulfur to Cu^{2+} ion was also proposed by Hasnain *et al.*[81] based on XAFS data and by Di Natale *et al.* based on potentiometric and spectroscopic data.[92] The involvement of thioether sulfur is likely in very acidic conditions but around neutral pH the thioether donor must compete with much stronger donors, such as the amide nitrogens adjacent to *N*-terminal amino group or His-111. The thermodynamic studies supported by spectroscopic data including NMR have not confirmed the involvement of a sulfur donor in the direct coordination to Cu^{2+} ions.[93,94] The NMR data clearly indicated the coordination of *N*-terminal amino nitrogen, His-111 imidazole nitrogen and adjacent amide nitrogen(s). The sequence of PrP106–126 peptide consists of two distinct domains, the *N*-terminal, polar and Cu^{2+}–ion binding domain comprising amino acids 106–111 (KTNMKH) and *C*-terminal hydrophobic chain from Met-112 to Gly-126 (MAGAAAAGAVVGGLG). The latter sequence is basic for the aggregation process. The hydrophobic chain does not interact with Cu^{2+} ion but metal–ion binding to *N*-terminal peptide fragment may influence distinctly the structure of the hydrophobic domain (Figure 11)[95] which is free to interact with the other peptide hydrophobic chain.

The NMR study of Zn^{2+} interactions with PrP106–126 have shown only negligible effects on NMR parameters, while Mn^{2+} has shown strong interactions with carbonyl oxygens of Gly-124 and Leu-125 and indirectly with His-111 imidazole.[95] The interactions of Mn^{2+} with Gly carbonyls could be of importance for the neurotoxicity mechanism. The glycine residues have a critical impact on the peptide flexibility and the aggregation ability. The substitution of Gly-114 and Gly-119 by more rigid Ala has reduced flexibility of this prion fragment increasing peptide solubility and decreasing ability to aggregate, although the mutated peptide is still highly neurotoxic even if it does not aggregate.[96] Thus, the aggregation process might not be a critical requirement for the peptide neurotoxicity.

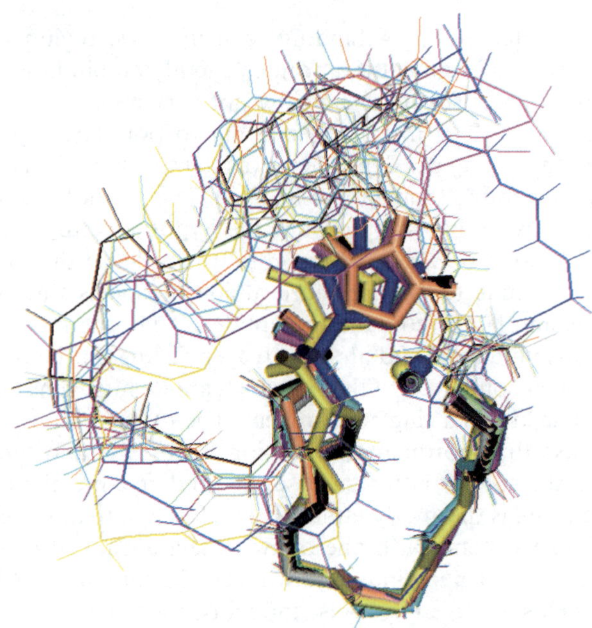

Figure 11   *NMR structure of Cu(II) complex of prion neurotioxic peptide*

9.5.1 Involvement of His-96 in the Interaction of Cu(II) with the Neurotoxic Peptide Fragment

On the *N*-terminal side of neurotoxic peptide PrP106–126 is located another His residue able to potentially interact with metal ions at position 96 (human PrP, hPrP). His-96 can act as an anchoring-binding site for Cu^{2+} ion as it does His-111. Two coordination modes were proposed for Cu^{2+} to PrP fragment containing both His-96 and His-111; two metal ions bind to His-96 and His-111 and the adjacent amides independently of each other,[97,98] or one Cu^{2+} bound by both His residues (Figure 12).[82,99] Both models may not exclude each other, as in case of binding of one metal ion involvement of two His residues is very likely, although their impact on the complex stability could be very different. In case of two metal ions bound in this region both His must be coordinated to different metal ions to make the binding possible.

Recent work of Viles group, which proposed earlier binding of two His to one Cu^{2+} ion,[82] based on the studies with Ni^{2+} probe has suggested that each His residue may coordinate one metal ion.[100] This finding again is chemically clear as in the case of binding of two metal ions; each His residue must be anchoring one of the metal ions or ligand-free metal ion would undergo hydrolysis. The comparative studies have shown that His-111 is much more effective in Cu^{2+} binding that His-96.[100,101]

9.5.2 The Comparison of the Binding Abilities of Octameric and Neurotoxic Regions Toward Cu^{2+} Ions

Several papers devoted to Cu^{2+} binding to neurotoxic region of PrP try to suggest that PrP91–126 fragment is a primary metal–ion binding site in PrP^C.[82] This question is not precise and chemically not very clear and that is why the answers are confusing. The metal–ion binding to poly-His peptides has two distinct features, the strong pH dependence (as to most peptides) and the strong dependence on the peptide to metal molar ratio. The physiologically important pH region, especially taking into account the hypothesis about copper transport by PrP^C, is located between pH 6 and 8. It is believed that in the high pH range when Cu^{2+} binds to two or more amide nitrogens the more effective binding site in a simple His-containing peptide is that one in which Cu^{2+} ion coordinates to His imidazole and amide nitrogens being on the *N*-terminal side of His residue, when compared to that in which amide nitrogens are situated on the *C*-terminal side like in a single octarepeat. The reason could be very simple, in the former case the coordinated Cu^{2+} ion forms six- and five-memebered chelate rings, while in the latter case seven- and five-membered rings. The seven-membered ring is thermodynamically less favored than the six-membered ring and for simple systems with one or two chelate rings, it could be a case. However, in the case of Cu^{2+} binding to whole or major part of PrP meaning of such question is elusive. Even if we use model peptides the comparison should be done between the whole tetrameric octarepeat unit with four His residues and PrP91–120, which contains two coordinating His residues (His-96 and

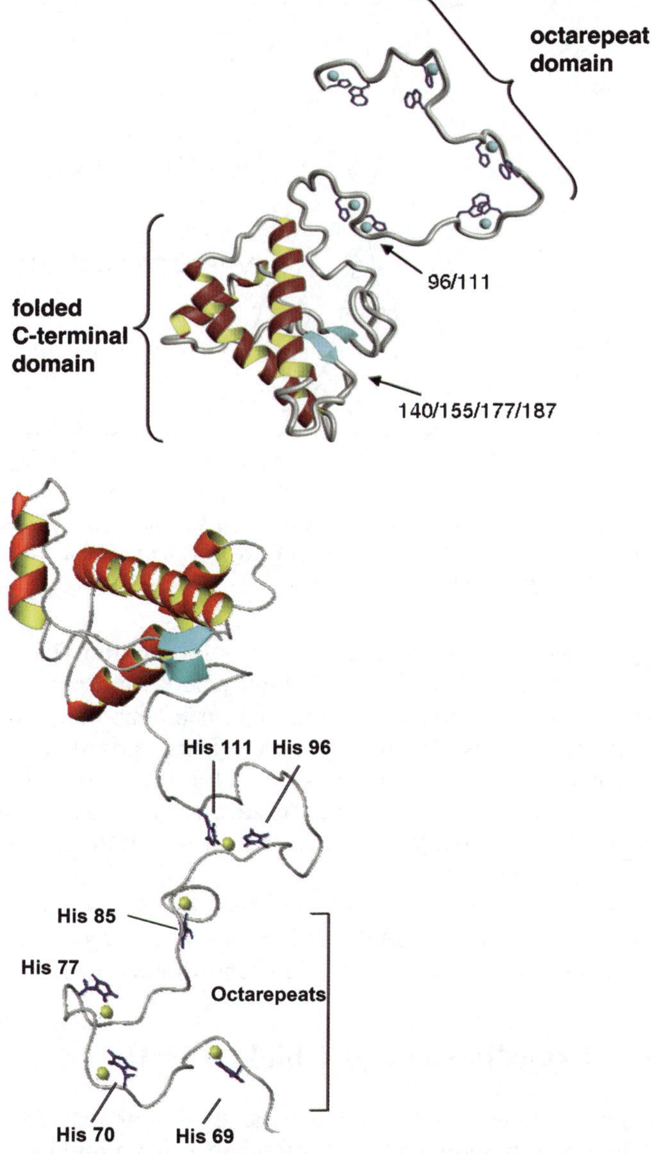

Figure 12 *Models of three-dimensional structure of PrP(61-231) with coppers included* (Reproduced with permission from Ref. 82 and 97)

His-111). Assuming the binding of one Cu^{2+} ion to protected 4-Oct and to the PrP91–120, the simple hypothetical plot expressing stability constants (Figure 13) clearly shows that majority of Cu^{2+} is bound to octarepeat peptide below pH 7.2, while PrP91–120 is a major binding site close to pH 8.[77,101]

Although this plot is representing graphically the stability constants rather than the situation in real solution it clearly shows that the involvement of the

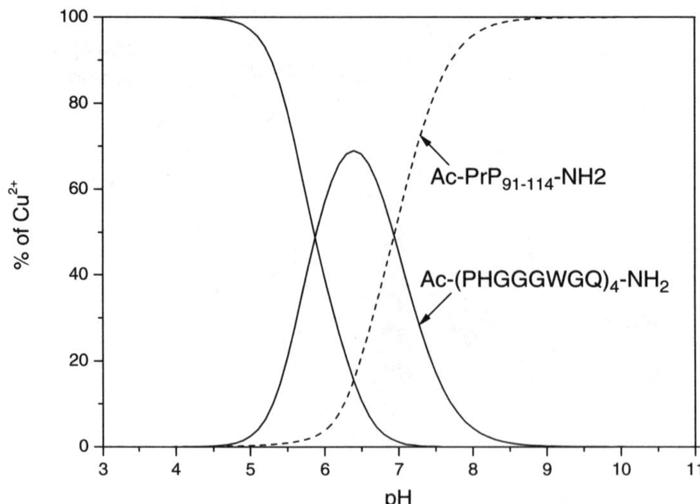

Figure 13 *Distribution profiles of competition between human tetra-repeated octapeptide (PHGGGWGQ)₄ (4-Oct) and neurotoxic fragment of human PrP (Ac-PrP₉₁₋₁₁₄–NH₂) in coordination of one Cu²⁺*

amide nitrogens occurring above pH 7.5 makes His-96 and His-111 the competing site for the Cu^{2+} ion at relatively high pH. Within pH range 5–7.5 at which single Cu^{2+} ion can be coordinated by four imidazoles of 4-Oct peptide the octarepeat fragment is a distinctly more efficient coordination site. This competition could be different but it is not[84] for the whole PrP protein. It should be also mentioned that the side chains around His anchoring sites could also be critical for Cu^{2+} complex stability and the direction (*N*- or *C*-terminal) of metal ion-bound amides.[101]

For binding of more than one Cu^{2+} ion, the positive co-operative effect is suggested, thus, the power of metal–ion binding by octarepeat domain could increase for the higher number of Cu^{2+} ions coordinated to PrP.

9.6 Cu^{2+} Coordination to Chicken PrP

The chicken prion protein is rather phylogenically remote PrP from the human protein.[102] There is only around 30% identity between mammalian and chicken protein, but 90% amino acid identity within bird species. (The amino acid identity within the mammalian species is also close to 90%.) The chicken PrP is strongly homologous to mammalian protein in the central region but diverges considerably in the *N*-terminal domain including tandem repeats.[103] While mammalian PrP consists of octapeptide repeats at the *N*-terminal, chicken PrP (chPrP) contains hexapeptide repeats (Pro-His-Asn-Pro-Gly-Tyr). Whereas it is well documented that mammalian PrP causes neurodegenerative diseases there is no evidence that birds may have any prion-related disorder.[104] However, Cu^{2+} ions are able to induce internalization of the chPrP inside the cell *via* endosome mechanism exactly as they do it for the mammalian PrP.[105] Similar to mammalian PrP, chPrP also exhibits the antioxidant activity probably due to Cu^{2+} binding.[34] Thus, Cu^{2+} interacts with chPrP causing the specific biological responses. The interactions of chPrP with Cu^{2+} ions may also allow understanding the differences between the octa- and hexa-repeat peptides and the effect of the metal–ion binding on the structure of PrP and biological implications.

ChPrP with its rather low amino acid identity to mammalian PrP has surprisingly a very similar molecular structure to that of the mammalian protein[106] with a flexible unstructured *N*-terminal tail and the globular *C*-terminal domain (Figure 14). The *C*-terminal domain consists of three α-helices, one short 3_{10}-helix and a short anti-parallel β-sheet. The tandem hexapeptide repeats $(PHNPGY)_n$ are located at the flexible *N*-terminal (residues 53 to 94).

Earlier work of Hornshaw *et al.* has shown that the *N*-terminal tandem repeats of chPrP binds Cu^{2+} ions in a concentration dependent manner.[27] The binding of a Cu^{2+} ion by a single hexapeptide unit is quite specific and different than that found in the case of mammalian octapeptide.[107,108] The His residue acts as an anchor and then His amide nitrogen enters the coordination.[109] The absorption and CD spectra suggested some involvement of the Tyr phenolate in the metal ion coordination around pH 7.[107] However, sophisticated NMR studies have shown that the involvement of the Tyr side in the Cu^{2+} ion coordination does occur but only in the specific peptide isomer.[110] Pro-residues involving their imino groups in the amide bonds form *cis-* and *trans-*isomers of

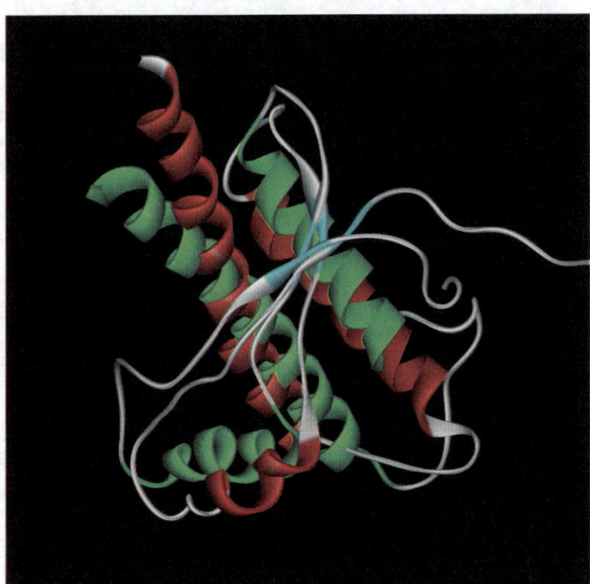

Figure 14 *The similarity of tertiary structures of mammalian (green) and chicken (red) PrP*

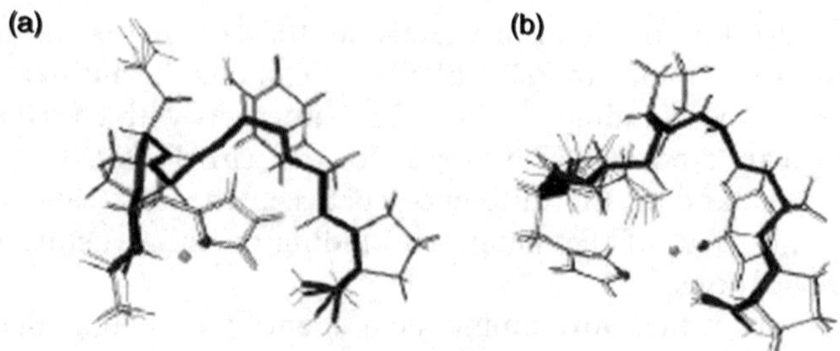

Figure 15 *Best five structures of Cu(II)Ch-PrP hexapeptide trans-trans (a) and cis-trans (b) conformers*

the peptide bond. When two Pro are within the peptide sequence *trans/trans* and *cis/trans* isomers are likely. The simultaneous binding of His and Tyr$^-$ is possible only in the *cis/trans* isomer (Figure 15).[110]

In the longer hexarepeats, *e.g.* dimer, Ac(HNPGYP)$_2$NH$_2$ to tetramer Ac(HNPGYP)$_4$NH$_2$ units similar to octapeptide fragments, the multi-imidazole inter-repeat coordination is observed.[107,109] In the equimolar solutions the coordination by three or four imidazoles to a trimeric (Figure 16) or tetrameric hexa-peptide fragment is a dominant binding mode within the pH range 6–8 as it was the case in mammalian PrP fragments (*vide supra*). This coordination mode could be biologically relevant both for the coordination pattern of Cu^{2+} to PrP as well as SOD enzyme activity. Comparing the binding power between monomeric units of chicken hexapeptide and mammalian octapeptide (Figure 17) the mammalian peptide is distinctly more efficient ligand for Cu^{2+} than the chicken analogue.

The kind of a cooperative effect observed for mammalian tandem repeats is also seen for the chicken hexapeptides (Figure 18). In the case of tetrameric repeats below pH 6 4-Hex dominates over the 4-Oct, while above this pH 4-Oct is more powerful ligand than the chicken tandem repeat (Figure 19). Above pH 6, the amide nitrogen donors become critical for the complex stability and octarepeats using two or three amides become more powerful ligands.[77,107,109]

9.7 Copper Mediated PrP Internalization

Two factors have been suggested that are important for the internalization of PrP^C. The first of these is an *N*-terminal sequence of the protein not including the octameric repeat region.[111–113] Internalization driven by the *N*-terminal causes PrP^C to move out of lipid raft domains to membrane regions that will form coated pits.[114] This response seems to be critically dependent on the basic amino acids KKRPKP at the *N*-terminal. It is possible that this domain interacts with other proteins,[115] but there is also evidence that this domain interacts with another internal domain within the hydrophobic region.[116]

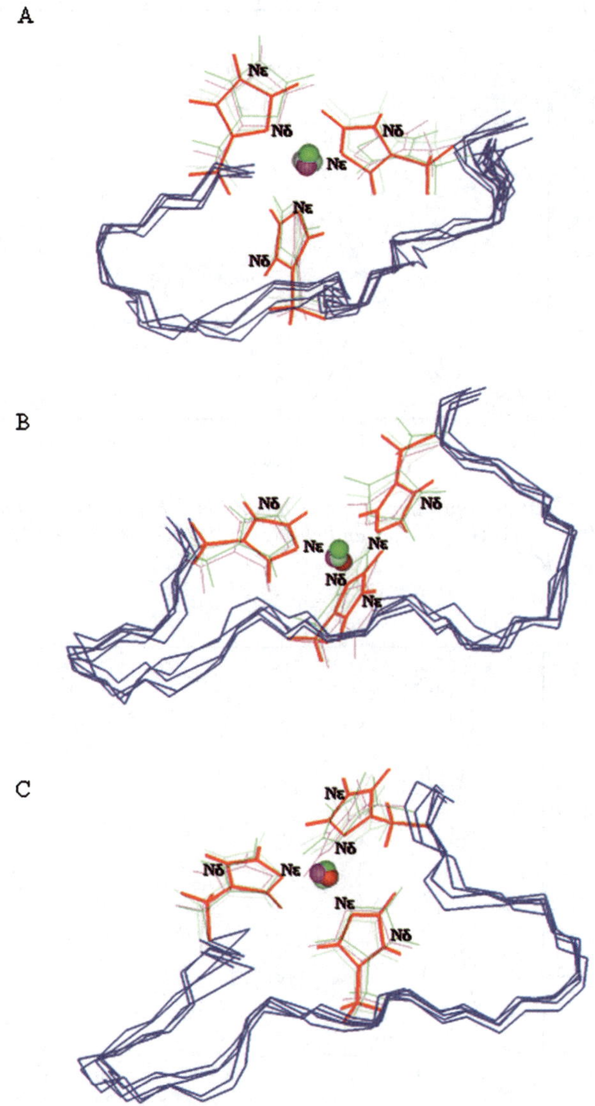

Figure 16 *Superimposition of Cu(II)-ChPrP(54-71) structures from energy minimization (magenta) and after 12 ps molecular dynamics (MD) calculation (red), the other coloured structures represent the snapshots from the MD trajectory. (A) {N_δ His-54, N_ε His-60, His-66} Cu(II) donor set; the RMSD of the structures was 0.71 ± 0.25 Å for backbone atoms and 0.88 ± 0.25 Å for heavy atoms (B) {N_δ His-60, N_ε His-54, His-66} Cu(II) donor set; the RMSD of the structures was 0.72 ± 0.21 Å for backbone atoms and 0.93 ± 0.25 Å for heavy atoms (C) {N_δ His-66, N_ε His-54, His-60} Cu(II) donor set; the RMSD of the structures was 0.75 ± 0.23 Å for backbone atoms and 0.89 ± 0.25 Å for heavy atoms*

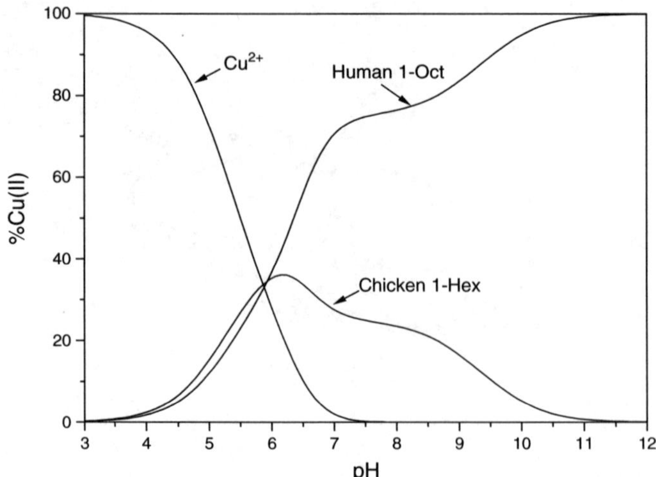

Figure 17 *Distribution profiles of competition between human octapeptide (PHGGGWGQ) 1-Oct and chicken (HNPGYP) 1-Hex in coordination of one Cu²⁺*

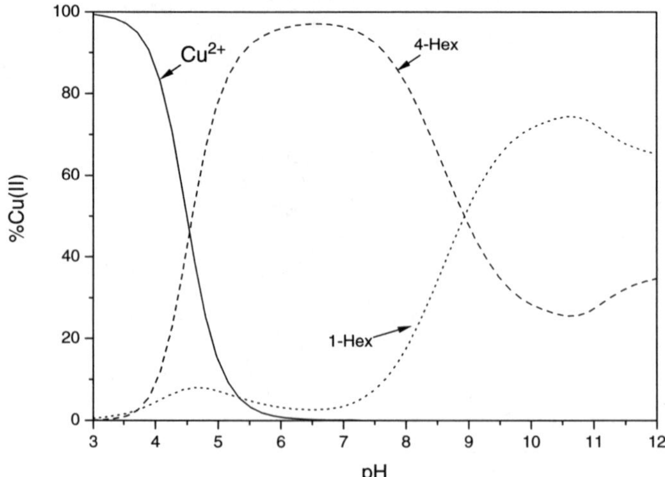

Figure 18 *Distribution profiles of competition between 1-Hex (dot line) and 4-Hex (dash line) in coordination of one Cu²⁺*

As indicated above, the second factor suggested to regulate internalization is binding of metal ions.[36,111] However, internalization of xenopus PrP^C (a PrP^C that has been suggested not to bind Cu) occurs in a similar way to mammalian PrP^c,[113] which would imply that internalization may not be Cu dependent, though this remains unproven. Cu binding to PrP^C at the cell surface causes the protein to enter endosomes.[105] This would clearly deliver Cu into the cell. A further suggestion is that internalization may be simply a propensity of

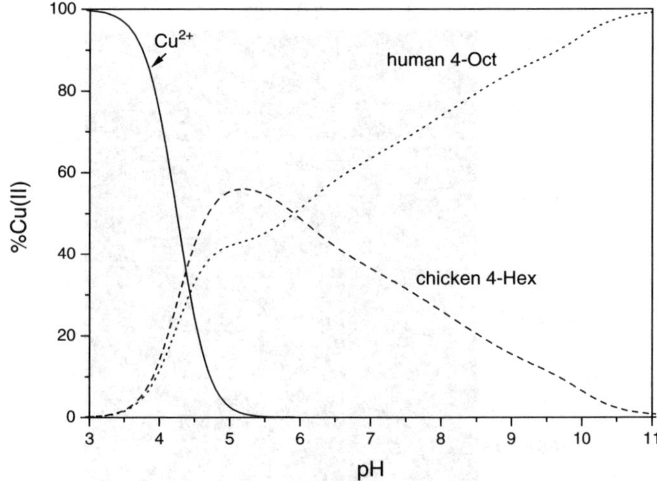

Figure 19 *Distribution profiles of competition between human tetra-repeated octapeptide* *(PHGGGWGQ)$_4$ (4-Oct) and chicken tetra-repeated hexapeptide (HNP-* *GYP)$_4$ (4-Hex) in coordination of one Cu^{2+}*

possessing a GPI anchor,[117] though given the evidence already considered this would seem unlikely.

As well as being internalized, there is some evidence that PrP^C is also secreted from the cell.[118] Possibly, PrP^C can be transferred between cells by this mechanism. There is evidence that metal ions such as Cu can induce this release[119] and that such shedding is a result of the action of metalloproteases.[120] However, there is also a metal independent shedding process. Inherited mutations in the *prnp* gene protein coding region can also modulate the membrane association of the protein,[21,121,122] and can prevent the protein from reaching the plasma membrane at all.[22]

A definitive answer to the question of what regulates internalization has come from a recent paper.[123] In this work Cu^{2+} was applied in a chelated form with glycine. Under these conditions as little as 100 nM Cu^{2+} was able to stimulate significant internalization of PrP^C. This concentration of Cu is equivalent to that found in cultured medium.[124] The implication is that under normal culture conditions internalization is driven by Cu^{2+} binding at the cell surface (Figure 20). This is clearly in contrast to previous work suggesting that binding of another protein to the very *N*-terminal of the protein is necessary for internalization. In this study, a mutant protein, lacking amino acid residues 23–38 of the mouse sequence could still be internalized by addition of Cu^{2+}.

In contrast, mutant proteins lacking either the octameric repeat region (amino acid residues 51–89) or the hydrophobic palindrome with the sequence AGAAAAGA (amino acid residues 112–119) showed no internalization of PrP^C and no response to Cu^{2+}. However, the expression of a single octameric repeat was sufficient to restore internalization in response to Cu. The implication of these findings is that Cu^{2+} binding to the octameric repeat region

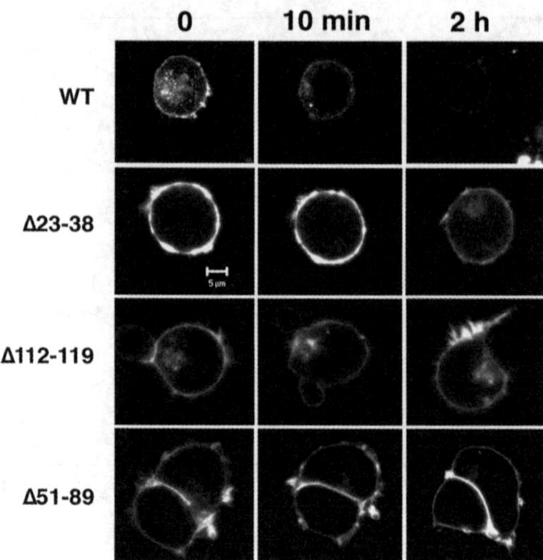

Figure 20 *Copper mediated internalisation of PrP. Cells were transfected with DNA constructs that allowed expression of PrP and mutants fused to GFP. GFP has no effect on the localisation of the protein. The cells expressing wild-type (WT) GFP-PrP were then treated with 100 μM Cu complexed with glycine. The same cells were visualised at time of addition (0) or at 10 min or 2 h later. Decreased GFP signal indicated internalisation of the protein and its rapid degradation. Cells expressing mutant PrPs were similarly treated Δ23-38, lacking the very N-terminus of the protein still responded to gluatmate while the mutants Δ51-89 (lacking the copper binding site) or Δ112-119 (lacking the hydrophobic palendrome) showed no response to copper, indicating that these domains were essential for internalisation of the protein in response to copper*

regulates internalization of PrP^C.[123] As conversion of PrP^C to PrP^{Sc} requires internalization of the PrP^C – PrP^{Sc} complex, then this suggests that availability of Cu to bind to PrP might limit protein conversion during disease.

9.8   Copper Transport

Experiments designed to look at the possible function of PrP^C in Cu transport employed the use of Cu^{67}. The studies involved three strains of mice: those over-expressing PrP^C, those deficient in PrP^C and wild-type controls. Transport of both chelated and unchelated Cu^{67} by cerebellar cells was studied in all three strains. Kinetic studies showed that only histidine chelated Cu^{67} was taken up at a rate proportional to PrP^C expression.[125] Kinetic parameters for Cu^{67} transport were also determined and the V_{max} values increased with higher expression of PrP^C while the K_m values (in the nM range) were not greatly different. These results suggested that there might be an increase in the number of Cu-binding sites within the cerebellar cells, which can be related to PrP^C expression. Further experiments in which cytosolic enzyme SOD1 was immuno-precipitated from cells loaded with Cu^{67} showed that the Cu could be incorporated into SOD1 in proportion with the level of PrP^C expression in the cells studied. Thus, PrP^C may participate in the incorporation of Cu into other cellular proteins.[126]

In the brain, the highest concentrations of PrP^C are found at the synapses and that synaptosomes of PrP-knockout mice demonstrated a strong reduction in Cu concentration. Also, Cu binding in the synaptic cleft showed a significant influence on synaptic transmission. Cu is released from the synapse during quantal release and it has been proposed that PrP^C may have a function in regulating Cu at levels at neuronal synapses. Several studies have been performed to understand this phenomenon. In one set of experiments, neuronal cells were loaded with Cu^{67}, allowed to release Cu spontaneously and subsequently treated with the depolarising agent, veratridine, which blocks Na^+ channels in neurons. Veratridine released Cu from cells at levels corresponding to the expression of PrP^C in those cells. Cells expressing no PrP demonstrated almost no veratridine induced Cu release while cells over-expressing PrP^C released higher quantities of Cu compared to wild-type.[126] Furthermore, electrophysiological experiments indicated that Cu applied to cerebellar slices inhibited the amplitude and frequency of inhibitory currents measured on Purkinje cells of PrP^C-deficient cells, but not wild-type cells suggesting that some protection against Cu is absent from the synapses lacking PrP^C. One proposal has been that PrP^C itself performs this protective function.[127]

9.9   PrP as an Antioxidant

The broadest range of evidence supporting a function for PrP^C suggests that it is an antioxidant. There are now a number of publications that shows oxidative damage as part of prion disease. Sections of brains from mice infected with mouse scrapie have been studied for markers of oxidative damage.[128] Additionally, biochemical studies to detect markers of protein and lipid oxidation have shown that such changes do occur as a result of prion disease.[129] During the conversion process from PrP^C to PrP^{Sc} the protein will lose any function that requires the native conformation. If PrP^C is an antioxidant in this form, then the loss of this activity due to conversion could contribute to the oxidative damage present in the brains of animals with a TSE. Mice lacking expression of PrP^C also show signs of oxidative damage.[130] In addition, exposure of mice to agents that cause oxidative damage in the brain produce far more damage in the brains of PrP-knockout mice than in wild-type mice indicating that they are more susceptible to oxidative damage.[131] The same PrP-knockout mice are less able to recover from sleep deprivation than wild-type mice.[132] This response to sleep deprivation is closely linked to a decreased ability of some brain regions to recover from oxidative damage. Similarly, the brains of PrP-knockout show altered electrophysiological parameters, which are likely to be due to inability to deal with oxidative stress with increased age.[133] PrP^C is also up-regulated in conditions tightly associated with stress. These include cerebral ischemia[134] or hypoxia-induced brain damage,[135] ageing,[136] AD[137] and in mouse ALS models.[138] PrP-knockout mice also show changes in other antioxidants.[139,140]

At the cellular level, cells lacking the expression of PrP^C are less able to deal with oxidative stress.[29,141,142] When cells are treated with agents that cause oxidative stress PrP^C is up-regulated.[143–148] Simple expression of the protein is not sufficient to bring about the changes that protect cells from oxidative damage. The protein must be tethered correctly to the cell membrane and be able to bind Cu^{2+}.[149] Also, cells that have been infected with prions are less able to respond to oxidative stress[150] with increased expression of antioxidants. Increased expression of prions causes an increased cellular resistance to oxidative stress.[143,151] This data collectively demonstrates the tight association between the expression of PrP^C and enhanced protection from oxidative damage. This implies that PrP^C is an antioxidant.

It has been demonstrated that both recombinant and brain-derived PrP^C have SOD-like activity when bound to Cu.[34,124] The depletion of PrP^C from cell extracts results in a lower SOD activity in the extract[152] and also when PrP^C converts to PrP^{Sc}, its SOD function is abolished.[16] Thus, it is possible that the normal function of PrP^C is to act as a SOD-like enzyme and control oxidative stress. It has, been however, demonstrated that PrP^C is different from cellular Cu/Zn-SOD, especially in the way that it forms its complex with Cu.[60] In addition, a very current question concerns oxidative stress as a factor causing neuronal damage in prion diseases and it is therefore of considerable interest to consider the SOD-like function of PrP in this context. Recombinant PrP^C, purified from *E. coli* expression systems and refolded with Cu^{2+}, can be used in quantitative assays designed to measure SOD activity. Experiments showed that both chicken and mouse rPrP^C could catalyze the dismutation of the superoxide radical at a rate equivalent to one-tenth of SOD1. SOD1 is a very potent enzyme, which catalyzes the reaction at around 100,000 times the spontaneous rate of superoxide degradation.[153] It was therefore concluded that PrP^C had significant SOD activity and this was also confirmed for native protein immuno-precipitated from mouse brain.[34] Recent evidence from D.R. Brown laboratory has also shown that PrP will generate hydrogen peroxide when Cu is bound. The generation of hydrogen peroxide is the standard product generated by the dismutation of superoxide. This new evidence confirms the possible enzymatic activity of PrP on superoxide.

Mutational analyses have been carried out to assess what components of the protein are necessary for the SOD-like activity.[154] In particular, the hydrophobic domain central in the protein was shown to be necessary for this activity. The implication is that two domains of the protein must interact for the activity. Finally the globular domain of the protein appears to play a role in regeneration of the protein to an active form. It is of particular interest to note that the two domains identified in this study as being important to antioxidant activity are identical to those found to be important to internalization. Therefore, from two different perspectives, the AGAAAAG palindrome and the octameric repeat regions are emerging as the regions with the highest functional significance.

There are currently three known SODs in mammals[155]: the cytosolic Cu/Zn SOD1, mitochondrial Mn-SOD or SOD2 and extracellular EC-SOD or SOD3. The former two are found in all cells at varying concentrations and often show increased expression under conditions of oxidative stress. EC-SOD exists in three different isoforms, binds one atom of Cu per molecule and is either released into the extracellular matrix or remains bound at the cell surface. In brain tissue, the expression of SOD3 is very low,[156] although it is elevated in PrP-knockout mice.[29] It is interesting to correlate these observations with the *in vivo* expression of PrP^C, considering its SOD activity. The expression of PrP^C is highest in the brain and particularly abundant at synapses. It is also present at neuromuscular junctions.[157] Thus, it has been proposed that PrP^C may serve as a synaptic SOD and may be released during transmission at the synapse. Superoxide is known to inhibit synaptic transmission and the presence of SOD activity in these regions of the nervous system may have a protective role. The possible protective role of PrP^C against oxidative stress was confirmed in cell culture using PC12 rat tumor cells, which can be differentiated into neurons using nerve growth factor (NGF). It was also observed that PrP^C expression increased in PC12 cell cultures on exposure to oxidative stress.[127]

The role of Cu^{2+} in the SOD mechanism of PrP is not yet clear. In classical Cu-containing SOD1 the dismutase reaction involves Cu^{2+}/Cu^+ redox cycle. For the effective and fast reaction the coordination pattern should be easily accommodated by both Cu^{2+} and Cu^+ ions. The donors like amide nitrogens and tetragonal geometry are favored by Cu^{2+} ions, while Cu^+ prefers trigonal or tetrahedral symmetry and perhaps instead of amide nitrogen donors the imidazole nitrogens, like it is the case in SOD1. Cu^{2+} bound to octarepeat region undergoes quite effective reduction to Cu^+. The most effective is the species having one Cu^{2+} ion bound by four imidazoles and the electron donor is the Trp side-chain indole ring.[75] Although the distance between Trp and Cu^{2+} ion in 1:1 Cu^{2+}–4-Oct species is not so well established as in the case of Cu^{2+}–1-Oct system (*vide supra*) the basic role of Trp in this redox process is well supported. The kinetic studies have shown that 4-Oct tandem repeat is essential for the peptide mediated reduction of Cu^{2+}[75] and the most effective pH (~ 6.5–7.0) is that at which Cu^{2+} forms complex with involvement of the imidazole donors only.[77] Trp mutation to Ala reduces the power of the peptide ability to produce Cu^+ almost completely.[74,75] This reduction process involving Trp could be useful in the copper transport and transfer into chaperons or other proteins. It shows, however, that Cu^{2+}/Cu^+ redox cycle is easily available only in the species in which one copper ion coordinates four imidazole nitrogens. This is in fact the main species in the Cu^{2+}–4-Oct system around pH 7 in equimolar solution.[77] If the process described above occurs PrP^C acts in biology also as copper reductase.

9.10   Manganese Binding

Interaction between PrP and other metals has also been studied but to a lesser degree. In addition to metal binding, it has also been suggested that interactions between metals can also increase the aggregation and conformational change of PrP. In particular, it has been suggested that Cu and Mn can accelerate the formation of aggregated, protease resistant protein.[17,158–160] Metals have a high potential to interact with β-sheet structural elements and therefore have the potential to exacerbate aggregation. It was found that Cu has the potential to enhance the infective potency of scrapie infection.[161] In contrast, treatment with penicilamine, a Cu chelator, lengthens the incubation period of scrapie, indicating a protective effect.[162] It has been found that hydrogen peroxide causes oxidation of methionine residues in PrP.[163] In addition, hydrogen peroxide can cause the cleavage of PrP at an alternative site in the presence of Cu[164] but hydrogen peroxide can also be generated when PrP or fragments of PrP interact with metal, such as Cu and Fe.[165] These data indicate that interactions between PrP and metals need not be beneficial.

Although most studies have examined the binding of Cu^{2+} to PrP, a small number of studies have also examined the potential of the protein to bind other metals. Studies with peptides have shown very little binding of any other metal besides Cu^{2+}.[27,58,108] Studies with full-length recombinant protein have suggested that three other metals could bind to the protein, manganese, nickel and zinc.[28] Of these Ni^{2+} ions may have in some cases the greatest similarity to Cu^{2+}, but measurements of its affinity in comparison to Cu^{2+} have suggested that this affinity is very poor. There has been more interest in zinc and PrP but again there is no strong evidence for metal substitution, only some suggestions that the expression of PrP^C at the cell surface might alter zinc entry into the cell.[166] However, initial studies indicated that manganese could compete equally with Cu for binding to PrP.[17] Since then, further studies have verified that manganese can bind to PrP.[82,167] It appears, however, that the exact binding site of Mn^{2+} to PrP is different to that of Cu^{2+}, with a preference for the proposed 5th metal binding site in the region of the two histidines at amino acid residues His-96 and His-111.[95] Mn^{2+} as a "hard" metal ion prefers oxygen rather than nitrogen donors and it was found to bind effectively to carbonyl oxygens of Gly-124 and Leu-125 residues in the neurotoxic peptide region.[95] In contrast to studies with full length PrP[17] studies with peptide suggest that Mn^{2+} ions do not interact with the octarepeat region effectively and are not able to substitute Cu^{2+} ions coordinated there.[108]

The binding of Mn^{2+} has one fundamental difference to that of Cu^{2+}. The binding of manganese to the protein leads to conversion of the protein to a protease resistant isoform that is a different conformation.[17] Acquisition of this β-sheet rich conformation requires time but could occur within minutes with exposure to infra-red.[167] We observed that PrP with manganese bound to it could form extensive fibrils when exposed to a broad range of the near infra-red spectrum. Additionally, PrP expressing astrocytes were also found to express a protease resistant form of PrP when grown in a high concentration of manganese. PrP with manganese bound was also shown to be capable of catalyzing superoxide dismutation indicating that the metal bond could undergo redox cycling.[34] However, this activity was lost following the conformational change from α-helix to β-sheet. Further analysis indicated that when Cu^{2+} is bound to the protein that the copper is largely protected from the interaction with water molecules, indicating that the redox cycling of Cu is tightly regulated.[167] The structure of Cu^{2+} bound PrP was found to be very stable, more stable than the protein without metal ion bond. However, PrP with Mn^{2+} bound was very unstable increasing the probability that the Mn would heavily oxidize the protein. This oxidation is likely to be responsible for the conformational change and aggregation of the protein. Manganese bound PrP converted to certain conformations is able to catalyze conversion of further PrP molecules to polymerize.[168] This polymerization was not auto-aggregation, as the substrate protein did not form aggregates, even after several weeks of incubation. This indicates that Mn-PrP was truly able to act as a seed for PrP polymerization. Therefore, the investigation of manganese-mediated PrP polymerization is a potentially important model for understanding the conversion process that generates PrP^{Sc}.

9.11 Cell Death and Metals

Currently, the majority of research suggests that neurodegeneration results from the innate toxicity of PrP^{Sc}. As this toxicity cannot occur without neuronal production of PrP^{C}, then clearly the neurotoxic mechanism requires two species of PrP. PrP^{Sc} is potentially necessary but not sufficient for neuronal death, similarly, loss of PrP^{C} function is necessary but not sufficient. The majority of experimental evidence suggests that PrP^{C} a Cu^{2+}-binding protein[17] that can function to transport Cu^{35} and/or PrP^{C} is an antioxidant requiring Cu binding for its activity.[34] Further consistent evidence from research suggests that oxidative damage is a feature of prion diseases[128] implying that break down of oxidative defense might contribute to neuronal death in prion disease. Generation of PrP by the brain might be sufficient to inactivate compensatory mechanisms that might come into play when PrP expression is impossible because of genetic ablation of the gene or temporal inactivation of gene expression. Therefore, expression of PrP that is either functionally inactive or can be converted to PrP^{Sc} is necessary for the mechanism of neurodegeneration.

The majority of research examining the mechanism of neurodegeneration that might be relevant to prion diseases has been carried out using cell culture techniques. The assumption used in these studies is that exposure of pure or mixed cell populations to PrP^{Sc} results in the induction of cell death either by apoptosis or some other mechanism. A considerable percentage of this research has focused on a single peptide fragment known as PrP106–126 because it is based on amino acid residues 106–126 of the human PrP sequence. The first study

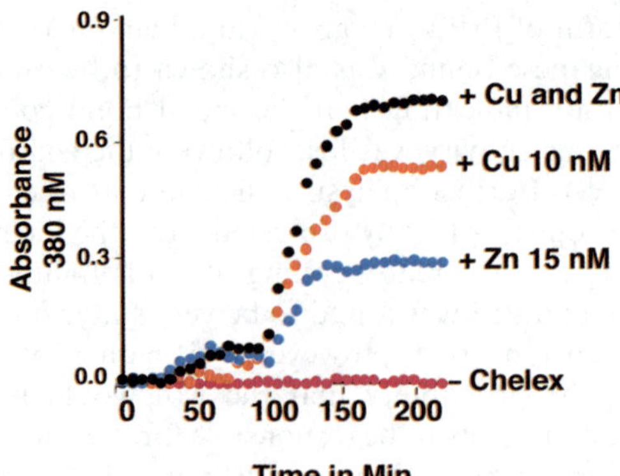

Figure 21 *Metal mediated prion protein peptide aggregation. A spectrophotometric assay was used to monitor the aggregation of the peptide PrP106-126. PrP106-126 was prepared in a buffer contain 10 mM Tis pH7.4 after treatment of the buffer with chelex medium which depletes the solution of metal ions. Chloride metal salts were then added and turbidity of the solution measured over time. Cu caused aggregation of the peptide*

using this peptide[38] established that this fragment is more toxic than other fragments and induces apoptosis in neurons. Since then, several other studies have examined different fragments either overlapping this region[169] or from elsewhere in the protein sequence.[170] Some of these fragments have the advantage of being more soluble than PrP106–126. However, the relative insolubility of PrP106–126 can be overcome by preparing the peptide in a different solvent such as DMSO and adding it to the culture in this form.[171] Whereas some PrP peptides are toxic, others have been found to be protective. In particular, peptides related to the octameric repeat region of the *N*-terminal have been shown to protect neurons from Cu toxicity by their potential to bind Cu.[172,173] This in contrast to PrP106–126 where there is some evidence that the toxic effects of the peptide are mediated by or require an interaction with Cu ions (Figure 21).[87,91,165]

Analyses of the toxicity of PrP106–126 have shown that the mechanism involves two basic effects. These effects will be discussed in detail. Both effects are necessary for the toxic effect.

9.11.1   Direct Effects

PrP106–126 interacts directly with cells and alters the cells' response to toxic substances produced by other cells or present in the extracellular environment. The peptide also changes the cellular response to non-toxic signals received in chemical form from other cells. These effects require the neuron to express PrP^C. As mentioned above, without continued cellular expression of PrP^C neither PrP^{Sc} nor PrP106–126 has a toxic effect. Possibly, these effects require a direct interaction between PrP106–126 and PrP^C. There could be other direct effects PrP106–126 has on cells that exacerbate the toxic effects, but these are secondary for the requirement of PrP^C expression.

9.11.2 Indirect Effects

PrP106–126 can also interact with other cells besides neurons. As glia vastly outnumber neurons in the brain it is quite likely that PrP^{Sc} will encounter glia in an animal with prion disease. Glial responses to either PrP106–126 or PrP^{Sc} have been assessed. Activation of astrocytes results in reduced activity of glutamate and/or Cu clearance by uptake mechanisms. This can result in increased levels of toxic substances in the vicinity of the neurons. Similarly, microglia activated by PrP106–126 or PrP^{Sc} result in the production of radicals, especially superoxide and possibly nitrogen radicals, but also inflammatory cytokines. In this way neurons are exposed to increased levels of stress agents or potentially toxic molecules.

9.11.3 Combined Effects

On their own these changes are probably insufficient to initiate cell death. However, reduced neuronal resistance to the oxidative stress combined with an increase in oxidative stress in the vicinity are potentially the two factors necessary to activate apoptosis.

9.11.4 Age Effects

One of the central curiosities of human prion diseases is the age of onset. In particular, both GSS and sCJD have clear age dependence. GSS, although an inherited disease, remains dormant until the patients reach their 50s. This clearly implicates age-dependent changes in the nervous system. One possibility is the increased inability of the aging nervous system to deal with oxidative stress. As oxidative stress is already implicated in prion diseases, then this might be the deciding factor in what initiates sporadic diseases such as CJD.

Binding of PrP106–126 to PrP^{C} has several effects. It causes direct inhibition of the antioxidant activity of the protein.[87] Treatment of neurons with PrP106–126 causes a marked reduction in the resistance of neurons to oxidative stress.[127] There is also a decrease in the activity of other antioxidant enzymes such as Cu/Zn SOD.[127,140] In addition, the peptide causes a decrease in the uptake of Cu by neuronal cells[133] and causes a decrease of Cu incorporation into Cu/Zn SOD.[126] It is unclear how these changes are brought about. However, there is evidence that PrP106–126 can enter cells[174] and might interact with intracellular proteins such as microtubules.[175] There is also evidence that aggregates of PrP106–126 can cause the shedding of PrP^{C}, which then becomes trapped in the aggregates.[133] Interestingly a subset of antibodies to PrP can also cause apoptosis suggesting that possibly inhibiting either the breakdown of the protein, its ability to bind Cu or its interaction with other proteins might be sufficient to trigger cell death.[172] Recently it has been confirmed that similar antibodies injected *in vivo* can also cause neuronal apoptosis.[176] As antibodies to PrP increase the toxicity of Cu to cells then it is quite possible that interfering in the protein's role in Cu metabolism might be central to the ability of PrP106–126 to initiate cell death. However, there is sufficient evidence to suggest that direct effects of PrP106–126 on neurons, mediated through PrP^C might be necessary to induce cell death but are insufficient for its execution. PrP106–126 has the effect of compromising the neurons ability to deal with stressful conditions[177] and in this way gives neurons a phenotype like that observed for PrP-deficient neurones.[29] Execution of cell death then comes about as a result of this compromised phenotype and one of a number of different stress events such as the production of superoxide.[177]

9.12   Metal Changes in TSEs

If loss of prion protein function has consequences for disease progression in TSEs, then one would expect that the earliest changes in prion disease would be seen at the synapse. Recent studies of changes in neurons in experimental prion disease have identified loss of dendritic spines occurring before any other change in prion disease.[47] However, such changes, although fitting with the hypothesis that loss of prion protein function contributes to neurodegeneration in prion disease, do not prove the connection.

As already mentioned PrP^C expression is necessary if not sufficient for prion disease.[4] Animals lacking PrP^C expression do not develop a spontaneous form of prion disease. Nevertheless, such animals do have a phenotype indicative of a disturbance and neurons lacking PrP^C expression in particular are more sensitive to oxidative stress.[29] PrP^C deficient cells have diminished cellular activity of SOD and diminished Cu content.

Recently studies of transition metals in prion diseases have begun to emerge. Studies of the brain of CJD patients have shown that the levels of Cu in their brains are decreased when compared to controls, which do not have CJD.[15] In addition, there was a striking elevation in manganese. The severity of these changes appeared to change with the *prnp* genotype of the patients. Those patients homozygous for methionine at codon 129 showed the largest changes. Accompanying the metal perturbations in CJD were changes to the levels of activity of antioxidant proteins. Although Cu/Zn-SOD was only mildly reduced in activity, Mn-SOD showed a threefold increase in activity in CJD patients. Also the brains of CJD patients showed large increases in the level of lipid peroxidation and carbonylation of protein that could be detected. These changes indicate that the brains of CJD patients show signs of ongoing oxidative stress and oxidative damage.

An immuno-affinity technique has been developed to isolate PrP^C from brain.[124] The same technique can be used to isolate PrP from brains of patients with CJD. This isolated protein contains both PrP^C and PrP^{Sc} but the majority of the protein is in the form of PrP^{Sc}. Metal analysis of PrP^{Sc} isolated from the brain of CJD patients showed that this protein lacked significant Cu-binding and that substitution with manganese and zinc had occurred.[15] Other researchers have also suggested that PrP in CJD patients' brains might bind metals other than Cu.[178] The SOD-like antioxidant activity associated with PrP^C was lost completely from the purified PrP^{Sc}. This finding confirms the notion that prion disease causes a loss of PrP^{C} function, which is directly related to the ability of the protein to bind Cu.

Brain tissue from CJD patients has a disadvantage in which it is the end stage of the disease. It is thus difficult to determine from studying such tissue what changes lead up to the final state described above. As tissue cannot be biopsied from the brains of living patients other models must be found. Experimental mouse scrapie is an effective model to study prion disease during its time course. With such a model it is possible to study changes during the asymptomatic incubation period of the disease before the accumulation of PrP^{Sc} in the brain. There are many strains of scrapie as defined by the sheep-derived innoculum used to infect the mice originally. These strains have names such as RML, ME7 and 79A. The RML-induced prion disease has been studied by many groups and the time course of the disease is well characterized. A series of mice were infected with RML. At various points during the incubation period samples of brain, liver, muscle and blood from RML-infected and control mice of similar age were collected. Mass spectroscopy techniques were used to determine the metals in those tissues.[16] There was a decrease in Cu in the brains of RML infected mice that reached a maximum at the onset of clinical signs. At the same time, there was an increase in the level of manganese but other metals did not change. In the liver there was only an increase in Cu but no change in any other metal examined. In blood, there was a small increase in blood Cu after onset of clinical signs but there was an elevation of manganese that occurred within the first 30 and 60 days after the intracerebrallar injection with the RML inoculum. Muscle showed some elevation in manganese but no other changes. These variations in metal ion content were accompanied by a decrease in the activity of Cu/Zn-SOD in the brains. However, this change only occurred after the onset of clinical signs suggesting it was secondary to the changes in the metal ions. These changes in metal ions for scrapie infected mice are similar to the findings in CJD brains but suggest that changes in metal ion content occur in parallel with changes in the levels of PrP^{Sc}.

The SOD-like activity of PrP purified from the brains of the infected mice was also examined and compared to that prepared from control mice.[16] There was a considerable loss of SOD activity from the protein after 60 days post-infection. This loss of activity was accompanied by a change in the metal occupancy of the purified protein with Cu being lost from the protein and a substitution with manganese occurring. These results show that changes in antioxidant defense may occur during prion disease. However, in particular, changes in the metabolism of the metals Cu and Mn may lie at the heart of these diseases.

Maintaining functional PrP^{C} is clearly advantageous and there is evidence to suggest that it can protect against prion disease PrP-knockout mice that have been modified to express hamster PrP *via* a GFAP promoter express PrP^{C} only in astrocytes. These mice are susceptible to infection with hamster scrapie and develop prion disease.[179] Wild-type mice are highly resistant to hamster scrapie because of specific differences between the protein sequence of hamster and mouse PrP^C. However, if wild-type mice are made transgenic to express hamster PrPc in astrocytes, they cannot be infected with hamster scrapie. The implication of this is that mouse PrP^C, which cannot be converted to mouse PrP^{Sc} by hamster PrP^{Sc}, protects against prion disease. This suggests that where there is sufficient functional PrP^C, then neurons may be protected from neuronal death caused by prion disease. In years to come strategies that protect or restore the normal Cu-dependent functions of PrP^C might be useful therapeutics to treat or prevent prion disease.

9.13 Copper and Mutant Prions

Many mutations in the *prnp* gene are known to be associated with inherited forms of prion diseases such as familial CJD and GSS. All the disease related mutations lie within the main part of the protein (amino acid residues 90–231). The mechanism by which these mutations can induce prion disease remains unclear but several mechanisms have been suggested.[8] It was recently suggested that a Cu^{2+}-binding site exists outside the octameric repeat region and in the globular domain of the protein (amino acid residues 121–231).[59] This binding site is different to those previously described.[28,55] The suggested site of Cu binding is His-187, although this finding seems to arise from a process of elimination rather than direct identification.[180] Evidence for a further two sites in the *C*-terminal are not substantiated. Analysis of whether inherited point mutations in this region (E200K, F198S, D178N) altered Cu binding to the protein has been carried out. Although, altered binding in the *C*-terminal domain for F198S and E200K mutants was identified, there was no clear link between this altered binding and the possible role of the point mutations in causing prion disease. In contrast, studies of the effect of these mutations on the neurotoxicity of PrP did show that these mutations do alter toxicity.[170] This effect is possibly a result of changes in intramolecular interactions. Evidence from an alternative source suggests that the mutations might alter loose interactions of the *C*-terminal domain of the protein and metals such as Fe and Cu.[165] It was found that a *C*-terminal fragment (121–231) carrying these mutations was able to generate hydroxyl radicals. The amount of hydroxyl radicals generated, as indicated by EPR, increased when the protein carried any of the three point mutations (E200K, F198S, D178N). In addition, catalase, an enzyme that breaks down hydrogen peroxide, a product of hydroxyl radicals and water, was able to inhibit the toxicity of the PrP121–231 peptides carrying these point mutations. This suggests that generation of toxic hydrogen peroxide as a result of metal-PrP interactions might be involved in inherited forms of prion disease.

9.14 Conclusions

This chapter has focused on the metal binding activities of the PrP. Abnormal binding of metals can cause conformation conversion and possibly play a role in disease progression. Binding of Cu^{2+} to the normal isoform potential endows it with one or more functions at the biochemical level. Other researchers have suggested that the function of PrP^c is not directly related to Cu-binding activity. In particular, it has been suggested the protein is involved in cell signaling or adhesion. It should be noted that any protein whose expression alters cellular viability will alter both cellular adhesion and intracellular signaling. This does not imply that the protein is a signaling protein or an adhesion factor. With the wealth of data supporting an antioxidant activity for the protein, then considerable further evidence would be necessary to validate the claims that the protein plays a direct role in adhesion of signaling. It is quite reasonable that PrP^c could have a number of other functions as well as its Cu dependent functions. However, further research related to metal binding is likely to focus on how PrP^c links up with other aspects of Cu metabolism and how alterations in metal metabolism could trigger protein conversion. In parallel, further studies are likely to assess the potential for metal mediated oxidative damage to play a role in the pathology of TSEs.

References

1. M. Bolton, P. McKinley and S.B. Prusiner, *Science*, 1982, **218**, 1309.
2. S.B. Prusiner, *Science*, 1982, **216**, 136.
3. H. Büeler *et al.*, *Nature*, 1992, **356**, 577.
4. H. Büeler *et al.*, *Cell*, 1993, **73**, 1339.
5. S.B. Prusiner, *Proc. Natl. Acad. Sci. USA*, 1998, **95**, 13363.
6. G. Legname *et al.*, *Science*, 2004, **305**, 673.
7. B. Ghetti *et al.*, *Brain Pathol.*, 1996, **6**, 127.
8. D.R. Brown, *Mol. Neurobiol.*, 2002, **25**, 287.
9. D.C. Gajdusek and C.J. Gibbs Jr., *Nature*, 1971, **230**, 588.
10. J. Collinge, *Hum. Mol. Genet.*, 1997, **6**, 1699.
11. M.E. Bruce *et al.*, *Nature*, 1997, **389**, 498.
12. F.E. Cohen and S.B. Prusiner, *Annu. Rev. Biochem.*, 1998, **67**, 793.
13. M. Purdey, *Med. Hypoth.*, 2000, **54**, 278.
14. D.R. Brown, *Medical Hypotheses*, 2001, **57**, 555.
15. B.-S. Wong *et al.*, *J. Neurochem.*, 2001, **78**, 1400.
16. A.M. Thackray *et al.*, *Biochem. J.*, 2002, **362**, 253.
17. D.R. Brown *et al.*, *EMBO J.*, 2000, **19**, 1180.
18. K. Basler *et al.*, *Cell*, 1986, **46**, 417.
19. N. Stahl *et al.*, *Biochemistry*, 1990, **29**, 5405.
20. R.S. Hegde *et al.*, *Science*, 1998, **279**, 827.
21. R.S. Stewart and D.A. Harris, *J. Biol. Chem.*, 2000, **276**, 2212.
22. A. Holme *et al.*, *Eur. J. Neurosci.*, 2003, **18**, 571.
23. R.K. Meyer *et al.*, *J. Biol. Chem.*, 2000, **275**, 38081.
24. X. Roucou *et al.*, *J. Biol. Chem.*, 2003, **278**, 40877.
25. A. Mange *et al.*, *J. Cell Sci.*, 2004, **117**, 2411.
26. N. Salès *et al.*, *Eur. J. Neurosci.*, 1998, **10**, 2464.
27. M.P. Hornshaw *et al.*, *Biochem. Biophys. Res. Commun.*, 1995, **207**, 621.

28. D.R. Brown *et al.*, *Nature*, 1997, **390**, 684–687.
29. D.R. Brown *et al.*, *J. Neurosci. Res.*, 2002, **67**, 211.
30. J. Collinge *et al.*, *Nature*, 1994, **370**, 295.
31. I. Tobler *et al.*, *Nature*, 1996, **380**, 639.
32. E. Graner *et al.*, *FEBS Lett.*, 2000, **482**, 257.
33. L. Chiarini *et al.*, *EMBO J.*, 2002, **21**, 3317.
34. D.R. Brown *et al.*, *Biochem. J.*, 1999, **344**, 1.
35. D.R. Brown, *J. Neurosci. Res.*, 1999, **58**, 717.
36. P.C. Pauly and D.A. Harris, *J. Biol. Chem.*, 1998, **273**, 33107.
37. W.E. Müller *et al.*, *Eur. J. Pharmacol.*, 1993, **246**, 261.
38. G. Forloni *et al.*, *Nature*, 1993, **362**, 543.
39. D.R. Brown, J. Herms and H.A. Kretzschmar, *Neuroreport*, 1994, **5**, 2057.
40. S. Brandner *et al.*, *Nature*, 1996, **379**, 339–343.
41. G. Mallucci *et al.*, *Science*, 2003, **302**, 871.
42. M.E. Bruce and H. Fraser, *Curr. Top. Microbiol. Immunol.*, 1991, **172**, 125.
43. A.E. Williams *et al.*, *Neuropathol. Appl. Neurobiol.*, 1994, **20**, 47.
44. A. Giese *et al.*, *Brain Pathol.*, 1995, **5**, 213.
45. S. Betmouni, V.H. Perry and J.L. Gordon, *Neuroscience*, 1996, **74**, 1.
46. A. Giese *et al.*, *Brain Pathol.*, 1998, **8**, 449.
47. P.V. Belichenko *et al.*, *Neuropathol. Appl. Neurobiol.*, 2000, **26**, 143.
48. J.F. Bazan *et al.*, *Protein Eng.*, 1987, **1**, 125.
49. B.-S. Wong *et al.*, *Biochem. Biophys. Res. Comm.*, 1999, **259**, 352.
50. R.M. Whittal *et al.*, *Protein Sci.*, 2000, **9**, 332–343.
51. T. Miura *et al.*, *Biochemistry*, 1999, **38**, 11560–11569.
52. J.H. Viles *et al.*, *Proc. Natl. Acad. Sci. USA*, 1999, **96**, 2042.
53. E. Aronoff-Spencer *et al.*, *Biochemistry*, 2000, **39**, 13760–13771.
54. J. Stöckel *et al.*, *Biochemistry*, 1998, **37**, 7185.
55. G.S. Jackson *et al.*, *Proc. Natl. Acad. Sci. USA*, 2001, **98**, 8531.
56. M.L. Kramer *et al.*, *J. Biol. Chem.*, 2001, **276**, 16711.
57. K.M. Pan, N. Stahl and S.B. Prusiner, *Protein Sci.*, 1992, **1**, 1343.
58. M.P. Hornshaw *et al.*, *Biochem. Biophys. Res. Comm.*, 1995, **214**, 993.
59. G.M. Cereghetti *et al.*, *Biophys. J.*, 2001, **81**, 516.
60. S. Van Doorslaer *et al.*, *J. Phys. Chem. B*, 2001, **105**, 1631–1639.
61. R. Riek *et al.*, *Nature*, 1996, **382**, 180.
62. R. Zahn *et al.*, *Proc. Natl. Acad. Sci. USA*, 2000, **97**, 145.
63. R. Riek *et al.*, *FEBS Letters*, 1997, **413**, 282.
64. F.L. Garcia *et al.*, *Proc. Natl. Acad. Sci. USA*, 2000, **97**, 8334.
65. D.G. Donne *et al.*, *Proc. Natl. Acad. Sci. USA*, 1997, **94**, 13452.
66. R. Zahn, *J. Mol. Biol.*, 2003, **334**, 477.
67. K.J. Knaus *et al.*, *Nat. Struct.Biol.*, 2001, **8**, 770.
68. L.F. Haire *et al.*, *J. Mol. Biol.*, 2004, **336**, 1175.
69. D.R. Brown and H. Kozlowski, *Dalton Trans.*, 2004, 1907.
70. H. Kozlowski *et al.*, *Coord. Chem. Rev.*, 1999, **184**, 319.
71. M. Luczkowski *et al.*, *J. Chem. Soc. Dalton Trans.*, 2002, 2269.

72. R.P. Bonomo *et al.*, *Chem. Eur. J.*, 2000, **6**, 4195.

73. C.S. Burns *et al.*, *Biochemistry*, 2002, **41**, 3991.

74. F.H. Ruiz *et al.*, *Biochem. Biophys. Res. Commun.*, 2000, **269**, 491.

75. T. Muira *et al.*, *Biochemistry*, 2005, **44**, 8712.

76. M. Luczkowski *et al.*, *Dalton Trans.*, 2003, 619.

77. D. Valensin *et al.*, *Dalton Trans.*, 2004, **9**, 4.

78. L. Redecke *et al.*, *J. Biol. Chem.*, 2005, **280**, 13987.

79. M. Chattopadhyay *et al.*, *J. Am. Chem. Soc.*, 2005, **127**, 12647.

80. K. Qin *et al.*, *J. Biol. Chem.*, 2002, **277**, 1981.

81. S.S. Hasnain *et al.*, *J. Mol. Biol.*, 2001, **311**, 467.

82. C.E. Jones *et al.*, *J. Biol. Chem.*, 2004, **279**, 32018.

83. C.E. Jones *et al.*, *J. Mol. Biol.*, 2005, **346**, 1393.

84. A. Thompsett *et al.*, *J. Biol. Chem.*, 2005, **280**, 42750.

85. C. Selvaggini *et al.*, *Biochem. Biophys. Res. Commun.*, 1993, **194**, 1380.

86. M. Ettaiche *et al.*, *J. Biol. Chem.*, 2000, **275**, 36487.

87. D.R. Brown, *Biochem. J.*, 2000, **352**, 511–518.

88. V. Meske, F. Albert and T. G. Ohm, *Acta Neuropathol.*, 2002, **104**, 560.

89. M. Perez *et al.*, *Biochem. J.*, 2003, **372**, 129.

90. C.S. Atwood *et al.*, *J. Biol. Chem.*, 1998, **273**, 12817.

91. M.F. Jobling *et al.*, *Biochemistry*, 2001, **40**, 8073.

92. G. Di Natale *et al.*, *Inorg. Chem.*, 2005, **44**, 7214–7225.

93. B. Belosi *et al.*, *ChemBioChem*, 2004, **5**, 349.

94. M. Remelli *et al.*, *Dalton Trans.*, 2005, 2876.

95. E. Gaggelli *et al.*, *J. Am. Chem. Soc.*, 2005, **127**, 996.

96. T. Florio *et al.*, *J. Neurochem.*, 2003, **85**, 62.

97. C.S. Burns *et al.*, *Biochemistry*, 2003, **42**, 6794.

98. G.L. Millhauser, *Acc. Chem. Res.*, 2004, **37**, 79.

99. G.S. Jackson *et al.*, *Proc. Natl. Acad. Sci. USA*, 2001, **98**, 8531.

100. C.E. Jones *et al.*, *J. Mol. Biol.*, 2005, **346**, 1393.

101. E. Gaggelli *et al.*, (unpublished data).

102. F. Wopfner *et al.*, *J. Mol. Biol.*, 1999, **289**, 1163.

103. E.M. Marcotte and D. Eisenberg, *Biochemistry*, 1999, **38**, 667.

104. D. Matthews and B.C. Cook, *Rev. Sci. Tech.*, 2003, **22**, 283.

105. L.R. Brown and D.A. Harris, *J. Neurochem.*, 2003, **87**, 353.

106. L. Calzolai *et al.*, *Proc. Natl. Acad. Sci. USA*, 2005, **102**, 651.

107. P. Stanczak *et al.*, *Dalton Trans.*, 2004, 2102.

108. A.P. Garnett and J.H. Viles, *J. Biol. Chem.*, 2003, **278**, 6795.

109. P. Stanczak *et al.*, *Biochemistry*, 2005, **44**, 12940.

110. P. Stanczak *et al.*, *Chem. Commun.*, 2005, **14**, 3298.

111. K.S. Lee *et al.*, *J. Neurochem.*, 2001, **79**, 79.

112. A.R. Walmsley, F. Zeng and N.M. Hooper, *J. Biol. Chem.*, 2003, **278**, 37241.

113. M. Nunziante, S. Gilch and H.M. Schätzl, *J. Biol. Chem.*, 2003, **278**, 3726.

114. C. Sunyach *et al.*, *EMBO J.*, 2003, **22**, 3591.

115. N. Madore *et al.*, *EMBO J.*, 1999, **18**, 6917.

116. Y. Yao, J. Ren and I.M. Jones, *J. Neurochem.*, 2003, **87**, 1057.

117. A. Magalhães *et al.*, *J. Biol. Chem.*, 2002, **277**, 33311.
118. D.R. Borchelt *et al.*, *Glycobiology*, 1993, **3**, 319.
119. D.R. Brown, *Neurobiol. Dis.*, 2004, **15**, 534.
120. E.T. Parkin *et al.*, *J. Biol. Chem.*, 2004, **279**, 11170.
121. W.S. Perera and N.M. Hooper, *Curr. Biol.*, 2001, **11**, 519.
122. R.S. Mishra *et al.*, *J. Biol. Chem.*, 2002, **277**, 24554.
123. C.L. Haigh, K. Edwards and D.R. Brown, *Mol. Cell Neurosci.*, 2005, **30**, 186–196.
124. D.R. Brown, C. Clive and S.J. Haswell, *J. Neurochem.*, 2001, **76**, 69.
125. D.R. Brown, *J. Neurosci. Res.*, 1999, **58**, 717.
126. D.R. Brown and A. Besinger, *Biochem. J.*, 1998, **334**, 423.
127. D.R. Brown *et al.*, *Exp. Neurol.*, 1997, **146**, 104.
128. M. Guentchev *et al.*, *Neurobiol. Dis.*, 2000, **7**, 270.
129. B.-S. Wong *et al.*, *J. Neurochem.*, 2001, **79**, 689.
130. B.-S. Wong *et al.*, *J. Neurochem.*, 2001, **76**, 565.
131. R. Walz *et al.*, *Epilepsia*, 1999, **40**, 1679–1682.
132. R. Huber, T. Deboe and I. Tobler, *Neuroreport.*, 2002, **13**, 1.
133. J. Curtis *et al.*, *Neurobiol. Dis.*, 2003, **13**, 55.
134. J. Weise *et al.*, *Neurosci. Lett.*, 2004, **372**, 146.
135. N.F. McLennan *et al.*, *Am. J. Pathol.*, 2004, **165**, 227.
136. W.M. Williams, E.R. Stadtman and J. Moskovitz, *Neuropathol. Appl. Neurobiol.*, 2004, **30**, 161.
137. T. Voigtlander *et al.*, *Acta Neuropathol.*, 2001, **101**, 417.
138. L. Dupuis *et al.*, *Mol. Cell Neurosci.*, 2002, **19**, 216.
139. F. Klamt *et al.*, *Free Readic. Biol. Med.*, 2002, **30**, 1137.
140. A.R. White *et al.*, *Am. J. Pathol.*, 1999, **155**, 1723.
141. C. Kuwahara *et al.*, *Nature*, 1999, **400**, 225.
142. B.H. Kim *et al.*, *Mol. Brain Res.*, 2004, **124**, 40.
143. D.R. Brown, B. Schmidt and H.A. Kretzschmar, *Int. J. Dev. Neurosci.*, 1997, **15**, 961.
144. H. Sauer *et al.*, *Free Radic. Biol. Med.*, 1999, **27**, 1276.
145. H. Sauer *et al.*, *Free Radic. Biol. Med.*, 2003, **35**, 586.
146. P.H. Frederikse *et al.*, *Curr. Eye Res.*, 2000, **20**, 137.
147. W.C. Shyu *et al.*, *Cell. Mol. Neurobiol.*, 2004, **24**, 257.
148. A. Senator *et al.*, *Free Radic. Biol. Med.*, 2004, **37**, 1224.
149. F. Zeng *et al.*, *J. Neurochem.*, 2003, **84**, 480.
150. O. Milhavet *et al.*, *Proc. Natl. Acad. Sci. USA*, 2000, **97**, 13937.
151. A. Sukudo *et al.*, *Biochem. Biophys. Res. Commun.*, 2004, **313**, 850.
152. B.-S. Wong *et al.*, *Biochem. Biophys. Res. Commun.*, 2000, **273**, 136–139.
153. I. Fridovich, Superoxide dismutases, *Ann. Rev. Biochem.*, 1975, **44**, 147–159.
154. T. Cui *et al.*, *Eur. J. Biochem.*, 2003, **270**, 3368.
155. I. Fridovich, *J. Biol. Chem.*, 1997, **272**, 18515.
156. T. Ookawara *et al.*, *Am. J. Physiol.*, 1998, **275**, C840.
157. C. Gohel *et al.*, *J. Neurosci. Res.*, 1999, **55**, 261.
158. K. Qin *et al.*, *J. Biol. Chem.*, 2000, **275**, 19121.
159. E. Quaglio, R. Chiesa and D.A. Harris, *J. Biol. Chem.*, 2001, **276**, 11432.

160. A. Giese *et al.*, *Biochem. Biophys. Res. Comm.*, 2004, **320**, 1240.
161. D. McKenzie *et al.*, *J. Biol. Chem.*, 1998, **273**, 25545.
162. E.M. Sigurdsson *et al.*, *J. Biol. Chem.*, 2003, **278**, 46199.
163. J.R. Requena *et al.*, *Arch. Biochem. Biophys.*, 2004, **432**, 188.
164. H.E. McMahon *et al.*, *J. Biol. Chem.*, 2001, **276**, 2286.
165. S. Turnbull *et al.*, *Biochemistry*, 2003, **42**, 7675.
166. N.T. Watt and N.M. Hooper, *Trends Biochem. Sci.*, 2003, **28**, 406.
167. R.N. Tsenkova *et al.*, *Biochem. Biophys. Res. Comm.*, 2004, **325**, 1005.
168. T. Lekishvili *et al.*, *Exp. Neurol.*, 2004, **190**, 233.
169. T. Pillot *et al.*, *J. Neurochem.*, 2000, **75**, 2298.
170. M. Daniels, G.M. Cereghetti and D.R. Brown, *Eur. J. Biochem.*, 2001, **268**, 6155.
171. D.R. Brown, *Mol. Cell Neurosci.*, 2000, **15**, 66.
172. D.R. Brown, B. Schmidt and H.A. Kretzschmar, *J. Neurochem.*, 1998, **70**, 1686.
173. M.A. Chacon *et al.*, *Mol. Psychiatry*, 2003, **8**, 853.
174. S.J. McHattie, D.R. Brown and M.M. Bird, *J. Neurocytol.*, 1999, **28**, 145–155.
175. D.R. Brown, B. Schmidt and H.A. Kretzschmar, *J. Neurosci. Res.*, 1998, **52**, 260.
176. L. Solforosi *et al.*, *Science*, 2004, **303**, 1514.
177. D.R. Brown, B. Schmidt and H.A. Kretzschmar, *Nature*, 1996, **380**, 345.
178. J.D.F. Wadsworth *et al.*, *Nature Cell Biol.*, 1999, **1**, 55–59.
179. A. Raeber *et al.*, *EMBO J.*, 1997, **16**, 6057.
180. G.M. Cereghetti *et al.*, *Biophys. J.*, 2003, **84**, 1985.

Are Metals Involved in Cu–Zn Superoxide Dismutase-Related Familial Amyotrophic Lateral Sclerosis

Amyotrophic lateral sclerosis (ALS), also known as motor neuron disease or Lou Gehrig's disease, is among the common adult motor neuron diseases (Figure 1).[1] First described in 1869 by the French neurobiologist and physician Jean–Martin Charcot, the disease's primary hallmark is the selective dysfunction and death of the neurons in the motor pathways. This leads to spasticity, hyper-reflexia (upper motor neurons), generalized weakness, muscle atrophy and paralysis (lower motor neurons).

Failure of the respiratory muscles is generally the fatal event, occurring within one to five years of onset. This neuromuscular disease is relatively rare with an incidence rate of 1–2 in 100,000. The degeneration of motor neurons causes muscular weakness, but, like in PD, intellect and personality are often unaffected. The National Institute of Neurological Disorders and Stroke reports that only 5–10% of all ALS cases can be traced to genetics, particularly to a mutation related to the superoxide dismutase (SOD) 1 enzyme. This leaves the vast majority of cases without a known etiology, with the potential for environmental association briefly outlined below. The disease has no racial, socioeconomic or ethnic boundaries. The life expectancy of ALS patients is usually 3–5 years after diagnosis. ALS is most commonly diagnosed in middle age and affects men more often than women.

As stated above, risk factors include an inherited genetic defect, which accounts for 5–10% of cases of familial ALS (FALS) in the United States. FALS is linked to a genetic defect on chromosome 21. This gene codes for the enzyme SOD, an antioxidant that protects motor neurons from free radical damage (*i.e.*, molecules introduced to the body, or produced by body processes that interact and cause cellular damage). More than 100 different mutations that cause SOD to lose its antioxidant properties have been found. However, only 20% of FALS cases are linked to SOD mutations, so there may be other unknown genetic defects involved.

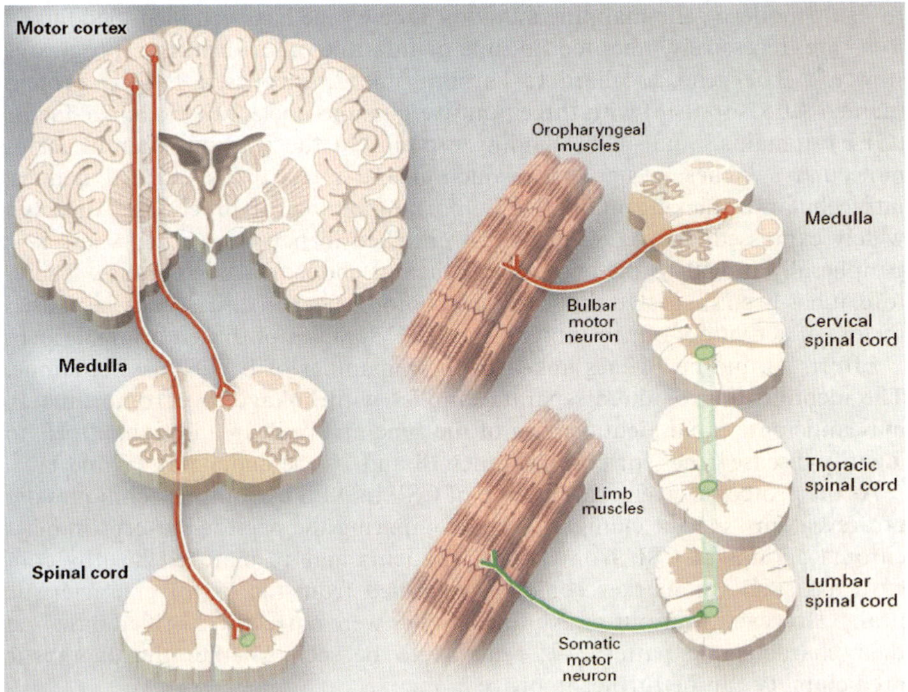

Figure 1 *Motor neurons selectively affected in ALS. Degeneration of motor neurons in the motor cortex leads to clinically apparent signs of upper motor neuron abnormalities: overactive tendon reflexes, Hoffmann signs, Babinski signs and clonus. Degeneration of motor neurons in the brain stem and spinal cord causes muscle atrophy, weakness and fasciculation.*
(Reprinted with permission·from ref. 1.)

In the United States, 90–95% of ALS cases are sporadic. Sporadic ALS (SALS) appears to be increasing worldwide. The causes are not clear, yet some evidence suggests that the immune system may be involved. Excessive levels of glutamate can over-stimulate motor neurons and cause them to die. Glutamate is one of the most important neurotransmitters for healthy brain function.

In Guamanian ALS, a dietary neurotoxin is the risk factor. The suspected neurotoxin is an amino acid found in the seed of the cycad *Cyas cirinalis*, a tropical plant found in Guam, which was used to make flour and was a major dietary component during the 1950s and the early 1960s, when this type of ALS had an exceptionally high incidence.

Although ALS was described more than 130 years ago, the mechanism underlying the characteristic selective degeneration and death of motor neurons in this disease has remained a mystery. Moreover, there is no effective remedy for this progressive, fatal disorder.[2] Few *Functional Genomics* studies have been performed with human ALS samples and little is known about global gene expression patterns.[3]

Using modern gene-mapping methods, a new gene has been identified linked to a rare, recessively inherited juvenile or infantile onset form that progresses slowly.[4,5] The gene, localized to chromosome 2, encodes a 184-kD protein (named ALS2 or alsin) with three putative guanine-nucleotide-exchange factor (GEF) domains. Small GTP-binding proteins of the Ras superfamily act as molecular switches in signal transduction and affect cytoskeletal dynamics, intracellular trafficking and other important biological processes. Although widely expressed, the ALS2 protein is enriched in nervous tissue, where it is peripherally bound to the cytoplasmic face of endosomal membranes. Therefore, it has been suggested that early-onset motor neuron disease may be caused by loss of activity of one or more of the GEF domains of this endosomal GEF.

Efforts to find new genes linked to the remainder of FALS cases continue. The identification of three separate families with linkage to chromosome 16 may anticipate rapid identification of the gene involved. In addition, efforts to identify disease genes for chromosomes 18 and 20 are under way (Table 1).[1]

At the proteome level, examination of CSF in ALS by routine techniques did not reveal any specific changes. When comparing the peptide patterns in mass chromatograms of CSF from 12 ALS patients and 10 matched healthy controls, no single biomarker could be identified from the list of characteristic peaks. However, four out of five test samples were correctly classified based on their characteristic pattern.[6] It remains to be seen whether the diagnostic predictability can be further improved.

By using the Kinetworks multi-immunoblotting technique to evaluate the expression of 78 protein kinases, 24 protein phosphatases and the phosphorylation states of 31 phospho-proteins in thoracic spinal cord tissue from control subjects and SALS patients, elevated expression and/or activation of many protein kinases was found in the SALS samples. These included different PKC isoforms and GSK3α/β, which may augment neuronal death in ALS, and CaMKK, PKBα, Rsk1, S6K and SAPK, which may be a response to neuronal injury that potentially can mitigate cell death.[7] As the human cellular signaling network includes genes for just around 500 kinases and 150 phosphatases, developing a blot with antibodies directed against all kinases and phosphatases

Table 1 *Genetics of ALS-related diseases*[a]

Disease	Inheritance	Linkage	Gene/protein
SALS	None	None	Unknown
FALS			
ALS	Dominant	21q22.1	SOD1
ALS	Dominant	16	?
ALS	Dominant	18	?
ALS	Dominant	20	?
ALS with dementia, Parkinsonism	Dominant	17q21	Tau
Juvenile-type 1	Recessive	15q15–22	?
Juvenile-type 2	Recessive	2q33	ALS2

[a] Modified from ref. 1.

and their activated forms (by using phosphorylation-dependent antibodies) is likely to be a fruitful approach.

The discovery of mutations in Cu/Zn-SOD immediately prompted hypotheses that toxicity is caused by oxidative damage and aberrant chemistry of the active copper and zinc sites of the misfolded enzyme.[8] There is also evidence that mitochondria may be primary targets for SOD1 mutant-mediated damage, based on studies in transgenic animals.[9] Despite these findings, the evidence that mitochondria are important targets for damage common to SOD1 mutants with different biochemical characters remains contradictory. Mitochondrial pathology has not been found in other rodent models that develop motor neuron disease from expression of lower levels of the same mutants or in any of the models that develop motor neuron disease from expression of mutants without dismutase activity.[1] The finding that SOD1 mutants impair slow axonal transport months prior to disease onset led to the conclusion that diminished transport correlated with the development of motor neuron disease.[10]

Furthermore, in ALS tissues, there is strong activation and proliferation of microglia in regions of motor neuron loss. Based on previous promising studies in mice the anti-inflammatory drugs minocylcine (Formula 10.1) and celecoxib (Formula 10.2) are currently studied in human clinical trials.[1] Despite the suggestion that trophic factors may be important in ALS, human trials with neurotrophic factors have been disappointing.[1] Furthermore, in a large European study, which had been designed to determine whether alterations in the vascular endothelial cell growth factor (VEGF) gene may be linked to human ALS, three single nucleotide polymorphisms in the promoter region of the VEGF gene were identified, implicating VEGF as a risk factor in the disease.[11]

Minocycline

Celecoxib

In FALS, the disease results from an acquired toxicity of mutant SOD1 that affects both neurons and glial cells. The exact nature of this toxicity is uncertain, but in neurons it likely disrupts several basic cellular functions including protein breakdown by the ubiquitin-proteasome system, slow anterograde transport, fast retrograde axonal transport, calcium homeostasis, mitochondrial function and maintenance of the cytoskeletal architecture.[1] The transcriptomic studies described above strengthen the previous findings of a role for oxidative damage and mitochondrial dysfunction, neuroinflammation, the proteasome system, growth control and the cytoskeleton, and further point at genes involved in transcriptional control and neuronal differentiation.

Metals may play a role in the development of ALS. Some studies have observed an association with occupation in welding or soldering, but not all have found metals to be related to ALS. More specifically, associations have been observed with exposure to lead,[12,13] but no association was observed between ALS and lead levels in various tissues or toenails; however, these studies had limited numbers of study participants. No association was observed between exposure to zinc and ALS, and the evidence from biomarker studies is inconclusive for the association observed for levels in brain tissue or toenails compared with controls. However, these studies may have had limited power based on the size of the study population. Although one epidemiologic study showed no association between exposure to copper and ALS,[14] there was decreased copper concentration observed in both cerebrospinal fluid and blood among patients with ALS versus controls.[15]

Case-control studies examining biomarkers of iron, manganese and Al and risk of ALS were found. Increased iron levels have been observed in brain tissue,[16] although not in blood or toenails. An increase of manganese was observed in cervical cords, both not confirmed in blood. An increase was observed in Al in CNS tissue[17] and cerebrospinal fluid.[18]

As already stated, pathologically, ALS is characterized by extensive loss of lower motor neurons in the spinal cord and brain stem, atrophy of ventral roots, degeneration of upper motor neurons in the motor cortex and corticospinal tract, somatic and axonal inclusions of aberrant neurofilament proteins and reactive astrocytosis. Pathological hallmarks of degenerating motor neurons are Bunina bodies, round eosinophilic inclusions containing a homogeneous granular matrix surrounded by vesicular and tubular structures[19] and skein-like inclusions that contain ubiquitin and are comprised of bundles of 15–20 nm thick neurofilaments.[20]

Although the precise molecular pathways that cause the death of motor neurons in ALS remain unknown, possible primary mechanisms include the toxic effects of mutant SOD1, including abnormal protein aggregation; the disorganization of intermediate filaments; glutamate-mediated excitotoxicity and other abnormalities of intracellular calcium regulation in a process that may involve mitochondrial abnormalities and apoptosis (Figure 2)[1].

Similar to other neurodegenerative disorders with incompletely defined etiology, such as AD and PD, ALS appears to be a syndrome originating from diverse pathogenic processes. ALS is predominantly a sporadic disorder

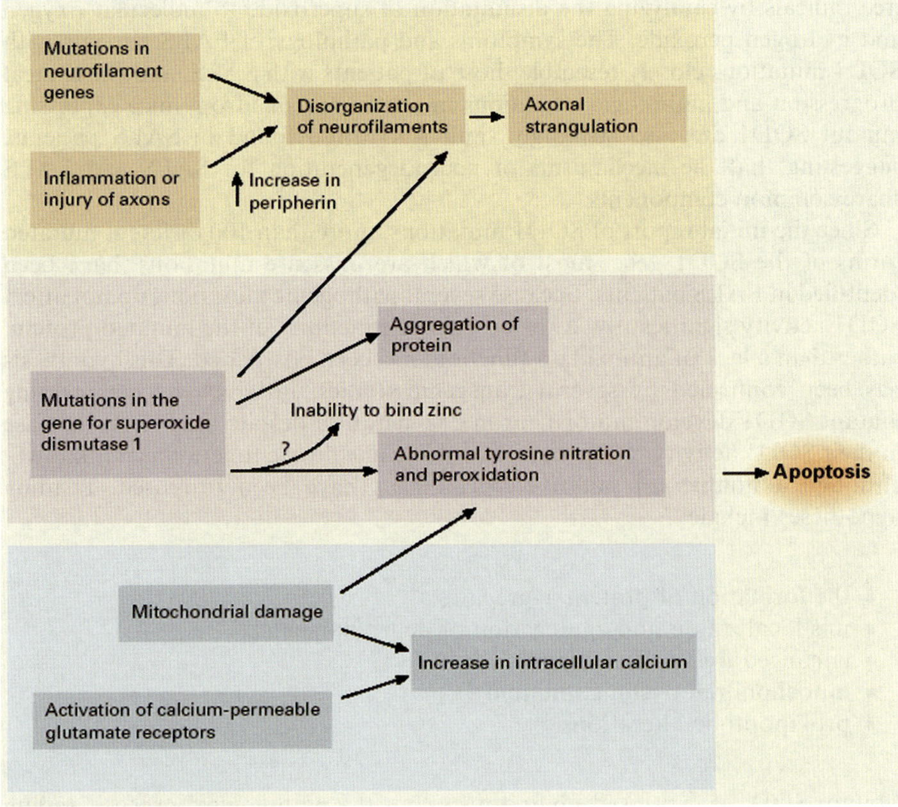

Figure 2 *Mechanisms that may contribute to the degeneration of motor neurons in ALS* (Reprinted with permission from ref. 1.)

(SALS), but approximately 10% of cases are familial (FALS). While the causes of SALS are probably multiple and still not deciphered, our current knowledge on the disease at the molecular level derives mostly from studies of FALS models. In the last decade, we have benefited from valuable transgenic animal and cellular models of FALS that re-capitulate the main clinical and pathological features of the disease. Such models have provided crucial tools not only for studying the pathogenesis of FALS, but also for testing a wide range of therapeutic approaches.[1] Significant advances have been made in understanding the development and the role of mitochondrial dysfunction in ALS using these transgenic models. Whether the findings obtained in these FALS models can be generalized to the more common forms of SALS remains an open question.

As already stated, approximately 10% of FALS cases are due to mutations in the gene encoding SOD1 (SOD1; Cu, Zn, dismutase; MIM147450).[21] In eukaryotic cells, superoxide is a normal byproduct of aerobic respiration and it is produced mostly by oxidative phosphorylation in the mitochondria.[22] SOD1 is a ubiquitous metalloprotein that prevents damage by oxygen-mediated free radicals by catalyzing the dismutation of superoxide to molecular oxygen and hydrogen peroxide. The symptoms and pathology of FALS patients with SOD1 mutations closely resemble those of patients with SALS, and the clinical progression and pathologic alterations in motor neurons from mice expressing mutant SOD1 are also strikingly similar to those found in SALS patients, suggesting that the mechanisms of neurodegeneration for SALS and FALS share common components.

Since the initial report of SOD1 mutations, more than 100 different mutated forms of the SOD1 gene, most of which are missense mutations, have been identified in FALS patients. Because several pathogenic mutations do not affect SOD1 activity significantly, a toxic "gain of function" of the mutated protein, rather than a lack of antioxidant function, has been postulated. This hypothesis has been confirmed by several transgenic studies, in which mice-expressing mutant SOD1 develop motor neuron degeneration despite an overall increase in their SOD1 activity.[23] The nature of this gained toxic function is not known, although a number of putative mechanisms have been proposed. Leading hypotheses include:

- the formation of protein aggregates;[24]
- mis-localization and aggregation of neurofilaments;[25]
- enhanced free radical generation;[26]
- mitochondrial dysfunction; and[27]
- pro-apoptotic alterations.[28,29]

Mutant SOD1 is expressed ubiquitously, but the pathological process leading to the disease strike motor neurons most severely. The tissue selectivity of SOD1 toxicity is a puzzling problem that has yet to be fully understood. While mouse models that express mutant SOD1 ubiquitously (similar to what happens in humans) develop motor neuron degeneration and ALS, models that express mutant SOD1 exclusively either in the motor neurons or in astrocytes do not develop the disease and pathologic phenotype, suggesting that the pathogenesis of FALS requires contribution and interaction from both motor neurons and the surrounding non-neuronal cells. Consistent with this view, in chimeric mice that contain a mixture of SOD1 mutant and WT cells, motor neurons with WT SOD1 develop signs of degeneration, whereas non-neuronal cells with WT genes appear to attenuate the degeneration of motor neurons with the mutant SOD1 gene. These findings suggest that FALS is caused by the failure in multiple, interacting cell types.

Morphological and ultrastructural abnormalities of mitochondria in SALS were initially observed in autopsies. Sub-sarcolemmal aggregates of abnormal mitochondria were found in skeletal muscle and in intra-muscular nerves. Mitochondrial morphological abnormalities were also detected in proximal axons and in the anterior horns of the spinal cord in SALS patients. Increased mitochondrial volume and elevated calcium levels within the mitochondria were found in muscle biopsies of ALS patients.[28]

Several studies have measured the function of mitochondrial respiratory chain, which is constituted by a series of enzymatic complexes that provide the energy necessary for ATP synthesis. Deficits in the activities of mitochondrial respiratory chain complex I and IV have been identified in the skeletal muscle and in the spinal cord of SALS patients.[30] Several other studies suggested that an impairment of the mitochondrial respiratory chain might be a significant occurrence in the pathogenesis of SALS.

The availability of the mutant SOD1 transgenic mice has provided an excellent platform to investigate more in depth the role of mitochondrial dysfunction in the pathogenesis of FALS. Although it is not yet clear whether the findings in this FALS model are applicable to other FALS and SALS cases, these studies have increased our understanding of mitochondrial involvement in ALS, raised new questions and generated new hypotheses that can be tested in the future. In the following sections of this review article, we will focus on the evidence for the involvement of mitochondria in the pathogenic process associated with motor neuron degeneration in models of FALS caused by mutations in SOD1 gene.

Several common forms of SOD exist: they are proteins co-factored with copper and zinc, or manganese, or iron:

- The cytosols of virtually all eukaryotic cells contain an SOD enzyme with copper and zinc (Cu/Zn-SOD). (The Cu/Zn-SOD available commercially is normally that found in the bovine erythrocyte: PDB 1SXA, EC 1.15.1.1). The Cu–Zn enzyme is a homodimer of molecular weight 32,500. The two sub-units are joined primarily by hydrophobic interactions.
- Chicken liver (and nearly all other) mitochondria, and many bacteria (such as *E. coli*) contain a form with manganese (Mn-SOD).
- *E. coli* and many other bacteria also contain a form of the enzyme with iron (Fe-SOD); some bacteria contain Fe-SOD, others Mn-SOD, and some contain both.

In humans, three types are recognized. SOD1 is located in the cytoplasm, SOD2 in the mitochondria and SOD3 is extracellular. The first is a dimer, while the others are tetramers. SOD1 contains copper and zinc, SOD3 contains Cu only, while SOD2 has manganese in its reactive center. The genes are located on chromosomes 21, 6 and 4, respectively (21q22.1, 6q25.3 and 4p15.3–p15.1).

The Cu/Zn-SOD1 (MW *ca.* 2×16 kDa) is found in all eukaryotic species where it catalyzes the conversion of the superoxide anion to hydrogen peroxide and oxygen. SOD1 also has pro-oxidant activities, including peroxidation, the generation of hydroxyl radicals, and the nitration of tyrosine (Figure 3)[1].

The superoxide anion radical (O_2^-) spontaneously dismutes to O_2 and H_2O_2 quite rapidly. However, SOD has the highest catalytic rate of any known enzyme with specific activity 10^8–10^9 times greater than the spontaneous rate of reaction. In fact, its rate is diffusion-limited. Thus, under real-world

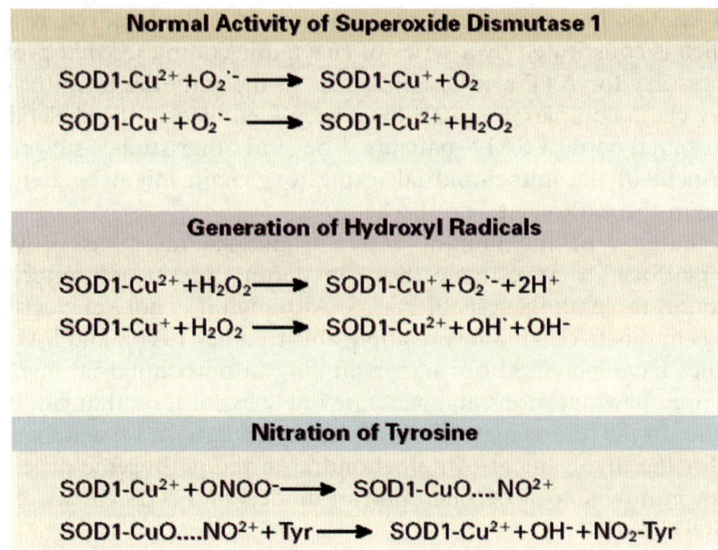

Figure 3 *Copper-Mediated Oxidative Reactions Catalyzed by SOD1 SOD1 normally catalyzes the conversion of toxic superoxide anions to hydrogen peroxide (H_2O_2) (top). Mutations in the gene for SOD1 may reverse this reaction, leading to the production of toxic hydroxyl radicals (middle), or promote the use of other abnormal substrates such as peroxynitrite, ultimately leading to the aberrant nitration of tyrosine residues in proteins (bottom)*
(Reprinted with permission from ref. 1.)

intracellular conditions, SOD greatly reduces the ambient level of the dangerous superoxide radical.

The presence of SOD has been shown to help protect many types of cells from the free radical damage that is important in aging, senescence and ischemic tissue damage. SOD also helps protect cells from DNA damage, lipid peroxidation, ionizing radiation damage, protein denaturation and many other forms of progressive cell degradation. Superoxide is a potent, but chemically selective, reactive oxygen species produced in all aerobic cells. Sometimes it functions as a signaling agent, but it can also cause oxidative damage, particularly at excessive concentrations or when it reaches the wrong cellular locations. One of the best-documented examples of superoxide-mediated damage occurs when it attacks and destroys labile iron–sulfur cluster co-factors of aconitase and other similar iron-containing proteins.[31]

The protein has been extensively investigated and characterized.[32,33] The two sub-units are identical, each containing 153 amino acids, one copper and one zinc ion. They are orientated such that their active sites face away from each other. The entire length of the chain is folded into a β barrel structure consisting of eight anti-parallel twisted β sheets (Figure 4). The non-sheet elements (random coil) are convoluted chains that connect one β sheet to the other one.

In the oxidized enzyme, Cu(II) is bound by four His residues and one water molecule in a distorted square planar arrangement. One of the His ligands

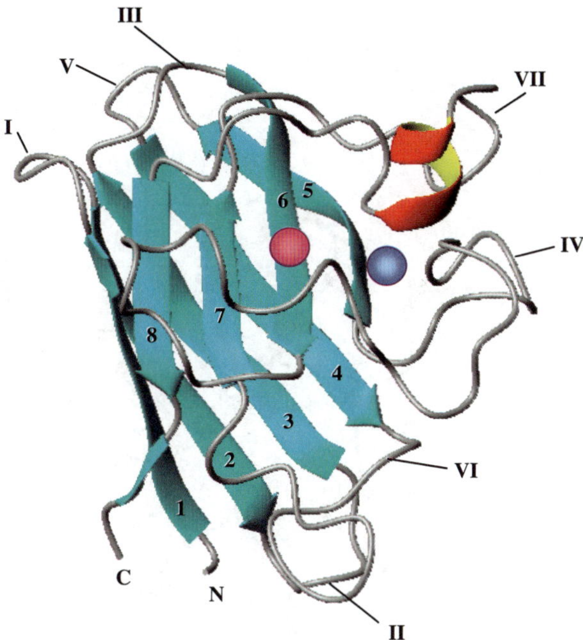

Figure 4 *Substructural motifs in copper/zinc-SOD1 β-strands 1-8 are numbered in the order they appear in the β-barrel starting from the N-terminus while loops are labelled with roman numerals I-VII in the order they appear in the primary sequence of the protein. Loop IV is the "Zn-binding" loop. Loop VII is the "electrostatic" loop*

bridges Cu(II) to Zn(II), which is further coordinated by two His and one Asp in tetrahedral coordination (Figure 5). Upon reduction, the bridging His and copper undergo a conformational re-arrangement due to breakage of the His–Cu bond with the metal moving *ca.* 0.17 nm away from His N_ε that becomes protonated.[34]

The results obtained on SOD1 mutants have recently been reviewed.[35] Over 100 distinct SOD1 mutations have been identified in FALS patients,[36,37] as summarized in Figure 6.[38]

Most of the individual mutations result in substitution of one single amino acid by another; such substitutions have been identified at over one-third of the 153 amino acid residues of the WT Cu/Zn-SOD protein. As shown in the Figure 6, the majority of the mutations are clustered on the top and bottom of the β barrel, in the dimer interface or along one of the loops that forms part of the Zn-binding pocket of SOD. The only parts of the SOD molecule where mutations have not been found are in the Zn-binding groups and on the residues facing out on backside of the β barrel, opposite the active site. In addition to the individual amino acid substitutions, there are also a smaller group of mutations resulting in amino acid deletions and truncations. A major part of the *C*-terminal region can be entirely deleted in a few individuals.

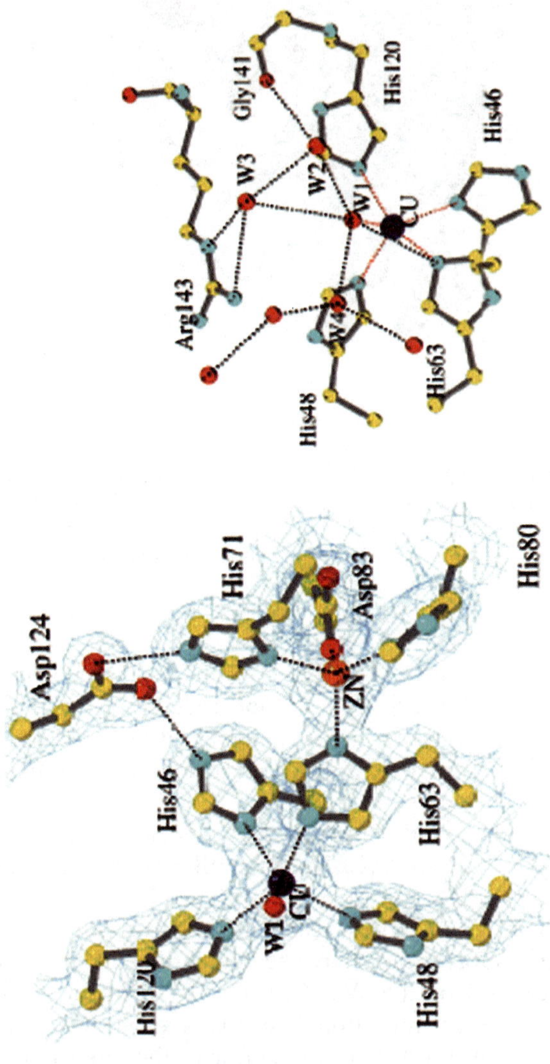

Figure 5 *The electron density map (left panel) and water network (right panel) of the Cu–Zn site of wild type SOD1* (Reprinted with permission from Strange et al., J. Mol. Biol. 2003, vol 328, pp. 877–891.)

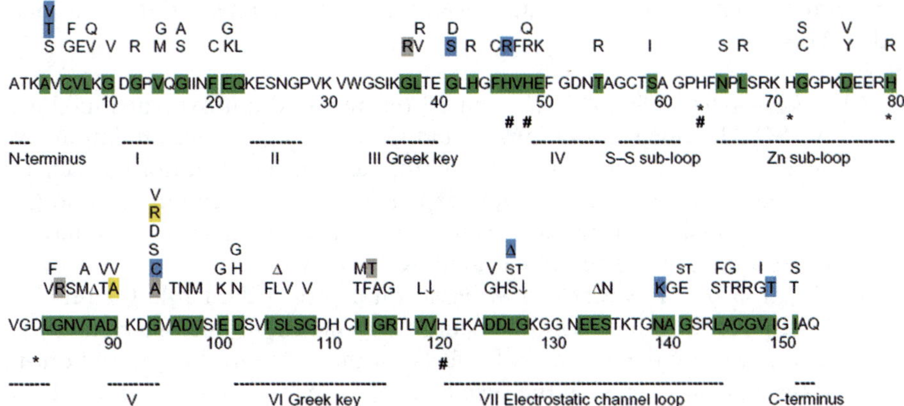

Figure 6 *Mutations of SOD1 observed in patients with FALS. More than 100 point mutations of the gene coding for SOD1 have been found in families with FALS. Most mutations are dominant, with the exception of those causing the substitutions D90A and N86S, which act as recessive. Mutations are distributed among all five exons of the gene and result in the alteration of amino acids throughout the SOD1 homodimeric β-barrel structure; whereas some mutations affect the active site, others are positioned at the dimer interface or inside β strands or connecting loops. Mutations found in patients are indicated above the main sequence (wild type). Copper-binding residues are indicated by #, zinc-binding residues are indicated by *. ↓ indicates a point of the sequence where an insertion has been detected in some patients; Δ indicates a deletion; ST indicates a stop codon, generating a truncated protein. Highlights: green, positions mutated in patients with FALS; blue, mutations introduced in cultured cells; yellow: mutations in transgenic rodents; gray: mutations introduced both in cell culture models and transgenic rodents.*
(Reprinted with permission from ref. 38.)

ALS mutant Cu/Zn-SOD proteins may be divided into two groups:[39]

(i) A group named WT-like:[40] copper and zinc levels are very similar to those found in the WT protein, *i.e.*, high in zinc but with variable levels of copper.[40] The SOD activities and the spectroscopic characteristics of this class of ALS mutant Cu/Zn-SOD proteins are virtually identical to those of WT Cu/Zn-SOD.[40]

(ii) A second class, named "metal-binding-region" mutants.[40] These include mutations in the metal-binding ligands themselves or in elements intimately associated with metal binding.[40] The side chain of Asp-124 is particularly important, since it contributes to the stabilization of both the copper- and zinc-binding sites by simultaneously H-bonding to the non-liganding imidazole nitrogen atoms of copper ligand His-46 and zinc ligand His-71. These mutants are isolated from the expression systems with very low zinc but also with very low copper. This class of ALS mutant Cu/Zn-SOD proteins is very likely to exist *in vivo* in zinc- *and* copper-deficient forms.

Several hypotheses have been advanced that would relate SOD1 mutations with FALS:

(i) *A loss-of-function mechanism* based on impaired antioxidant functions of SOD1. Such mutations could lead to toxic accumulation of superoxide radicals, but (i) no motor neuron degeneration is seen in transgenic mice in which SOD1 expression has been eliminated and (ii) over-expression of mutant SOD1 in transgenic mice causes motor neuron disease in spite of elevated SOD1 activity.

(ii) *A gain-of-function mechanism* based on a pro-oxidant role for mutant SOD1 contributing to motor neuron degeneration. Such mechanism is challenged by the absence of effects on motor neuron degeneration in transgenic mice lacking the specific copper chaperone for SOD1, which deprives SOD1 of copper and eliminates enzymatic activity.[41]

(iii) *An aggregation mechanism* based on the possible deleterious effects of accumulating aggregates of mutant SOD1, as supported by the observation that intracellular inclusions are found in motor neurons in murine models of mutant SOD1-mediated disease.[42]

Although a variety of inclusions have been described in sporadic cases of ALS, there is poor evidence for deposition of SOD1 in these inclusions and no convincing evidence that aggregation contributes to the pathogenesis of SALS.

An aberrant redox chemistry of mutant SOD1 may arise from modification of active copper and zinc sites and yield enhanced peroxidase activity,[43] and superoxide production.[44] However, the requirement of copper ion for toxicity is not universally accepted, since ALS symptoms are observed in mice expressing the SOD1 mutant in which all of the copper ligands are mutated.[45]

The thermodynamic and kinetic stability of SOD largely depends upon copper and zinc ions such that the loss of these ions may facilitate partial or global unfolding transitions. These changes may lead to misfolding and aggregation. As a matter of fact, the metals have been shown to contribute significantly to the kinetic stability of the protein, with apo-SOD showing acid-induced unfolding rates about 60-fold greater than the holo-protein.[46] However, the unfolding rates of WT and mutant SOD were similar to each other in both the holo and apo states, indicating that regardless of the effect of mutation on thermodynamic stability, the kinetic barrier toward SOD unfolding is dependent on the presence of metals. Thus, pathogenic SOD mutations that do not significantly alter the stability of the protein may still lead to SOD aggregation by compromising its ability to bind or retain its metals and thereby decrease its kinetic stability.

As for the other hypothesis for a toxic function based on an increased propensity for cytoplasmic aggregation of mutant SOD1, ALS mutations have been considered to induce protein misfolding and/or destabilization; in fact, ALS mutations provoke a decrease of 1–6°C in the melting temperature of SOD1,[47] while WT SOD1 remains active after treatment at 80°C. While it remains unclear how SOD1 acquires zinc ions, the copper chaperone for SOD1, controls not only copper acquisition but also disulfide formation in apo-SOD1.[48] ALS mutants are more susceptible to disulfide reduction compared to the WT protein,[49] and, hence, adventitious reductions may constitute an important step in the disease. It has in fact been shown that uptake of the SOD1 molecule into the intermembrane space of the mitochondria is dependent on the status of the disulfide bond.[50] The reduced form of SOD1 is imported through the mitochondrial outer membrane, but the disulfide-bonded apo-SOD1, the Zn(II)-loaded SOD1, and the holo-form or fully mature form of SOD1 are not readily transferred from the cytosol into the intermembrane space of the mitochondria. Since the effects of disulfide reduction on the SOD1 structure are so relevant to the intracellular localization and stability of the SOD1 molecule, it has been shown that, even after removal of both copper and zinc ions from the active and mature form of hSOD1, the dimeric state still persists. However, upon reduction of the disulfide bond, the protein can readily dissociate to the monomer form.[51] Zn(II) addition to the reduced apo-hSOD1 restores the dimeric state, indicating that only the most immature form of hSOD1, before any post-translational modifications, favors the monomeric state.[48]

A study on post-translational modifications influencing the stability of both WT and ALS-mutant SOD1 has recently been recently reported.[52] Consideration of the four post-translational events (Cu-binding, Zn-binding, disulfidesulphide formation, dimerization) preceding SOD1 activation makes it possible for the protein to adopt over 44 states. Results showed that ALS mutations have the greatest effect on the most immature form of SOD1, destabilizing the metal-free and disulfide-reduced polypeptide to the point that it is unfolded at physiological temperature. Moreover, the immature states of ALS-mutant, but not WT, SOD1 were readily forming oligomers at physiological concentrations. The oligomers were found to be more susceptible to mild oxidative stress, promoting incorrect disulfide crosslinks between conserved cysteines and driving aggregation. It was therefore concluded that it is the earliest disulfide-reduced polypeptides in the SOD1 assembly pathway that are most de-stabilized with respect to unfolding and oxidative aggregation by ALS-causing mutations.

Since the loss of copper and zinc ions decreases the thermodynamic stability of SOD, this would probably also decrease its kinetic stability and favor partial or global unfolding transitions leading to misfolding and aggregation. Five FALS-related mutants (G37R, H46R, G85R, D90A and L144F) were used to demonstrate that the unfolding rates were very similar to WT SOD in the metalated as well as in the demetalated states,[53] with demetalated SODs showing unfolding rates about 60-fold faster than the metalated proteins. As a consequence, pathogenic SOD mutations may lead to SOD aggregation by affecting its ability to bind or retain metals and it may also be hypothesized that the loss of metals in WT SOD is involved in non-familial forms of ALS.

The A4V mutation is responsible for a rapidly progressive disease course and is particularly prone to aggregation when expressed in *E. coli*: it remains soluble when expressed at 18°C, but aggregates in inclusion bodies when expressed at 23°C or above.[54] Since the SOD aggregates dissolve with 4 M urea, inter-molecular hydrophobic interactions are predominantly responsible for making SOD insoluble. Many of the urea-solubilized subunits were cross-linked *via* disulfide bridges.

The H43R and A4V mutations, when examined in the context of the C6A/ C111S HSOD (HSOD-AS) parent, which retains the WT fold and activity, showed nearly WT structures; however, an overall architectural destabilization was demonstrated propagating from the point mutations and promoting the formation of fibril-like aggregates.[55] The cysteine-independent aggregation of FALS mutants was 7–80-fold enhanced compared to the parent HSOD-AS, and aggregation severity could be suggested to be enhanced by the presence of free cysteine residues, which can form disulfide–sulfide bonds to covalently lock the fibrous aggregates. The fibrils were visualized for the first time by electron microscopy and atomic force microscopy and resembled those found in post-mortem studies of FALS patients.

Cu/Zn-SOD mutants with modifications near the metal binding sites show altered SOD activities and spectroscopic properties. Moreover, some of them, specifically metal-deficient S134N and H46R SOD1, crystallize in filament-like arrays with abnormal contacts between the individual protein dimers.[56] To test hypothesis that the abnormal protein–protein contacts in the crystal structures might also be responsible for the earliest steps of the protein oligomerization in solution, NMR studies were carried out on S134N SOD. These studies demonstrated a dramatically different behavior between the fully metalated mutant and the fully metalated WT SOD1, with a region of subnanosecond mobility located close to the site of the mutation.[57] In addition, concentration dependent chemical shifts were providing evidence of abnormal intermolecular contacts in solution.

According to one gain-of-function theory, a mutation in *SOD1* alters the enzyme in a way that enhances its reactivity with abnormal substrates (Figure 3). For example, abnormal tyrosine nitration could damage proteins if the radical peroxynitrite is used as a substrate of SOD1.[8] Spinal cord levels of free nitrotyrosine are elevated in patients with SALS and in those with FALS, as well as in *SOD1*-knockout mice, but specific targets of nitration have not been identified so far.

Since mutations in *SOD1* may cause oxidative damage by impairing the ability of the enzyme to bind zinc, once deprived of zinc, both mutant and WT SOD1 behave as less efficient superoxide scavengers, and the rate of tyrosine nitration increases.[58]

Possible targets of SOD1-induced toxicity include the neurofilament proteins, composed of heavy, medium and light subunits. These proteins play a role in axonal transport and in determining the shape of cells. Large-caliber, neurofilament-rich motor axons are preferentially affected in human ALS, and the level of neurofilaments may be important in selective neuronal vulnerability.[1]

In both SALS and FALS patients, neurofilaments accumulate in the cells and proximal axons of motor neurons. Abnormalities in neurofilaments could be either causal or a byproduct of neuronal degeneration.[59]

The direct involvement of neurofilaments in ALS pathogenesis is ratified by:

(i) over-expression of mutant subunits in mice yields the dysfunction of motor neurons and the degeneration of axons;

    (ii) the same over-expression determines neurofilament swellings similar to those detected in ALS patients;

   (iii) mutations in the gene for the heavy subunit of neurofilaments are found in both SALS and FALS patients; and

   (iv) a mutation in the gene for the light subunit of neurofilaments was found in another motor neuron disorder, the neuronal form of Charcot–Marie–Tooth disease.[60]

The way in which the aberrant expression of neurofilaments causes the degeneration of motor neurons is unclear. Disorganized neurofilaments could impede the axonal transport of molecules necessary for the maintenance of axons ("axonal strangulation") (Figure 2).[1] Such abnormalities in neurofilaments may result from the toxic effects of mutant SOD1. In mice with a mutation in *SOD1*, elimination of the expression of the light subunit or overexpression of the heavy subunit of neurofilaments ameliorated the motor neuron disease. Axonal neurofilaments may be targets of the toxic effects of mutant SOD1, which could explain why reducing the number of axonal neurofilaments is protective. Alternatively, the accumulation of neurofilaments in motor neuron cells could protect against SOD1-mediated injury by buffering calcium,[61] or diminishing zinc binding.

Peripherin – another intermediate filament – is found with neurofilaments in the neuronal inclusions of patients with SALS and mice with *SOD1* mutations. Peripherin is normally expressed in motor neurons, but levels of peripherin increase in response to cellular injury or inflammatory cytokines. Increased expression of peripherin after neuronal injury or inflammation could cause motor neuron disease through an interaction with the medium and heavy subunits of neurofilaments in the absence of the light subunits, leading to the formation of toxic aggregates. This could explain why the over-expression of peripherin kills only motor neurons, which contain high levels of neurofilaments, and not sensory neurons, which do not express neurofilaments.

There is much evidence to indicate that ALS involves a derangement of intracellular free calcium. Abnormal calcium homeostasis activates a train of events that ultimately triggers cell death. In patients with ALS and in mice with mutant SOD1, the resistance of particular motor neurons (*e.g.*, oculomotor neurons) may be related to the presence of calcium-binding proteins that protect against the toxic effects of high intracellular calcium levels.

The mechanism of excitotoxic injury of neurons involves excessive entry of extracellular calcium through the inappropriate activation of glutamate receptors. Glutamate, the chief excitatory neurotransmitter in the CNS, acts through two classes of receptors: the G protein-coupled receptor, which, when activated, leads to the release of intracellular calcium stores, and the glutamate-gated ion channels, which are distinguished by their sensitivity (or insensitivity) to *N*-methyl-D-aspartic acid (NMDA).

The NMDA-receptor channel is calcium-permeable, whereas the permeability of the non-NMDA-receptor channel (activated by the selective agonists kainate and α-amino-3-hydroxy-5-methyl-4-isoxazole propionic acid [AMPA]) varies with the subunit composition of the receptor. If a particular subunit (named GluR2) is present, the channel is impermeable to calcium. In contrast, AMPA receptors that lack GluR2 are calcium-permeable. The selective vulnerability of motor neurons to AMPA,[62] could be explained either by the fact that the expression of GluR2 in motor neurons is normally lower than in other neurons or by an impairment in the editing of GluR2 mRNA in patients with ALS.[63] Either mechanism would lead to the expression of calcium-permeable AMPA receptors.

The possibility of glutamate excitotoxicity in patients with ALS was suggested by the finding of increased glutamate levels in cerebrospinal fluid in patients with SALS.[64] High levels of glutamate could be excitotoxic, increasing levels of free calcium through the direct activation of calcium-permeable receptors or voltage-gated calcium channels.

The increased levels of glutamate in cerebrospinal fluid could also result from impaired glutamate transport in the CNS. The synaptic activity of glutamate is normally terminated by reuptake of the neurotransmitter by excitatory amino acid transporters (EAATs). In patients with FALS, mutant SOD1 could lead to excitotoxic neuronal injury by catalyzing the inactivation of EAAT2, as it does in the presence of hydrogen peroxide.[65] This process would represent another link between FALS and SALS.

Mutant SOD1 may also affect intracellular calcium levels through a direct toxic effect on mitochondria, which are essential for calcium homeostasis. The high metabolic load of motor neurons and the consequent dependence of these cells on oxidative phosphorylation may make them particularly vulnerable to the loss of mitochondrial function.

The many possible triggers of ALS could perturb diverse cellular functions essential for the survival of motor neurons. In SOD1-mediated ALS, motor neurons have been suggested to die as a result of apoptosis,[66] although other reports have disputed this point.[67] Apoptosis involves the activation of the caspase proteases. In mice with the G93A mutation in *SOD1*, the expression of anti-apoptotic Bcl-2 delayed the onset of motor neuron disease and prolonged life. An inhibitor of the caspase, interleukin-1β-converting enzyme, also slowed progression and extended survival, as did the intra-cerebroventricular administration of zVAD-fmk, a broad caspase inhibitor. Although apoptosis is a late event in the degeneration of motor neurons, inhibition of programmed cell death might ameliorate ALS.

It may be concluded that multiple theories have been proposed to explain the molecular pathogenesis of ALS. It is likely that more than one of these mechanisms contributes to human ALS. How these pathways interact remains to be explained.

References

1. L.P. Rowland and N.A. Shneider, *N. Engl. J. Med.*, 2001, **344**, 1688.
2. L.I. Bruijn *et al.*, *Annu. Rev. Neurosci.*, 2004, **27**, 723.
3. F. Hoerndli *et al.*, *Progr. Neurobiol.*, 2005, **76**, 169.

4. S. Hadano *et al.*, *Nat. Genet.*, 2001, **29**, 166.

5. Y. Yang *et al.*, *Nat. Genet.*, 2001, **29**, 160.

6. M. Ramstrom *et al.*, *Proteomics*, 2004, **4**, 4010.

7. J.H. Hu *et al.*, *J. Neurochem.*, 2003, **85**, 432.

8. J.S. Beckman *et al.*, *Nature*, 1993, **364**, 584.

9. M.C. Dal Canto and M.E. Gurney, *Am. J. Pathol.*, 1994, **145**, 1271.

10. T.L. Williamson *et al.*, *Proc. Natl. Acad. Sci. USA*, 1998, **95**, 9631.

11. D. Lambrechts *et al.*, *Nat. Genet.*, 2003, **34**, 383.

12. C. Armon *et al.*, *Neurology*, 1991, **41**, 1077.

13. F. Kamel *et al.*, *Epidemiology*, 2002, **13**, 311.

14. M. Vinceti *et al.*, *Amyotroph. Lateral Scler. Other Motor Neuron Disord.*, 2002, **3**, 208.

15. E. Kapaki *et al.*, *J Neurol Sci.*, 1997, **147**, 171.

16. E.J. Kasarskis *et al.*, *J Neurol Sci.*, 1995, **130**, 203.

17. M. Yasui *et al.*, *Neurotoxicology*, 1993, **14**, 445.

18. K. Sood *et al.*, *Indian J. Med. Res.*, 1990, **92**, 9.

19. M. Tomonaga *et al.*, *Acta Neuropathol.*, 1978, **42**, 81.

20. A. Migheli *et al.*, *Neurosci. Lett.*, 1990, **114**, 5.

21. D.R. Rosen *et al.*, *Nature*, 1993, **362**, 59.

22. G. Lenaz, *Biochim. Biophys. Acta*, 1998, **1366**, 53.

23. Z. Xu, *Amyotroph. Lateral Scler. Other Motor Neuron Disord.*, 2001, **1**, 225.

24. J.D. Wood *et al.*, *Neuropathol. Appl. Neurobiol.*, 2003, **29**, 529.

25. R.C. Lariviere and J.P. Julien, *J. Neurobiol.*, 2004, **58**, 131.

26. J.S. Beckman *et al.*, *Trends Neurosci.*, 2001, **24**, S15.

27. Z. Xu *et al.*, *J. Bioenerg. Biomembr.*, 2004, **36**, 395.

28. S. Przedborski, *Neurologist*, 2004, **10**, 1.

29. L. Siklos *et al.*, *Ann. Neurol.*, 1996, **39**, 203.

30. G.M. Borthwick *et al.*, *Ann. Neurol.*, 1999, **46**, 787.

31. J.S. Valentine *et al.*, *Curr. Opin. Chem. Biol.*, 1998, **2**, 253.

32. E.L. Shipp *et al.*, *Biochemistry*, 2003, **42**, 1890.

33. S.Z. Potter and J.S. Valentine, *J. Biol. Inorg. Chem.*, 2003, **8**, 373.

34. L. Banci *et al.*, *Eur. J. Biochem.*, 2002, **269**, 1905.

35. J.S. Valentine *et al.*, *Annu. Rev. Biochem.*, 2005, **74**, 563.

36. C. Guegan and S. Przedborski, *J. Clin. Invest.*, 2003, **111**, 153.

37. J.S. Beckman *et al.*, *Trends Neurosci.*, 2001, **24**, S15.

38. C. Bendotti and M.T. Carrí, *Trends Mol. Med.*, 2004, **10**, 393.

39. J.S. Valentine and P.J. Hart, *Proc. Natl. Acad. Sci. USA*, 2003, **100**, 3617.

40. L.J. Hayward *et al.*, *J. Biol. Chem.*, 2002, **277**, 15923.

41. J.R. Subramaniam *et al.*, *Nat. Neurosci.*, 2002, **5**, 301.

42. D.W. Cleveland and J. Liu, *Nature Med.*, 2000, **6**, 1320.

43. M. Wiedau-Pazos *et al.*, *Science*, 1996, **271**, 515.

44. A.G. Estevez *et al.*, *Science*, 1999, **286**, 2498.

45. J. Wang, *Hum. Mol. Genet.*, 2003, **12**, 2753.

46. S.M. Lynch *et al.*, *Biochemistry*, 2004, **43**, 16525.

47. J.A. Rodriguez *et al.*, *J. Biol. Chem.*, 2002, **277**, 15932.

48. Y. Furukawa *et al.*, *EMBO J.*, 2004, **23**, 2872.
49. A. Tiwari and L.J. Hayward, *J. Biol. Chem.*, 2003, **278**, 5984.
50. L.S. Field *et al.*, *J. Biol. Chem.*, 2003, **278**, 28052.
51. F. Arnesano *et al.*, *J. Biol. Chem.*, 2004, **279**, 47998.
52. Y. Furukawa and T.V. O'Halloran, *J. Biol. Chem.*, 2005, **280**, 17266.
53. T.M. Dawson and V.L. Dawson, *Science*, 2003, **302**, 819.
54. A. Siderowf and M. Stern, *Ann. Intern. Med.*, 2003, **138**, 651.
55. N. DiDonato *et al.*, *J. Mol. Biol.*, 2003, **332**, 601.
56. J.S. Elam *et al.*, *Nat. Struct. Biol.*, 2003, **10**, 461.
57. L. Banci *et al.*, *J. Biol. Chem.*, 2005, **280**, 35815.
58. L.B. Corson *et al.*, *Proc. Natl. Acad. Sci. USA*, 1998, **95**, 6361.
59. J.P. Julien and J.M. Beaulieu, *J. Neurol. Sci.*, 2000, **180**, 7.
60. I.V. Mersiyanova *et al.*, *Am. J. Hum. Genet.*, 2000, **67**, 37.
61. S. Lefebvre and W.E. Mushynski, *Biochemistry*, 1988, **27**, 8503.
62. F. Terro *et al.*, *Brain Res.*, 1998, **809**, 319.
63. H. Takuma *et al.*, *Ann. Neurol.*, 1999, **46**, 806.
64. J.D. Rothstein *et al.*, *Ann. Neurol.*, 1990, **28**, 18.
65. D. Trotti *et al.*, *Nat. Neurosci.*, 1992, **2**, 427.
66. L.J. Martin, *J. Neuropathol. Exp. Neurol.*, 1999, **58**, 459.
67. B.P. He and M.J. Strong, *Neuropathol. Appl. Neurobiol.*, 2000, **26**, 150.

CHAPTER 11

Parkinson's Disease: Any Role for Metals

11.1   Introduction

Parkinson's disease (PD) is a severe, progressive motor disorder caused by changes in the central nervous system. True PD is tightly linked to degeneration of neurons in an area of the ventral mid-brain or basal ganglia known as the *substantia nigra pars compacta*. The neurons affected are specifically ones that generate the compound dopamine (DA) as a neurotransmitter and are termed dopaminergic neurons. The disease was first described in 1817 by James Parkinson and was also termed shaky palsy because of the shaking movement made by the patients. The disease affects approximately 1 in every 500 people. Other diseases with similar symptoms are often described as "parkinsonian" because of the symptoms exhibited. One such disease is manganism. However, these diseases must be separated from true PD, which is specifically a disease resulting from the degeneration of dopaminergic neurons in the substantia nigra (SN). Around 50–70% of all the dopaminergic neurons are lost from this region before symptoms of the disease appear.[1,2] Like most neurodegenerative disorders, the true cause of the diseases remains uncertain. There is strong evidence that familial or inherited forms are linked to particular point mutations in certain genes such as α-synuclein or parkin. A common treatment of the disorder is to supply the lost neurotransmitter, L-DOPA.

The clinical symptoms of PD include resting tremors, muscle rigidity and bradykinesia, as well as extensive dopaminergic neuronal loss, the presence of Lewy bodies (LBs) containing α-synuclein fibrils in the SN (Figure 1) and other brain regions are characteristic of the disease.[1,3] Inclusions containing α-synuclein are also found in dementia with Lewy bodies (DLB), multiple system atrophy and the "Lewy body variant" of Alzheimer's disease (AD).[4]

It is likely that α-synuclein plays a critical role in the pathogenesis of these diseases since rare missense mutations in the *SNCA* gene (resulting in amino acid substitutions A30P, E46K, A53T) or duplication or triplication of the α-synuclein locus have been linked to familial forms of either PD[5–9] or DLB.[7] Furthermore, transgenic animals over-expressing wild-type or mutant human α-synuclein develop clinical and pathological features very similar to those

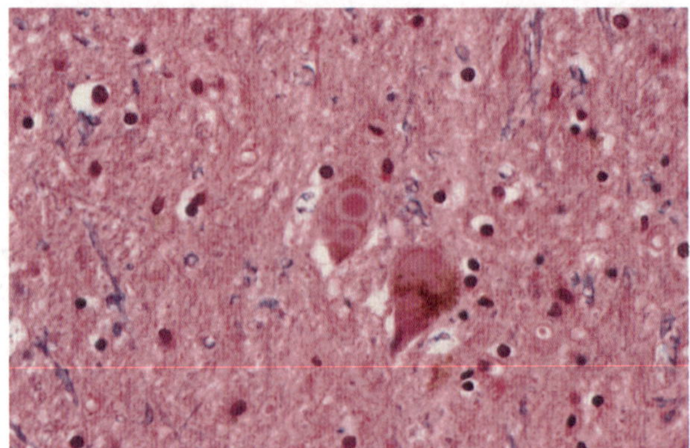

Figure 1   *Lewy bodies. A transverse section of the SN from a PD patient showing two Lewy bodies. These deposits are composed largely of α-synuclein*

observed in PD,[10,11] suggesting that the accumulation of aggregated forms of α-synuclein in the brain could be the underlying cause of neurodegeneration in PD and related disorders.

Most patients with PD have a sporadic form that has not been possible to link to mutations in any known gene. Around 15% of patients claim to have a family member who also had the disease.[12] A huge number of genetic linkage studies have been undertaken to attempt to find the gene associated with PD in the inherited forms. Paradoxically, there has been no single gene identified as the PD gene. Mutations in a large number of proteins have been found. These include α-synuclein, parkin, dardarin, DJ-1, Nurr-1 and series of genes in which the loci are known, but the protein product remains to be identified (*e.g.*, PARK10). Studies of the mutated proteins have provided insight into how the mutations might cause disease. It has been hoped that the same proteins might be somehow affected in the sporadic forms of the disease. However, evidence for this is scant. Looking for factors that might make people susceptible to PD has lead to a range of studies looking at candidate susceptibility genes. These studies have taken as their lead ideas provided by those genes for which mutations cause the inherited forms including genes related to DA synthesis or the ubiquitin–proteasome system or even oxidative stress. Although many of these studies provide interesting clues, the lack of reproducibility between linkage studies of susceptibility has meant that currently there is clearly no evident set of susceptibility genes and this field requires further development.

11.2   Cell Death in Parkinson's Disease

The search for a single mode of cell death in PD has been elusive given the contradictory results that have been obtained. There is evidence that some amount of the neuronal death is due to apoptotic mechanisms, but also autophagy.[13,14] There is evidence for the involvement of the tumor necrosis factor-α receptor[15] and caspase 8.[16] Studies on the execution of cell death have largely been contradictory, but the one which is consistent does not show anything highly specific about the cell-death mechanism. What is more important to consider is the nature of the factors that initiate cell death. Given the link between the diseases and alterations in neurotansmitters, it is highly likely that some form of excitotoxicity is involved. This hypothesis has been strengthened by the finding that complex I of the respiratory chain showed less activity in mitochondria from Parkinson's patients.[17] Also, in one of the main animal models of PD, inhibition of nitric oxide synthase has neuroprotective effects.[18] Models using animals rely on toxins such as MPTP (1-methyl-4-phenyl-1,2,3,6-tetrahydropyridine) and rotenone. The toxins have their effect by altering mitochondrial complex I and induce oxidative stress. As indicated above, the main characteristic of these diseases is the breakdown of the dopaminergic system. Dopamine and DA-breakdown products can possibly have oxidative effects. In addition, there is evidence that iron levels are elevated in PD again pointing to an oxidative mechanism. The other characteristic of these diseases is the presence of aggregated α-synuclein. It is possible that this protein is neurotoxic in its own right. Additionally, the formation of the aggregated form of the protein could result from alterations in the processing of cellular proteins. Two of the other proteins associated with PD are involved in the ubiquitin–proteasome system. This is the main protein degradation system in the cell and its function is necessary to rapidly clear proteins. Abnormalities in the function of this pathway could result in increased concentrations of proteins that have potential toxicity or that might aggregate.

The current state of the field does not allow a clear understanding as to which of these aspects are important and which are not or how the different components might fit together into a cohesive story. Whatever the mechanism involved, it must be one that is selective for dopaminergic neurons. Therefore, an understanding of this system requires an understanding of the changes to release this neurotransmitter. Naturally, proteins that clearly have a role in the disease, because mutations in their genes result in familial forms, must be studied in detail. One possible way to combine the various suggested neurotoxic mechanisms is the following: The main initiator of cell death is a change in the mitochondria, which results in the changes in intracellular iron and increased oxidative stress. The damage to the mitochondria results from the collection of high levels of proteins in the cytoplasm that damage mitochondria, including aggregated α-synuclein. This protein accumulates because it cannot be broken down. One major cause of this would be alteration in the ubiquitin–proteasome system. Alterations in this system are unlikely to be specific to dopaminergic neurons. However, these neurons must, for some currently unknown reason, be more sensitive to these changes. Unfortunately, the key to the exact nature of the trigger of cell death remains unknown.

11.3 Genetics of Parkinson's Disease

The genetic loci associated with PD have been given the designation PARK. The first of these to be identified was PARK1, which is associated with the protein α-synuclein and is found on human chromosome 4q21.[5] In this case, the disease either arises through missense point mutations (A53T, A30P or E46K) or through triplication of the gene. The latter demonstrates that simple increased expression of α-syncuclein could be sufficient to cause disease. These mutations are associated with early onset of the disease and the pathology includes Lewy bodies and is autosomal dominant.[19]

PARK2 is another early onset locus but is autosomal recessive and this form of disease lacks Lewy bodies. The protein associated with this locus is parkin. This is an E3 ubiquitin-ligase containing two RING finger domains.[20] Mutations in this gene are also associated with juvenile onset PD. Most of the mutations causing disease are found in exon 7.

PARK3 has been associated with chromosome 2p13 and is autosomal dominant, but the gene product remains unknown. PARK4 is no longer accepted as a separate locus for PD but has been identified as a triplication of the α-synuclein gene.[8] PARK5 is also autosomal dominant and associated with another ubiquitin-associated enzyme, ubiquitin *C*-terminal hydrolase-L1 (UCH-L1). A single known mutant of this protein has been identified.[21] UCHL-1 hydrolyses small *C*-terminal adducts of ubiquitin to generate the ubiquitin monomer. The protein is highly expressed in neurons.[22] This protein is often found in Lewy-bodies.[23] A polymorphism on exon 3 of the UCHL-1 gene has also been associated with reduced susceptibility to PD. The protein is a major target for oxidative damage and is found to be down regulated in expression in both AD and PD.[24] The finding that this protein is altered in PD further emphasizes that some alteration in the ubiquitin–proteasome system lies at the heart of this disease.

Both PARK6 and PARK7 are autosomal recessive loci found at chromosome 1p35 and 1p36 and both are associated with early onset PD. However, they code two different proteins, DJ-1 and PINK-1, respectively. DJ-1 is a ubiquitously expressed homodimeric protein. Eleven different mutations of the gene for DJ-1 are associated with PD.[25] The mutations are found throughout the gene. This suggests any change in the protein's function could cause the same problem in terms of altered metabolism. The protein stabilizes mRNA in association with *c-myc*. This protein has also been suggested to be an antioxidant or a sensor for oxidative stress.[26] Further studies also suggest a protein chaperone role. Studies with one mutant form of DJ-1 suggest that it could cause toxicity by overwhelming the protein degradation systems.[27] Once again DJ-1 is another protein that seems to play a role in protein folding or protein breakdown. PARK6 is one of the more common early onset loci in European families. Two different mutations in the gene for PTEN-induced putative kinase (PINK-1) were recently identified.[28] This protein is localized to the mitochondria. Mitochondrial dysfunction has been shown in PD and verification of this finding could have important consequences. Mitochondrial breakdown could be a mechanism of cell death for dopaminergic neurons.

Another recently identified protein is dardarin. This is the protein product of the PARK8 locus[29] associated with an autosomal dominant form of the disease usually associated with late onset. As this is a new finding, little is known about this protein or its cellular role. Less is known about PARK9 or PARK10. There are other loci that are associated with PD, but there are two other genes that are of note. These are nuclear receptor related-1 (Nurr1) and neurofilament-M. Nurr1 or NR4A2 is a protein associated with cellular differentiation and the maintenance of dopaminergic neurons. Owing to this function, it has already been suggested to be a possible candidate for a role in PD. Mutations in this gene have been found in 10% of familial PD cases in one study.[30] Also, a particular polymorphism is higher in patients with sporadic PD than in controls. Mutations in the Nurr1 gene alter expression of genes associated with DA transport.[31] Neurofilament M is one of the forms of the cytoskeletal proteins of neurons associated with the formation of the axons. A missense mutation in this protein has been found in one family with familial PD.[32] The importance of this remains unknown but it does suggest that a change in the cytoskeleton could play a role in the disease process.

11.4   The Proteins Associated with Parkinson's Disease

As indicated from the genetic studies, there are quite a number of proteins associated with PD by virtue of linkage studies and the evidence of mutations in specific genes. It is unclear at present how these mutations cause the disease. It is also unclear which of these proteins are the most important in pathology of PD. The weight of evidence points toward α-synuclein and parkin, but other protein should not be ruled out. Many of the other proteins have only recently been identified. However, the two well-characterized proteins are worthy of further consideration here.

11.4.1   Parkin

A form of inherited PD termed autosomal recessive juvenile PD, is the most frequent form of familial PD. Linkage studies first showed that this form of the disease was closely associated with mutations in the gene for the protein termed parkin.[33] The wild-type form of this protein is an ubiquitin ligase (Figure 2). The mutant forms of this protein do not have this activity.[20,34,35] Parkin forms a complex with other proteins known as F-box protein, cullin-1 and hSel-10. These proteins form a SCF-like E3 complex, named because it can degrade cyclin-E.[36] Thus, parkin may play a role in regulating the cell cycle by its possible action of the degradation of this cyclin. Parkin has also been associated with chaperones Hsp70 and CHIP.[37] Parkin also ubiquitinates polyglutamine proteins.[38] All these findings suggest a role for parkin in quality control of protein synthesis. Substrates for parkin are quite variable but include cyclin-E, synaptotagmin XI, synphilin-1, tubulin, CDCrel-1 and the Pael receptor.[39] As the ubiquitination pathway regulates degradation of proteins, excess levels of some of these proteins might lead to cell death. There is some evidence for this for the Pael receptor.[40] Also synaptotagmin XI is associated with exocytosis in the SN.[41] Accumulation of his protein might explain disorder of DA release.

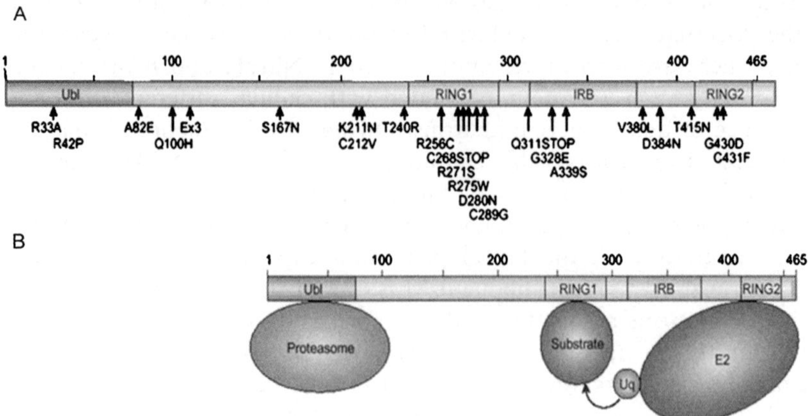

Figure 2 *Parkin. (A) Linear structure of the parkin protein. The known mutations in the protein that cause PD are shown below the sequence indicated by arrows. Shown are the Ubl (ubiquitin-like domain) and the two RING (really interesting new gene) domains. (B) An illustration of the ubiquitination activity of the protein. The substrate and the E2 ubiquitinase bind to the two RING domains. The proteasome binds to the Ubl domain via interaction with Rpn10*

Parkin-knockout mice have been produced.[42,43] These mice do not show loss of dopaminergic neurons. However, the mice have altered behavior associated with neuronal dysfunction. In particular, the striatum of these mice show increased levels of DA and DA metabolites. This suggests that parkin is necessary for maintaining synaptic function in dopaminergic and other neurons.

There is no evidence of Lewy body formation in parkin-associated PD. However, parkin is present in Lewy bodies.[44] The use of proteasomal inhibitors results in the formation of protein inclusions in cells. One of the proteins that will be bound into the inclusions is parkin.[45] Thus, sequestration of parkin in Lewy bodies could have a similar effect to mutations in the inherited forms and prevent its activity. It is not clear if loss of parkin activity plays a role in other forms of PD.

Oxidative stress could alter the activity of parkin. The protein's activity is dependent on cysteine residues. Hydrogen peroxide has been shown to cause misfolding of parkin.[46] In addition, three amino acid residues at the *C*-terminus are needed for the proper folding of the protein and one of the parkin mutations, W453Stop, which loses these, has both lost activity and is misfolded.[39] Therefore, both oxidative stress and mutation could alter the protein to prevent its activity. In addition, parkin could potentially protect against oxidative stress. Parkin over-expression has been shown to protect against various kinds of cell death associated with oxidative stress or mitochondrial dysfunction.[47] Astrocytes show a redistribution and increased expression of parkin during unfolded protein stress.[48]

Although it remains unclear as to how loss of parkin could be neurotoxic, there has been a recent study suggesting that parkin expression protects against manganese toxicity.[49] It is unknown as to whether parkin is a Mn-binding protein, but the potential for parkin to be involved in the regulation of Mn metabolism suggest that Mn toxicity could result from the loss of parkin's activity. Mn toxicity in the brain is known to produce parkinsonian symptoms in patients. This is currently a speculative idea but suggests yet another possible link between a neurodegenerative disease and alter metabolism of a specific metal.

11.4.2 α-Synuclein

α-synuclein is a small (14 kD), highly conserved, presynaptic protein of unknown function, expressed highly in specific brain regions.[50–53] The protein has a series of imperfect repeats (KTKEGV) at the *N*-terminus, as well as an acid *C*-terminal domain (amino acid residues 96–140) and appear from recombinant studies, to be a natively unfolded protein.[51–57] In addition, like most natively unfolded proteins, it has low overall hydrophobicity and a large net negative charge.[58] α-synuclein is one of a family of three proteins. The other two are β-synuclein and γ-synuclein. These proteins show 55–62% identity.[59] α- and β-synuclein have identical *N*-termini and both these proteins are concentrated in nerve terminals in the proximity of synaptic vesicles.[60] γ-synuclein is expressed throughout nerve terminals. α-synuclein became of interest to the study of neurodegenerative diseases after the discovery of the nature of what was then termed the "non-amyloid-beta component" (NAC) of plaques in AD.[61,62] The protein then termed NAC-precursor or phosphoneuro-protein-14 turned out to be a homologue of synuclein originally identified in the electric organ of the Pacific electric ray (*Torpedo californica*).[50] Since then, there has been no conformation of a role of NAC in AD but α-synuclein is now accepted as the main component of Lewy bodies as found in PD and Lewy body dementias.

α-Synuclein can bind to lipids membranes through its *N*-terminal repeat region[63] and can selectively inhibit phospholipase-D2.[57] This phospholipase is localized to the cell membrane where it is involved in signal-induced cytoskeleton regulation and endocytosis. It is therefore possible that α-synculein regulates vesicular transport processes. α-Synuclein appears to be phosphorylated[64] and this may have some consequences for the protein's function. There is some evidence that the protein interacts with synphilin-1,[65] another protein of unknown function, and this protein has also been identified in Lewy bodies.[66]

Knockout mice have been generated that do not express α-synuclein. These mice do not show any neuropathological changes suggesting that loss of function of the protein does not play a direct role in any form of cell death.[67] However, loss of the protein does result in abnormal activity of dopaminergic neurons in SN, with reduced levels of DA detected in the striatum. This implies that the protein could play a role in the regulation of neurotransmitter release. A second strain of α-synuclein knockout mice was also developed.[68] Certain toxins will induce parkinsonian changes in mice. One such compound

MPTP-induced degeneration and loss of dopaminergic neurons. This second line of knockout mice was proved to be resistant to the effect of MPTP. MPP+, the metabolic product of MPTP, acts on various elements of the synaptic machine. Again, these results suggest a role for α-synuclein in vesicular function. Recently, a double-knockout mouse has been generated lacking both α- and β-synuclein.[69] Again DA levels were found to be reduced in the brain but studies of neurons isolated in culture found no differences to wild-type mice. This suggests that synucleins are not essential components of the machinery that causes neurotransmitter release, but they may contribute to long-term regulation of presynaptic activities. Given the similarity between the synuclein, possibly a triple-knockout mouse is necessary to understand the function of these proteins in the CNS. It is more likely, however, that the role-played by α-synuclein in disease results from dysfunction due to its aggregation.

Studies examining the potential of α-synuclein to cause PD have used three models. The first uses animals. Animal models are either transgenic mice or rodents injected with compounds such as MPTP or rotenone, which also induced parkinsonian changes.[70] The second model uses cell-culture expression of α-synuclein, either in the wild-type form or expressing the human gene mutants (see above). The third model uses recombinant protein or peptides to study the aggregation properties of the protein under various conditions. Only in the animal models can the parkinsonian condition truly be assessed. However, these models have only really been effective in proving the causal relation between α-synuclein and the loss of dopaminergic neurons. The complexity of an animal model makes it difficult to assess a mechanism by which α-synuclein can act to cause cell death or the mechanism by which it switches from a normal monomeric protein to an aggregated and potentially toxic one.

There have been more than 10 groups that have generated α-synuclein transgenic mice.[71–80] The mice either expressed human wild-type α-synuclein or the human protein carrying one of the two main mutants (A53T or A30P). These mice differ in the level of expression of the protein and in the kind of promoter used to generate expression. Promoters used include those for PDFG-β, Thy-1, prion protein, tyrosine hydroxylase or an oligodendrocyte specific promoter. The results from these many experiments were quite variable. However, many of the mice produced accumulations of α-synuclein and showed changes in the dopaminergic system. In addition, many of the mice showed motor changes reminiscent of the parkinsonian tremor or altered locomotion or coordination. However, none of the mice showed neuronal loss no matter how high the expression level or the presence of mutations was. Expression within glial cells resulted in inclusions with a greater similarity to Lewy bodies. The failure of transgenic mice to result in a reliable model of PD is possibly due to the expression of a human protein in a mouse. Co-expression of mouse and human α-synuclein could alter the ability of the human protein to form a toxic molecule, as it is known that mixing human and mouse α-synuclein inhibits the ability of the protein to aggregate.[81] Therefore, the animal models based on neurotoxins such as MPTP and rotenone are more reliable and reproduce the disease more effectively than transgenic mice. Unfortunately, such models do not provide insight into how changes in α-synuclein cause disease as the disease is generated by another source.

Recently, a viral vector system was used to directly transfer the human α-synuclein into the SN of a rat.[82] The rAAV (recombinant adeno-associated viral) vector resulted in high expression of the human protein in the SN and after 13 weeks the researchers observed a 50% reduction in the number of dopaminergic neurons. Unlike other models, the progression of cell loss was slower, more like PD. In addition, there were other changes that were more similar to the human disease, including phosphorylation of α-synuclein at serine 129 and activation of caspase 9. This system possibly represents a better model of PD.

Considerable research has focused on the mechanism by which α-synuclein can aggregate. The limitation of these studies is that aggregation of α-synuclein has not been proven to be the cause of PD. The formation of Lewy bodies or other aggregates of this protein might be a result of the disease rather than the cause. However, as inherited mutations in the protein are associated with PD, then it is likely that the protein plays some role in the progression of the disease.

Both peptides and recombinant protein have been used to study the aggregation properties of α-synuclein. Filaments will form from the *N*-terminal domain of the protein with similar properties the protein aggregates extracted from the brains of Parkinson patients.[56,83–86] The kinetics of fibrillation is consistent with a nucleation mechanism.[87,88] The key step in the transformation of the protein to a form able to aggregate involves a partially folded intermediate.[56] The final transition of the protein results in a gain of β-sheet content. The fibrils that are formed are amyloid-like around 5–10 nm in length with a diameter of 4–8 nm. These can cluster together to form bundles of 50 nm and up to 1 mm in length.[52,85,89] Peptides from amino acid residues 1–18 behaved similarly to the full-length protein suggesting this domain is key to the aggregation. In comparison, a peptide based on amino acid residues 19–35 remained soluble and unstructured.[90] Further analysis suggested that the residues 8–16 are key to the formation of β-sheet. These findings are in contrast to another study suggested that the main site regulating protein aggregation is around amino acid residues 64–100.[91] Small peptides from this region (residues 69–72) have been shown to block aggregation of the full-length α-synuclein. It is possible that both these regions add to the fibril formation of the protein.

Of greater interest is the assessment of factors that could contribute to the aggregation of the protein. Phosphorylation of serine129 increases fibril formation.[92] Sequence modification is the most obvious cause of increased aggregation and mutations associated with pathology in particular.[93] In addition, oxidation and nitration can also increase the rate of conversion.[94]

Current research has shown that one of the main factors affecting self-oligomerization of the protein are metals, such as Cu and Fe. Cu has been suggested to be the most effective metal in terms of inducing oligomerization.[95] Cu chelators were shown to abolish aggregation. This oligomerization seemed to be mediated by interaction of Cu with the *C*-terminus of the protein. This was shown by limited proteolysis of the α-synuclein that cleaved off part of the *C*-terminus, either at residue 97 or 114. The shorter fragment showed no response to Cu in terms of oligomerization, while the 1–114 fragment did produce a limited amount of oligomerization.[95] Further studies showed that metals not only induce aggregation, but also induce conformational change. In this study, aluminum was found to be the most effective metal at induction of polymerization with Cu and Fe being similarly effective.[96] Analysis suggested that the mechanism of polymerization could either lead to amorphous aggregates or structured fibrils. Structural analyses also showed that the metals induced a switch from unstructured to a β-sheet structure.

Despite early studies suggesting that Cu causes aggregation of the protein *via* the *C*-terminus, a more recent study suggests that Cu can bind to the histidine at residue 50 (Figure 3). Further binding can occur at the *C*-terminus, but this is of much lower affinity than that at *N*-terminal site.[97] Binding at the high affinity site was shown to be sufficient to drive oligomerization of the protein. The high affinity site appears to be a type-II site with square-planar coordination.

However, the affinity for this site is suggested to be 0.1 μM. That for the second site was shown to 50 μM, which is inline with previous findings.[95,98] However, these affinities seem rather low for an intracellular protein where Cu would be bound by proteins with a much higher affinity. An intracellular increase in free Cu or Fe could potentially increase these metals to the concentration where they could interact with α-synuclein and trigger its aggregation. Our laboratory has also studied the affinity of Cu binding to α-synuclein (Figue 4). Our results indicate that Cu binds with a high affinity and the binding of the first Cu initiates dimerization of the protein before a further Cu atom will bind. The finding of a specific high affinity interaction between α-synuclein and redox active metals is the conclusive finding linking PD to the biology of metals.

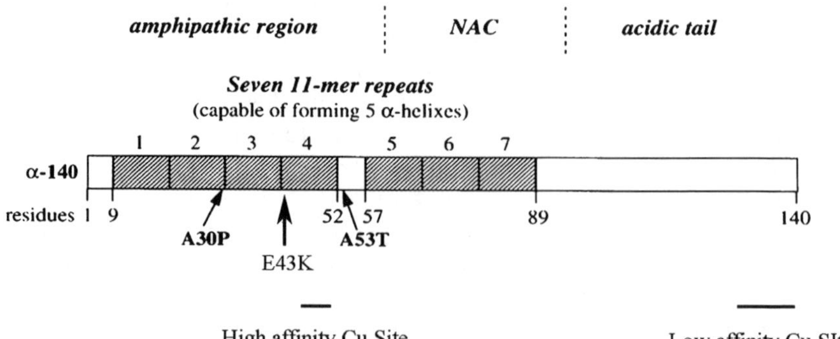

Figure 3   *α-synuclein. Linear representation of the α-synuclein protein. Shown are the location of the repeats in the N-terminus, the NAC domain (non-amyloid component) thought to be involved in aggregation and the proposed Cu-binding domains. Also shown are the location of the three mutations associated with inherited forms of PD*

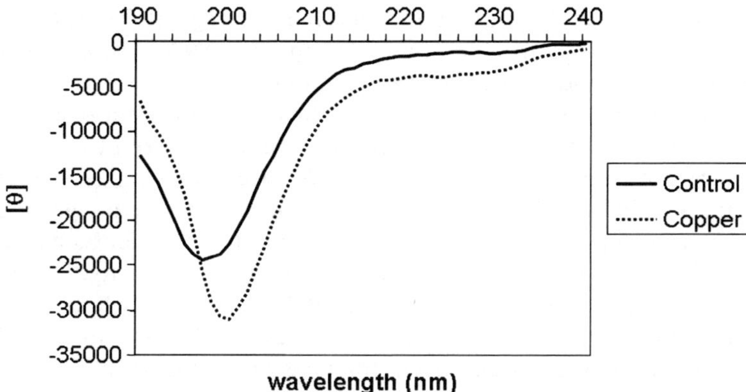

Figure 4   *Copper effect on α-synuclein structure. Recombinant α-synuclein was analyzed by circular dichroism spectroscopy. The protein with no metal bound shows a spectrum indicative of large amounts of random coil structure. Addition of Cu causes the protein to gain more structure as indicated by the shift of the spectrum to the right*

11.5   Metals in Parkinson's Disease

The possible involvement of metals in the cause of PD follows from epidemiological studies. In Michigan, it was found that industrial exposure to copper and iron was related to higher incidences of PD.[99] In a similar study in Quebec, an association was found between increased risk of PD and exposure to manganese, iron or aluminum.[100] In Detroit, exposure to copper or manganese was also found to be associated with increased risk of PD.[101] Analysis of Lewy bodies from parkinsonian substantiae nigrae showed there was increased concentrations of iron, zinc and aluminum compared to other brain regions.[102–104] The iron present showed increased level of Fe^{3+} and an increase in the iron-binding protein, ferritin. Experimental studies using $FeCl_3$ injected directly into the SN of rats resulted in a 95% reduction in striatal DA and altered behavior. This supports the notion that iron initiates dopaminergic degeneration in PD.[105]

As stated above, one possible source of the increased iron in PD could be from altered heme metabolism due to mitochondrial damage. Although there is evidence for alterations in heme oxigenase-1,[106] there is no evidence for alterations in heme synthesis or Fe incorporation. Therefore, it is possible that the altered Fe content of the brain is the result of import of Fe from outside the brain and that Fe itself could trigger the mechanism leading to cell death or α-synuclein aggregation. These changes in brain Fe could be due to alterations in Fe regulator mechanisms in the body.[107]

The main route of entry of Fe into cells is *via* the transferrin receptor. Iron deficiency results in the up-regulation of the transferrin receptor. The endosomes release captured Fe into the cytoplasm where it is captured by Fe-storage protein such as ferritin. In Fe deficiency, ferritin expression is down regulated allowing release of iron. The free Fe in the cell forms a labile iron pool.

Response to iron levels is achieved through the iron regulatory proteins (IRPs) that are able to bind to iron response elements in DNA. These proteins can alter both cellular and mitochondrial functions. Several proteins in the mitochondria can be affected by IRPs. These include the respiratory complex I, succinate dehydrogenase and mitochondrial aconitase. Current understanding of the biology of Fe does not provide a clear picture as to whether disregulation of Fe causes oxidative damage or results from oxidative damage. Increased levels of hydrogen peroxide can alter the activity of IRPs, increasing the levels of free Fe[108] and antioxidants can abolish increased IRP activity.[109] Therefore, oxidative stress can cause a disregulation of Fe metabolism but this in itself could cause further increased of oxidative stress in the cell. In particular, if somehow the normal feedback system for Fe is disrupted, high concentrations of Fe could develop. This disruption would be evidenced by increased levels of Fe in the absence of increased proteins such as ferritin. This has been found in one study of PD.[110] The data suggested that the IRP proteins were more active than required, even though there were low levels of ferritin compared to Fe levels, translation of ferritin by the IRP–IRE system was unaltered. A possible reason for the failure of inactivation of IRP would be compartmentalization of Fe by proteins binding it and sequestering it elsewhere. It has been suggested that IRP-2 maintains Fe homeostasis under normal conditions, but under oxidative conditions IRP-1 becomes aberrantly activated.

Fe causes an increase in the rate of α-synuclein aggregation[96] and Fe is deposited in Lewy bodies that are composed largely of this protein.[111] Fe in association of α-synuclein can generate hydrogen peroxide,[112] which suggests that this combination has the potential to cause oxidative damage. It is therefore possible that Lewy bodies form as a result of the elevated Fe concentrations in the brain. What is unclear is whether the Lewy bodies play any role in cell death or are merely markers of the disease process. It is possible that diversion of α-synuclein into Lewy bodies inhibits the protein-causing toxic effects elsewhere.[113]

Neuromelanin (NM), a complex biopolymer could be another molecule causing damage to neurons *via* interaction with Fe. The pigmented neurons of the SN contain the highest levels of NM. This molecule occurs as granules, possibly inactivated in lysosomes.[114] Energy-dispersive X-ray analysis has shown the presence of Fe in intraneuronal NM granules in the SN of Parkinson's patients but not in controls.[115] This suggests that these granules are a significant pool of Fe in the dopaminergic neurons.[116] The function of NM is unknown, but it could act as an iron chelator or a scavenger to remove Fe.[117] In this sense, it could be protecting against Fe-mediated damage or alternatively it could be trapping redox active Fe within the cell. The NM-Fe complex has been shown to potentiate oxidative damage to membranes.[118] Neuromelanin can also activate microglia inducing the release of cytotoxic factors that can cause neurodegeneration.[117]

Increased Fe in cells could have another effect, altering the breakdown of DA. Dopamine is broken down both enzymatically and non-enzymatically. Enhanced breakdown of DA could result in the production and retention in the

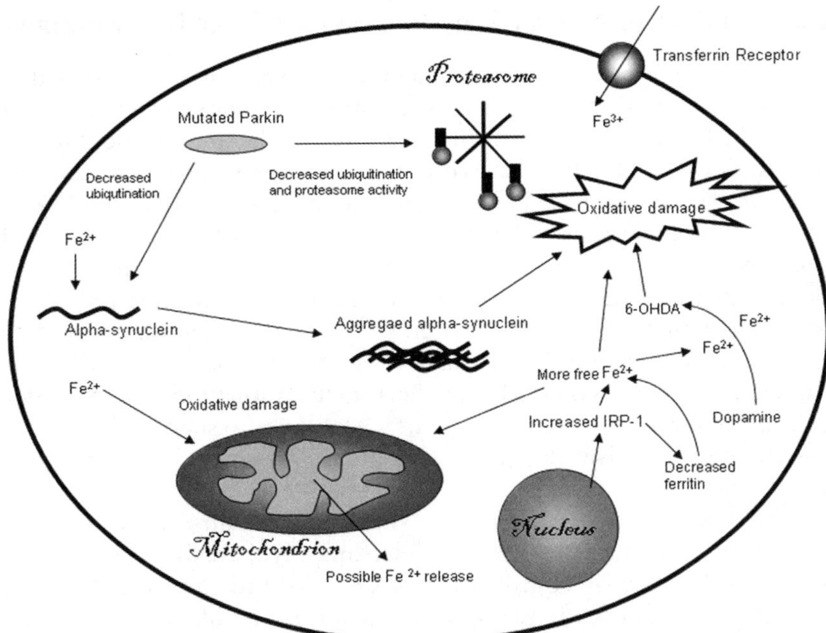

Figure 5 *Cellular changes in PD. Changes associated with PD include increased intracellular free Fe, alterations in the activity of Fe-binding proteins and increased levels of IRP-1. Increased levels of Fe could cause increased levels of aggregated α-synuclein or cause oxidation of the mitochondria or conversion of DA to 6-OHD. In addition mutations in Parkin can alter the activity of the proteasome and possibly decrease breakdown of α-synuclein resulting in aggregation of the protein*

cell of toxic products such as 6-hydroxydopamine (6-OHDA). Auto-oxidation of DA by radicals generated through Fe will increase 6-OHDA levels. Monoamine oxidase is the enzyme that breaks down DA levels and is expressed at high levels in the SN. Potentially, levels of monoamine oxidase B could be elevated by increased levels of oxidative stress, especially mediated by Fe. In this way, Fe could specifically increase oxidative damage that would be specific to dopaminergic neurons (Figure 5). However, currently this theory is not substantiated.

11.6 Bioinorganic Chemistry in Parkinson's Disease

Metals in PD may be involved in different processes. They can affect the protein behavior, induce the oxidative stress as in the other neurological disorders and interact with catecholamins, the neurotransmitters and effective ligands for metal ions like Fe^{3+} or Cu^{2+}. Metal ions, especially Fe^{3+} can be very effectively bound to NM present in high amounts in SN. Metals can also induce the aggregation of α-synuclein a major component of Lewy body.

11.6.1   Binding of Copper Ions to α-Synuclein and its Fragments

As mentioned above α-synuclein a major component of neuronal and glial inclusions (Lewy bodies) is rather a small protein comprising 140 amino acid residues, which can be divided in three regions: (i) the amphipatic *N*-terminal domain (residues 1–60), (ii) the hydrophobic self-aggregating domain, NAC (residues 61–95), and (iii) the acidic *C*-terminal domain (residues 96–140). The hydrophobic domain seems to be critical to start the fibrillation process.[119] The protein does not show any distinct secondary structure[54–57] although it may adopt an α-helical conformation upon interactions with membranes[120,121] and β-conformation in rigid amyloid-like fibrils.[122,123] The mechanism of the formation of amyloid structure is still unclear. However, there is good evidence indicating that metal ions accelerate the protein fibrillation and aggregation affecting most likely the protein structure.[95–96,124] It is also suggested that Cu and Fe may induce oxidative oligomerization and aggregation in the presence of H_2O_2.[125,126]

Earlier studies have suggested that α-synuclein is a multi-copper protein able to bind five or even more metal ions.[95,98] Chemically these findings were rather surprising as the effective coordination of Cu^{2+} ions to an unstructured short protein or peptide is usually realized *via* anchoring site like an imidazole side chain of His residue or nitrogen of the *N*-terminal amino group.[127]

In α-synuclein the role of anchoring sites could play the imidazole moiety of His-50 and the *N*-terminal amino group. The His-50 is indeed involved in the Cu^{2+} coordination contributing to a strong-binding site.[97] According to the spectroscopic data the metal-ion binding site is tetragonal and it involves two nitrogen donors (imidazole and deprotonated amide nitrogen) and two oxygen donor atoms. The carboxylate-rich *C*-terminal domain seems to be able to coordinate the second Cu^{2+} ion involving possibly three carboxylates (Asp-119, Asp-121 and Glu-123) but with much lower affinity than that at His-50 site.[97] However, the *N*-terminal amino group with Met-Asp-Val sequence has potentially better binding ability than that at the *C*-terminal domain or even His-50 site. The insertion of Asp-2 residue with its relatively effective β-carboxylate function causes that the $\{NH_2,N^-,\beta COO^-\}$coordination set may easily coordinate to Cu^{2+} ion forming very stable complex species.[128,129] The study has shown that the *N*-terminal fragments of α-synuclein (residues 1–17, 1–28 and 1–39) are really able to coordinate Cu^{2+} ion *via* $\{NH_2,N^-,\beta COO^-\}$ donor set quite effectively.[130] The formed species (CuH_2L) with this coordination mode is very stable in the broad pH range (Figure 6) including the physiological region.

The physiological concentration of Cu seems to be sufficient to form complexes with α-synuclein and promote the protein aggregation.[97]

11.7   Metal Ions and Catecholamines

Catecholamines (Figure 7) acting among others as neurotransmitters could be consider as very attractive ligands for metal ions serving efficient catechol functions.

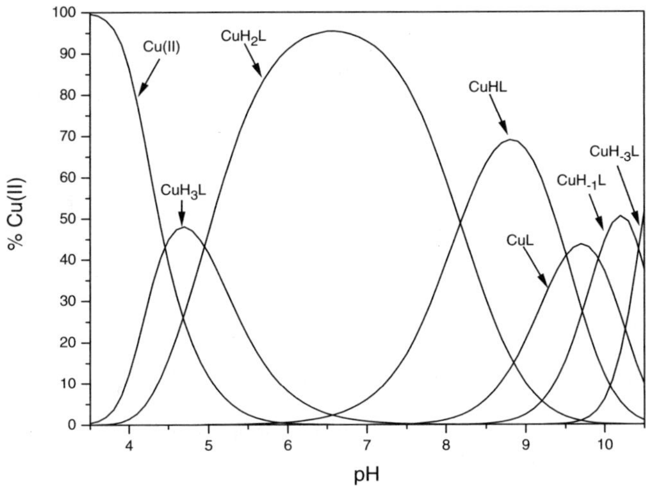

Figure 6 *Species distribution curves for Cu^{2+} complexes with 1–28 fragment of α-synuclein. CuH_2L complex involves $\{NH_2,N^-,\beta COO^-\}$ coordination mode at the N-terminus*

Figure 7 *The catecholamines*

Chelation by catechol oxygens is especially effective for Fe^{3+} ions. Besides, ability to coordinate metal ion the catechol moiety can be easily oxidized during two-electron process to quinone. The coordinated Fe^{3+} ion may oxidize catechol to semiquinone being reduced to Fe^{2+}. Second Fe^{3+} ion will oxidize semiquinone to quinine (Figure 8). Fe^{2+} can also induce catecholamine oxidation binding dioxygen molecule and then acting as catalyst.[131] Quinone species

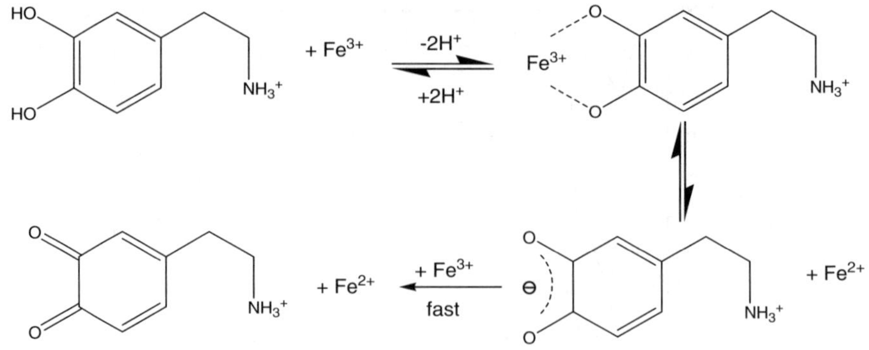

Figure 8 *The general mechanism at low pH for the oxidation of catecholamines by iron(III) ions. Initial formation of a mono-complex followed by internal electron transfer. The semiquinone formed reacts rapidly with another iron ion to form the quinine*

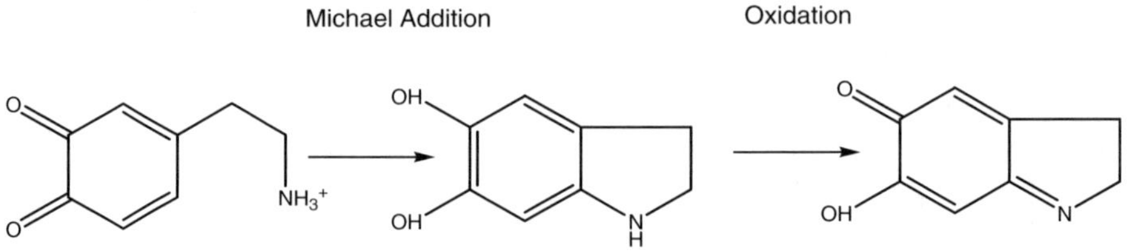

dopaminoquinone leucodopaminochrome dopaminochrome

Figure 9 *Ring closure of the side chain in dopaminoquinone yields leucodopaminochrome, which is further oxidized to dopaminochrome*

may, *via* Michael addition, form indole rings and recover the catechol donor set able to bind metal ion and undergo further oxidation with formation as in case of DA which forms dopaminochrome (Figure 9). The amino function serves the competitive metal-ion binding site, which is very effective when accompanied, *e.g.*, by hydroxyl group (adrenaline, noradrenaline). Cu^{2+} ions are able to bind to both catecholate and amino sites equally well leading to formation of the polymeric species.[132,133] The metal-ion coordination to the amino site may not have any direct biological implication although it may have a strong impact on the chemistry of catecholamines.

The *bis*-catecholamine complexes with Fe^{3+} behave differently, they are stable in respect to the electron transfer within the complex molecule and, in some cases they may bind dioxygen to oxidize ligand *via* catalytic mechanism.[131] The formation of the *tris*-complexes is also likely but their formation is not relevant biologically due to the high pH needed (>8).

The general mechanism of Fe^{3+} coordination and resulting oxidation reactions are similar for most catecholamines except two biologically basic derivatives, noradrenaline[134] and 6-hydroxydopamine (6-OHDA)[135,136] for which oxidation occurs without complex formation prior to ligand oxidation reaction. In the latter case the complex is a transient species hardly detected. The reaction between iron and 6-OHDA is critical for the cell chemistry in PD as 6-OHDA produced by Fenton reaction is able to reduce iron in ferritin and remove it.[137,138]

6-OHDA can enter the ferritin system and reduce iron,[139] which by use of external chelating agent including DA itself or other low-molecular biomolecules like citrate, amino acids or nucleotides can be remove from ferritin to act as the Fenton reaction catalyst.[131] This is very strong support for the notion that 6-OHDA plays a basic role as neurotoxin in progression of PD.

11.8 The Neuromelanin of Substantia Nigra and Metal Ions

The pigmented neurons of SN and the locus coeruleus have the highest levels of NM in the brain.[140] In PD, a selective loss of dopaminergic neurons containing NM is observed, while those unpigmented remain.[141] The biological function of NM is not clear. It could play a protective role owing to sequestration of redox-active metals and organic toxic compounds. NM can also produce free radicals by reacting with H_2O_2.[142] The latter process occurs when level of iron is too high to be effectively bound by NM. The produced free radicals increase the oxidative damage to the cell leading to its death. Cell death results in the release of NM which may activate microglia and release toxic compounds damaging other neurons and deteriorating the neurodegeneration process.[143]

NM is an insoluble complex polymeric pigment having multilayer 3D structure similar to some extent to that of synthetic or natural melanins.[144] The layers being 4.7 Å apart from each other are composed most likely of dihydroxyindole rings and benzothiazine residues. The presence of peptide component (15%) is characteristic for NM and also for other natural melanins.[145] Synthetic melanins incubated, *e.g.*, with homogenates of brain are able to bind peptides in similar manner as does it NM. It could indicate that the presence of the peptide structures within NM pigment is the consequence of the direct reactions between the melanic polymer and proteins. It is also likely that DA used for NM biosynthesis contains peptydyl moieties.[117,145] The major components found in NM are build up with DA and CysDA-derived units. This could suggest that NM synthesis plays a critical role for detoxification processes protecting neurons against accumulation of the toxic compounds derived from DA and its derivatives.[142,146]

The unusual structure of NM and the variety of interactions possible within this structure causes that NM may form effective binding with variety of organic molecules like peptides, lipids, pesticides and various toxic compounds as well as with different metal ions, especially iron. The comparison of the metal-ion concentrations for Fe, Zn and Cu in SN with those in NM isolated from the same sample clearly shows the very high accumulation ratio of iron and zinc in NM equal to 63 and 47, respectively. Accumulation of copper seems to be relatively low even if melanin is very effective in copper-ion binding.[147]

Ferritin, a storage-iron protein is present in oligodendrocytes and astrocytes but it is not detected in SN dopaminergic neurons.[140] Thus, NM is the only known storage system for iron in SN neurons. NM has high and low affinity binding sites for iron. The high affinity binding sites are able to bind iron efficiently enough to protect neurons against radical production.[148] NM is able to bind also other redox active metals like Cu, Cr or Mn as well as toxic metals like Cd, Hg and Pb.[149,150] NM is then the high capacity storage system for many toxic compounds including redox active and redox inactive metal ions. However, in the case of high levels of, *e.g.*, iron NM may involve also the low affinity binding sites, which form ferritin-like iron-hydroxy clusters.[114,151,152] This iron can be redox active able to produce reactive oxygen species causing the neurodegenerative disorders.

Iron in NM is coordinated usually *via* oxygen donors derived from very effective phenolates of catechole-like units or less effective carboxylate functions in the octahedral geometry.[153–155]

11.9   Conclusions

The studies described above suggest that in PD, like other conditions, a number of metals could be involved. In particular, either Cu or Fe may play some role. Until the finding that α-synuclein binds Cu, the evidence that links metals and PD was rather circumstantial. The finding is very recent and at the time of writing it is not clear what the true consequence of this finding is. Possibly, it might only relate to the normal function of the protein. However, given the potential of metals to play a role in the aggregation of the protein, it does possibly point to involvement in the disease process. There is more evidence for a role of Fe in PD. Oxidative stress mediated through alterations in the IRP-1 protein could be central to the mechanism. Oxidative damage *via* Fe could have effects such as increased breakdown of DA. The accumulation of toxic DA by-products in the cell could then result in specific DA-mediated cell death. This potential requires further investigation and many interesting studies are likely to eventuate in the coming years.

References

1. A. Barzilai and E. Melamed, *Trends Mol. Med.*, 2003, **9**, 126.
2. M.G. Spillantini *et al.*, *Nature*, 1997, **388**, 839.
3. K. Arima *et al.*, *Brain Res.*, 1998, **808**, 93.
4. M.C. Bennett, *Pharmacol. Ther.*, 2005, **105**, 311.
5. M.H. Polymeropoulos *et al.*, *Science*, 1997, **276**, 2045.
6. R. Krüger *et al.*, *Nat. Genet.*, 1998, **18**, 106.
7. J.J. Zarranz *et al.*, *Ann. Neurol.*, 2004, **55**, 164.
8. A.B. Singleton *et al.*, *Science*, 2003, **302**, 841.
9. M.C. Chartier-Harlin *et al.*, *Lancet*, 2004, **364**, 1167.
10. E. Masliah *et al.*, *Science*, 2002, **287**, 1265.

11. M.B. Feany and W.W.A. Bender, *Nature*, 2000, **404**, 394.
12. Y. Huang *et al.*, *Brain Res. Brain Res. Rev.*, 2004, **46**, 44.
13. P. Anglade *et al.*, *Histol. Histopathol.*, 1997, **12**, 25.
14. P. Anglade *et al.*, *Histol. Histopathol.*, 1997, **12**, 603.
15. M. Mogi *et al.*, *J. Neural. Transm.*, 2000, **107**, 335.
16. A. Hartmann *et al.*, *J. Neurosci.*, 2001, **21**, 2247.
17. A.H. Schapira *et al.*, *J. Neurochem.*, 1990, **54**, 823.
18. F. Beal *et al.*, *Brain Res.*, 1998, **783**, 109.
19. M. Vila and S. Przedborski, *Nat. Med.*, 2004, **10**(Suppl), S58.
20. H. Shimura *et al.*, *Nat. Genet.*, 2000, **25**, 302.
21. E. Leroy *et al.*, *Nature*, 1998, **395**, 451.
22. J.F. Doran *et al.*, *J. Neurochem.*, 1983, **40**, 1542.
23. K.D. Wilkinson *et al.*, *Science*, 1989, **246**, 670.
24. J. Choi *et al.*, *J. Biol. Chem.*, 2004, **279**, 13256.
25. V. Bonifati *et al.*, *Neurol. Sci.*, 2003, **24**, 159.
26. T. Taira *et al.*, *EMBO Rep.*, 2004, **5**, 213.
27. D.J. Moore *et al.*, *J. Neurochem.*, 2003, **87**, 1558.
28. E.M. Valente *et al.*, *Science*, 2004, **304**, 1158.
29. C. Paisan-Ruiz *et al.*, *Neuron*, 2004, **44**, 595.
30. W.D. Le *et al.*, *Nat. Genet.*, 2003, **33**, 85.
31. A.N. McEvoy *et al.*, *J. Immunol.*, 2002, **168**, 2979.
32. C. Lavedan *et al.*, *Neurosci. Lett.*, 2002, **322**, 57.
33. T. Kitada *et al.*, *Nature*, 1998, **392**, 605.
34. Y. Imai *et al.*, *J. Biol. Chem.*, 2000, **275**, 35661.
35. Y. Zhang *et al.*, *Proc. Natl. Acad. Sci. USA*, 2000, **97**, 13354.
36. J.F. Staropoli *et al.*, *Neuron*, 2003, **37**, 735.
37. Y. Imai *et al.*, *Mol. Cell*, 2002, **10**, 55.
38. Y.C. Tsai *et al.*, *J. Biol. Chem.*, 2003, **278**, 22044.
39. Y. Imai and R. Takahashi, *Curr. Opin. Neurobiol.*, 2004, **14**, 384.
40. Y. Yang *et al.*, *Neuron*, 2003, **37**, 911.
41. D.P. Huynh *et al.*, *Hum. Mol. Genet.*, 2003, **12**, 2587.
42. J.M. Itier *et al.*, *Hum. Mol. Genet.*, 2003, **12**, 2277.
43. M.S. Goldberg *et al.*, *J. Biol. Chem.*, 2003, **278**, 43628.
44. M.G. Schlossmacher *et al.*, *Am. J. Pathol.*, 2002, **160**, 1655.
45. M.M. Muqit *et al.*, *Hum. Mol. Genet.*, 2004, **13**, 117.
46. K.F. Winklhofer *et al.*, *J. Biol. Chem.*, 2003, **278**, 47199.
47. F. Darios *et al.*, *Hum. Mol. Genet.*, 2003, **12**, 517.
48. M.D. Ledesmav *et al.*, *J. Neurochem.*, 2002, **83**, 1431.
49. Y. Higashi *et al.*, *J. Neurochem.*, 2004, **89**, 1490.
50. L. Maroteaux *et al.*, *J. Neurosci.*, 1988, **8**, 2804.
51. R. Jakes, M.G. Spillantini and M. Goedert, *FEBS Lett.*, 1994, **345**, 27.
52. A. Iwai *et al.*, *Neuron*, 1995, **14**, 467.
53. J.M. George *et al.*, *Neuron*, 1995, **15**, 361.
54. M. Goedert, *Nature*, 1997, **388**, 232.
55. P.H. Weinreb *et al.*, *Biochemistry*, 1996, **35**, 13709.
56. V.N. Uversky, J. Li and A.L. Fink, *J. Biol. Chem.*, 2001, **276**, 10737.

57. D. Eliezer *et al.*, *J. Mol. Biol.*, 2001, **307**, 1061.
58. V.N. Uversky, J.R. Gillespie and A.L. Fink, *Proteins*, 2000, **41**, 415.
59. M. Goedert, *Nat. Rev. Neurosci.*, 2001, **2**, 492.
60. D.F. Clayton and J.M. George, *J. Neurosci. Res.*, 1999, **58**, 120.
61. T. Tobe *et al.*, *J. Neurochem.*, 1992, **59**, 1624.
62. K. Ueda *et al.*, *Proc. Natl. Acad. Sci. USA*, 1993, **90**, 11282.
63. W.S. Davidson *et al.*, *J. Biol. Chem.*, 1998, **273**, 9443.
64. A.N. Pronin *et al.*, *J. Biol. Chem.*, 2000, **275**, 26515.
65. S. Engelender *et al.*, *Nat. Genet.*, 1999, **22**, 110.
66. K. Wakabayashi *et al.*, *Ann. Neurol.*, 2000, **47**, 521.
67. A. Abeliovich *et al.*, *Neuron*, 2000, **25**, 239.
68. W. Dauer *et al.*, *Proc. Natl. Acad. Sci. USA*, 2002, **99**, 14524.
69. S. Chandra *et al.*, *Proc. Natl. Acad. Sci. USA*, 2004, **101**, 14966.
70. E. Maries *et al.*, *Nat. Rev. Neurosci.*, 2003, **4**, 727.
71. E. Masliah *et al.*, *Science*, 2000, **287**, 1265.
72. P.J. Kahle *et al.*, *EMBO Rep.*, 2002, **3**, 583.
73. P.J. Kahle *et al.*, *J. Neurosci.*, 2000, **20**, 6365.
74. Y. Matsuoka *et al.*, *Neurobiol. Dis.*, 2001, **8**, 535.
75. B.I. Giasson *et al.*, *Free Radical Biol. Med.*, 2002, **32**, 1264.
76. M.K. Lee *et al.*, *Proc. Natl. Acad. Sci. USA*, 2002, **99**, 8968.
77. E.K. Richfield *et al.*, *Exp. Neurol.*, 2002, **175**, 35.
78. E. Rockenstein *et al.*, *J. Neurosci. Res.*, 2001, **66**, 573.
79. T. Gomez-Isla *et al.*, *Neurobiol. Aging*, 2003, **24**, 245.
80. S. Gispert *et al.*, *Mol. Cell. Neurosci.*, 2003, **24**, 419.
81. J.C. Rochet *et al.*, *Biochemistry*, 2000, **39**, 10619.
82. M. Yamada *et al.*, *J. Neurochem.*, 2004, **91**, 451.
83. K.A. Conway *et al.*, *Nat. Med.*, 1998, **4**, 1318.
84. R.A. Crowther *et al.*, *FEBS Lett.*, 1998, **436**, 309.
85. O.M. El-Agnaf *et al.*, *FEBS Lett.*, 1998, **440**, 71.
86. B.I. Giasson *et al.*, *J. Biol. Chem.*, 1999, **274**, 7619.
87. S.J. Wood *et al.*, *J. Biol. Chem.*, 1999, **274**, 19509.
88. K.A. Conway *et al.*, *Proc. Natl. Acad. Sci. USA*, 2000, **97**, 571.
89. H. Han *et al.*, *Chem. Biol.*, 1995, **2**, 163.
90. A.M. Bodles *et al.*, *Eur. J. Biochem.*, 2000, **267**, 2186.
91. O.M. El-Agnaf *et al.*, *FASEB J.*, 2004, **18**, 1315–1317.
92. H. Fujiwara *et al.*, *Nat. Cell Biol.*, 2002, **4**, 160–164.
93. M.J. Volles *et al.*, *Biochemistry*, 2003, **42**, 7871.
94. B.I. Giasson *et al.*, *Science*, 2000, **290**, 985.
95. S.R. Paik *et al.*, *Biochem. J.*, 1999, **340**, 821.
96. V.N. Uversky *et al.*, *J. Biol. Chem.*, 2001, **276**, 44284.
97. R.M. Rasia *et al.*, *Proc. Natl. Acad. Sci. USA*, 2005, **102**, 4294.
98. E.N. Lee *et al.*, *J. Neurochem.*, 2003, **84**, 1128.
99. B.A. Rybicki *et al.*, *Mov. Disord.*, 1993, **8**, 87.
100. J. Zayed *et al.*, *Can. J. Neurol. Sci.*, 1990, **17**, 286.
101. J.M. Gorrell, D. DiMonte and D. Graham, *Environ. Health Perspect.*, 1996, **104**, 652.

102. E.C. Hirsch *et al.*, *J. Neurochem.*, 1991, **56**, 446.

103. P. Riederer *et al.*, *J. Neurochem.*, 1989, **52**, 515.

104. D.T. Dexter *et al.*, *Brain*, 1991, **114**, 1953.

105. M.B. Youdim *et al.*, *Eur. Neurol.*, 1991, **31**(Suppl. 1), 34–40.

106. H.M. Schipper, A. Liberman and E.G. Stopa, *Exp. Neurol.*, 1998, **150**, 60.

107. C. Borie *et al.*, *J. Neurol.*, 2002, **249**, 801.

108. E.A. Martins *et al.*, *Arch. Biochem. Biophys.*, 1995, **316**, 128.

109. C. Nunez-Millacurav *et al.*, *J. Neurochem.*, 2002, **82**, 240.

110. B.A. Faucheux *et al.*, *J. Neurochem.*, 2002, **83**, 320.

111. R.J. Castellani *et al.*, *Acta Neuropathol.*, 2000, **100**, 111.

112. S. Turnbull *et al.*, *Free Radical Biol. Med.*, 2001, **30**, 1163.

113. M.S. Goldberg *et al.*, *Nat. Cell Biol.*, 2000, **2**, E115.

114. M. Gerlach *et al.*, *J. Neurochem.*, 1995, **65**, 923.

115. L. Zecca *et al.*, *J. Neurochem.*, 2001, **76**, 1766.

116. K. Jellinger *et al.*, *J. Neurochem.*, 1992, **59**, 1168.

117. F.A. Zucca *et al.*, *Pigment Cell Res.*, 2004, **17**, 610.

118. K.L. Doublev *et al.*, *Drugs News Develop.*, 1999, **12**, 333.

119. B.I. Giassonv *et al.*, *J. Biol. Chem.*, 2001, **276**, 2380.

120. S. Chandra *et al.*, *J. Biol. Chem.*, 2003, **278**, 15313.

121. C.C. Jao *et al.*, *Proc. Natl. Acad. Sci. USA*, 2004, **101**, 8331.

122. L.C. Serpell *et al.*, *Proc. Natl. Acad. Sci. USA*, 2000, **97**, 4897.

123. A. Der-Sarkissian *et al.*, *J. Biol. Chem.*, 2003, **278**, 37530.

124. G. Yamin *et al.*, *J. Biol. Chem.*, 2003, **278**, 27630.

125. S.R. Paik, H.J. Shin and J.H. Lee, *Arch. Biochem. Biophys.*, 2000, **378**, 269.

126. M. Hashimoto *et al.*, *NeuroReport*, 1999, **10**, 717.

127. H. Kozłowski, T. Kowalik-Jankowska and M. Jeazowska-Bojczuk, *Coord. Chem. Rev.*, 2005, **249**, 2323.

128. W. Bal *et al.*, *J. Inorg. Biochem.*, 1993, **52**, 79.

129. J.F. Galey *et al.*, *J. Chem. Soc. Dalton Trans.*, 1991, 2281.

130. T. Kowalik-Jankowska *et al.*, *J. Inorg. Biochem.*, 2005, **99**, 2282.

131. W. Linert, G.N.L. Jameson, R.F. Jameson and K.A. Jellinger, *Neurodegenerative diseases and metal ions*, in: *Metal Ions in Life Sciences*, A. Sigel, H. Siegl and R.K.O. Sigel (eds), vol. 1, Wiley, Chichester, UK, 2006, p. 281.

132. J.E. Gorton and R.F. Jameson, *J. Chem. Soc. Dalton Trans.*, 1972, 304.

133. J.E. Gorton and R.F. Jameson, *J. Chem. Soc. Dalton Trans.*, 1972, 307.

134. U. El-Ayaan, R.F. Jameson and W. Linert, *J. Chem. Soc. Dalton Trans.*, 1998, 1315.

135. G.N.L. Jameson, A.B. Kudryavtsev and W. Linert, *J. Chem. Soc. Perkin Trans. 2*, 2001, **4**, 557.

136. G.N.L. Jameson and W. Linert, *J. Chem. Soc. Perkin Trans. 2*, 2001, **4**, 569.

137. Z. Maskos, J.D. Rush and W.H. Koppenol, *Arch. Biochem. Biophys.*, 1992, **296**, 521.

138. W. Linert *et al.*, *Biochim. Biophys. Acta*, 1999, **1454**, 143.

139. G.N.L. Jameson, R.F. Jameson and W. Linert, *Org. Biomol. Chem.*, 2004, **2**, 2346.

140. L. Zecca *et al.*, *Proc. Natl. Acad. Sci. USA*, 2004, **101**, 9843.

141. W.R. Gibb, *Brain Res.*, 1992, **581**, 283.
142. L. Zecca *et al.*, *Trends Neurosci.*, 2003, **26**, 578.
143. H. Wilms *et al.*, *FASEB J.*, 2003, **17**, 500.
144. J. Cheng, S.G. Moss and M. Eisner, *Cell Res.*, 1994, **7**, 263.
145. L. Zecca *et al.*, *J. Neurochem.*, 2000, **74**, 1758.
146. D. Szuler *et al.*, *Proc. Natl. Acad. Sci. USA*, 2000, **97**, 11869.
147. L. Zecca *et al.*, *J. Neural. Transm.*, 2002, **109**, 663.
148. K.L. Double *et al.*, *Biochem. Pharmacol.*, 2003, **66**, 489.
149. L. Zecca *et al.*, *J. Neurochem.*, 1994, **62**, 1097.
150. L. Zecca *et al.*, *Neuroscience*, 1996, **73**, 407.
151. L. Zecca *et al.*, *J. Neurochem.*, 2001, **76**, 1766.
152. L. Lopiano *et al.*, *Biochim. Biophys. Acta*, 2000, **1500**, 306.
153. T. Shima *et al.*, *Free Radical Biol. Med.*, 1997, **23**, 110.
154. A.J. Kropf *et al.*, *Biophys. J.*, 1998, **75**, 3135.
155. M.G. Bridelli, D. Tampellini and L. Zecca, *FEBS Lett.*, 1999, **457**, 18.

CHAPTER 12
Chelating Agents in Metal Neurotoxicity

Chelation therapy is a safe, effective and inexpensive alternative to drugs and surgeries and is used to treat illnesses such as heart disease, strokes diabetes, AD and adverse reactions to environmental pollutants. The word chelation is derived from the Greek word chele that means claw (like that of a scorpion or crab). The concept of chelation is based on the observation that when ethylenediaminetetraacetic acid (EDTA) comes in contact with certain positively charged metals, it grabs them (hence the chele or claw), and removes them. Chelation therapy is the process of removing from the body the undesirable ionic material by the infusion, or taking orally, of an organic compound which has suitable chelating properties.

EDTA is a synthetic amino acid first used in the 1940s for treatment of heavy metal poisoning. It is widely recognized as effective for that use as well as certain other conditions including emergency treatment of hyper-calcemia and the control of ventricular arrhythmias associated with digitalis toxicity. Studies by the National Academy of Sciences/National Research Council in the late 1960s indicated that EDTA was considered possibly effective in the treatment of occlusive vascular disorders caused by arteriosclerosis. EDTA grabs metallic cations such as lead or calcium from the body and forms a stable compound that is then excreted from the system. The stability of such binding is vital to success in chelation therapy. If the bonding is weak, other chemicals can break this bond to form their own metal-containing compounds.

EDTA was first used medically in the 1940s to treat workers from battery factories, who had developed lead poisoning. In the early 1950s, Dr. Norman Clarke, Sr. director of research at Providence Hospital in Detroit, Michigan, was using EDTA for lead poisoning and found that his patients also reported less pain from angina (chest pain due to blocked arteries). In addition, they noted improved memory; better sight, hearing and smell; and an increase in energy.

At the time, Clarke and other doctors postulated that it was EDTA's effect on calcium within the body that might account for these results. They theorized that, during chelation, the EDTA could be grabbing onto the calcium within the arterial plaque lining blood vessels and removing it, the way it removed lead in poisoning cases.

With the calcium gone, they felt, the plaque dissolved, circulation improved (not only to the heart, but also to the brain, eyes and other organs) and patients reported better feeling. One writer at the time even dubbed chelation therapy "a Roto-Rooter for the arteries". Unfortunately, X-rays and biopsies later showed that chelation had no effect at all on calcium within the arteries, and this early theory was discounted.

A second, more widely accepted theory – and one that continues to be popular – suggests that, by removing toxic metals, the EDTA also removed a significant source of destructive oxygen molecules known as free radicals. With free-radical production slowed, the arteries could then heal, shedding their plaque and lessening the symptoms of heart disease. Today antioxidant vitamins are thought to play a key role in "mopping up" free radicals throughout the body and, for this reason, large doses of antioxidants are typically administered along with the EDTA.

Chelation therapy is widely used for the treatment of atherosclerosis and other chronic degenerative diseases involving the circulatory system. It also has other benefits. Many scientists suggest that the beneficial effect of chelation treatment is from the removal of metallic catalysts that cause excessive free-radical proliferation. This reduces the oxidation of lipids, DNA, enzyme systems and lipoproteins. The chelation halts the bad effects and initiates the body's healing process, often reversing the damage. It removes the calcium and copper cations from the blood stream. The plaques lining the artery walls are made porous and brittle. Eventually, they may get dislodged. Even if only a microscopic layer of the plaque is removed, it, along with a smoothening of the artery wall due to the healing of the cells that line the arteries, can improve the blood flow to the artery muscles substantially. This can prevent artery spasm and minimize or prevent angina pain. Many patients who could not walk due to muscle pain or angina pain have reported that they can walk without pain after chelation therapy.

The majority of lipid peroxidation activity involves the presence of metal ions such as iron, copper or calcium. EDTA effectively locks onto these ions, preventing their destructive action. Proponents of chelation therapy claim that EDTA can reduce the production of free radicals by up to a million-fold!

Research over the past 30 years has confirmed the benefits of EDTA. This protective influence of EDTA would be enhanced by an appreciable presence of antioxidant nutrients, such as vitamins A, C and E selenium, and amino acid complexes such as glutathione peroxidase. These not only mop up free radicals but also assist in reinforcing the stability of cell membranes.

While many chelation studies have been done over the years, very few rigorous studies have ever been performed on humans. Critics of the therapy note that most studies showing its effectiveness have been done by physicians with a financial interest in the therapy. Proponents respond by saying that studies disproving chelation have typically been performed under the supervision of physicians with a financial interest in costly surgical procedures.

Despite the disagreements, interest in chelation as an alternative treatment for heart and vascular problems has continued to grow, with thousands of people seeking out the therapy annually as an option to coronary bypass surgery and balloon angioplasty.

Once the intravenous solution of EDTA enters the bloodstream, it is believed to attach itself to metal molecules. It then takes about 48 h for the body to excrete these substances through the urine. Because the therapy can also remove small amounts of zinc, copper, calcium, manganese and other essential minerals from the body, supplemental vitamins and minerals are often added to the EDTA infusion. This is a complicated process that should only be performed by a trained professional.

Chelation therapy is medically indicated when toxic levels of heavy metals, such as iron, arsenic, lead and mercury are present.

- *Lead toxicity* most commonly occurs with young children exposed to old houses with lead paint dust or chips. Occupational exposure (soldering, welders, smelters, battery reclamation) is also a risk. Lead screening for children has now become a standard procedure of a doctor's visit for children in many states.
- *Mercury toxicity* almost always occurs with high-risk occupational exposures including dental workers, manufacturers of batteries/thermometers, tannery work/taxidermy and contaminated seafood.
- *Arsenic poisoning* usually occurs from exposure to insecticides, herbicides, rodent poisons, veterinary parasitic medications or intentional poisoning.

Other heavy metals, mentioned only in passing because toxic exposure is extremely uncommon, include cadmium, manganese, aluminum, cobalt, zinc, nickel, copper and magnesium.

Heavy metal toxicity can cause a wide range of problems including severe injury to the body organs and the brain.

Common chelating agents include

- *Desferoxamine Mesylate* used for iron toxicity, intravenous preferred route of administration (Formula 1).

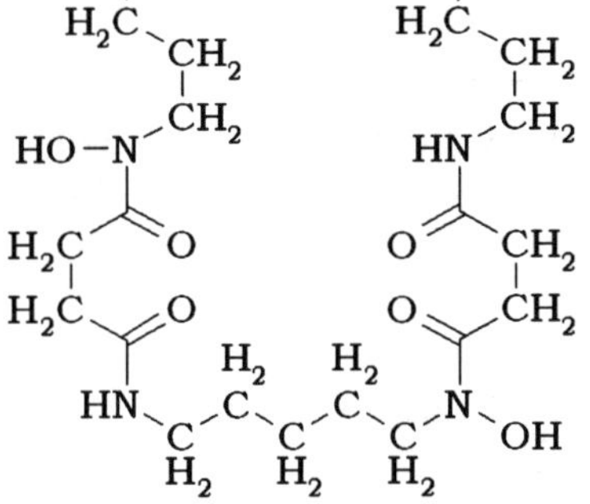

Desferoxamine mesylate Desferal, is an iron-chelating agent, available in vials for intramuscular, subcutaneous, and intravenous administration.

Desferoxamine mesylate is a white to off-white powder. It is freely soluble in water and slightly soluble in methanol.

Also called desferal, it is available in vials for intramuscular, subcutaneous and intravenous administration. It is a white to off-white powder. It is freely soluble in water and slightly soluble in methanol.

- *Dimercaprol (BAL)* lead, preferred agent for arsenic and mercury toxicity, given intramuscularly (Formula 2).

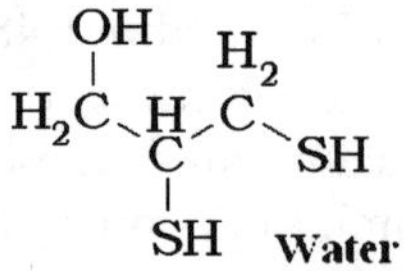

Water solubility is 1 in 20. Solubility in vegetable oils is 1 in 18 Dimercaprol is administered by deep intramuscular injection. The generally recommended doses are similar for arsenic, gold and inorganic mercury poisoning and range between 2.5 mg/kg to 5 mg/kg every 4 hours

It displays a water solubility of 1 in 20. Dimercaprol is administered by deep intra-muscular injection. The generally recommended doses are similar for arsenic, gold and inorganic mercury poisoning and range between 2.5 mg kg^{-1} and 5 mg kg^{-1} every 4 h.

- *Dimercaptosuccinic acid (DMSA) and DMPS*: *meso*-DMSA (Formula 3), rac-DMSA (Formula 4) and DMPS (Formula 5) are analogues of Dimercaprol for lead, mercury and arsenic poisoning.

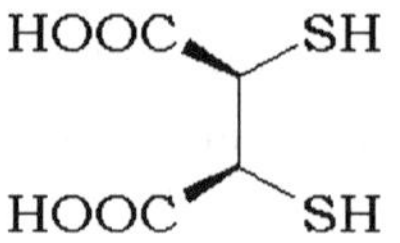

meso-DMSA

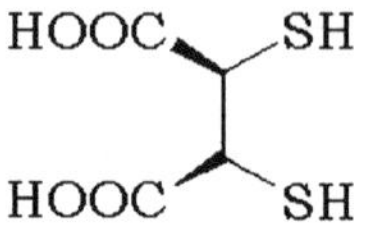

rac-DMSA

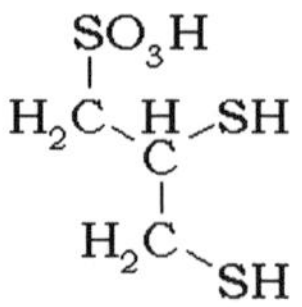

DMPS

- *D-penicillamine*: an oral chelating agent used for lead, arsenic or mercury poisoning (Formula 6). Much less expensive but not as effective as DMSA.

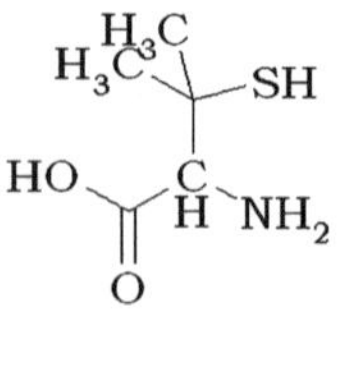

- *Calcium Disodium Versante (CaNa$_2$–EDTA)*: can be used in conjunction with BAL in lead toxicity (Formula 7).

Never used alone in treating lead toxicity because chelates only extracellular, not intra-cellular lead.

Diagnosis of heavy metal toxicity is serious and must be made by a physician based on clinical symptoms in conjunction with laboratory testing. Chelating agents are potentially toxic and should not be used unless absolutely indicated.

12.1   Copper

Angiogenesis is required for solid tumor growth, and may be an Achilles' heel of cancer. Many angiogenic promoters appear to be dependent on normal levels of copper. Lowering copper levels moderately with a drug such as tetrathiomolybdate (TTM) produces strong anti-angiogenesis and potent anti-cancer effects in several rodent models. The use of TTM in advanced and metastatic cancer in canine pets and in human patients has provided very encouraging results. Most recently, it has been discovered that fibrotic and inflammatory cytokines are also copper dependent, and animal model studies of pulmonary fibrosis and cirrhosis have revealed strongly protective effects of TTM therapy, suggesting that TTM will be helpful in many human diseases of this type.

Copper may be involved in producing neuronal oxidant damage in AD, may be involved in precipitation of amyloid β protein and may play a relevant role in other diseases of neurodegeneration, such as amyotrophic lateral sclerosis and prion diseases.

The most important of the genetic abnormalities involving copper in humans are Wilson's disease and Menkes' disease, caused by recessive defects in genes *ATP7B* and *ATP7A*, respectively. These genes are differentially expressed in tissues, their products account for much cellular copper trafficking, and as their functions are better understood, so are the phenotypes that result from their mutations.

Wilson's disease is an autosomal recessive genetic disorder, and is caused by disabling mutations in both copies of the ATP7B gene.[1] This gene functions in a pathway in the liver for biliary excretion of excess copper. With absent or reduced function, copper accumulates and causes progressive damage in the liver, and often in the brain.[2,3] New causative mutations are constantly being found, so that there are now over 200.[4] These vary among populations but, in almost all populations, no one mutation dominates to a level accounting for 50% or more of all mutations. It is not, therefore, practical to diagnosis Wilson's disease by mutation screening, although haplotype analysis in siblings of diagnosed patients can be used for genotyping. Attempts are being made to correlate phenotype and genotype, but this field is in its infancy, hampered by a large proportion of patients in many populations being compound heterozygotes.

Therapy of Wilson's disease continues to advance, with the addition of zinc as a US Food and Drug Administration approved maintenance drug in 1997.[5] Zinc's mechanism of action involves induction of intestinal cell metallothionein, which binds food copper and copper in gastrointestinal secretions, and prevents its absorption. Zinc offers efficacy and low toxicity, whereas the drug it is replacing, penicillamine, has a long list of toxicities. Penicillamine's mechanism of action involves reductive chelation of copper in the body, and enhanced excretion of copper in the urine. Penicillamine has been used most extensively for this purpose in the past, but has a high risk of making neurologically presenting patients permanently worse.[6] TTM's mechanism of action involves formation of a tripartite complex with protein and copper, which can be used to generate two anti-copper effects. Given with food, TTM prevents copper absorption. When given separated from food, TTM is absorbed and complexes the body's readily available copper with itself and albumin, rendering the copper unavailable. Progress has also been made in treating patients who present with some degree of renal failure, using a combination of zinc and trientine (Formula 8), a combination that may allow recovery through medical therapy in some patients who would otherwise require hepatic transplantation. Trientine's mechanism of action is similar to that of penicillamine, but it is a gentler, less toxic drug. Reviews of modern therapy, as well as discussion of recognition, diagnosis and other areas related to Wilson's disease, will be found in two recent books.[3,7]

$$H_2N-CH_2-CH_2-NH-CH_2-CH_2-NH-CH_2-CH_2-NH_2$$

Trientine

Menke's disease is an X-linked inherited disorder, and is caused by a mutation in the ATP7A gene.[8] The syndrome that results is one of severe pre- and post-natal copper deficiency in affected males, leading to brain damage and mental retardation, presence of a type of scalp hair called kinky or steel wool, fragile bones, aortic aneurisms and other manifestations of copper deficiency. Both the protein made by ATP7A and the one made by ATP7B are membrane-bound, copper-transporting ATPases. The expression profiles and functions are different, with ATP7B expressed primarily in the liver with functions as discussed above, whereas ATP7A is much more ubiquitously expressed and has important functions in several organs. ATP7A is crucial in causing copper efflux from cells. Failure of function of ATP7A in the intestine leads to a failure of copper efflux from intestinal cells, accumulation of excess copper in the intestine, a failure of copper absorption into the blood and generalized copper deficiency. Failure of function of ATP7A in the BBB leads to a failure of copper efflux from cells of this barrier, accumulation of copper in these cells, and a failure of copper uptake in the brain, even if circulating copper levels are normalized by parenteral copper therapy. Menke's disease generally results in severe pre-natal brain damage making post-natal attempts at copper therapy not useful. The exception is with mutations that allow some function of ATP7A to remain, and cause a milder syndrome called the occipital hormone syndrome; in this case, parenteral copper therapy may be helpful. A review that includes a good description of Menke's disease, the function of ATP7A in cellular copper trafficking, and various types of ATP7A mutations has been published.[9]

Besides Menke's and Wilson's diseases, copper chelation has also been shown to be possibly used in repressing the vascular response to injury.[10] The elucidation of the pathways involved in the regulation of the vascular response to injury is critical for the management of human diseases in which the pathology may be regulated by stress-induced endothelial and vascular smooth muscle cell responses. Although a variety of polypeptide growth factors and cytokines have been implicated as mediators of the vascular response to injury, a fundamental role has been demonstrated for the migration of peripheral blood mononuclear cells into sites of injury as a delivery system for these biological response modifiers as regulators of both the inflammatory and angiogenic responses.

Members of the IL-1 and fibroblast growth factor (FGF) gene families may significantly contribute to vessel-wall pathology in response to injury. Because it is well established that the IL-1 and FGF prototypes function in the extracellular compartment as ligands for high-affinity cell-surface receptors, the identification of the mechanism(s) used by the IL-1 and FGF prototypes for non-classical release could potentially yield new insight into pro-inflammatory and angiogenic disorders.

The appearance of IL-1α and FGF1 in the extracellular compartment is regulated by convergent, yet distinct non-classical export pathways induced by cellular stress.[11] Both IL-1α and FGF1 use intracellular Cu^{2+} to force the assembly of a multi-protein complex near the inner surface of the plasma membrane.[12,13] Although both IL-1α and FGF1 form Cu^{2+}-dependent heterotetrameric complexes with S100A13 to facilitate their release, the FGF1 release pathway also requires the function of the extravesicular domain of synaptotagmin (Syt)1 for export.[14] It is interesting to note that IL-1α, FGF1, Syt1 and S100A13 are all Cu^{2+}-binding proteins and are able to associate with phosphatidylserine, an acidic phospholipid that is known to translocate from the inner to the outer surface of the plasma membrane. It has been shown that, in both murine NIH 3T3 cells and human U937 cells, the stress-induced IL-1α and FGF1 release pathways are sensitive to inhibition by the Cu^{2+} chelator

TTM, suggesting that intra-cellular Cu^{2+} may play an important role in the regulation of these pathways *in vivo*. It is well established that Cu^{2+} is not only a potent inducer of angiogenesis *in vivo*, but it is also implicated as a promoter of the arterial response to injury.[15] The chelator TTM

   (i) has been suggested to be efficacious in the clinical management of human cancer;[16]

  (ii) is able to significantly inhibit NF-κB activity as a component of its anti-angiogenic activity *in vivo*;[17]

 (iii) since IL-1 is an important regulator of NF-κB activity, it can be used to limit the recruitment of mononuclear cells in response to vascular injury *in vivo* by restricting the export of IL-1α.[10]

12.2 Lead

One of the best-known toxic effects of lead is its interference with the heme biosynthesis. The ability of lead to produce encephalopathy and particularly the vulnerability of the developing brain to lead has also been recognized for many years.[18] Although several mechanisms have been proposed to explain the lead-induced toxicity, no mechanism has been defined explicitly. Oxidative stress has been suggested to be one of the important mechanisms of toxic effects of lead.[19,20] Oxidative stress has been implicated for its contribution to lead-associated tissue injury in the liver, kidneys, brain and other organs,[21] and a number of studies has confirmed the possible involvement of ROS in lead-induced toxicity.[22] The current approved treatment for lead poisoning is to administer chelating agents that form insoluble complexes with lead and remove the same from lead-burdened tissue. Chelation therapy using *meso*-2,3-dimercaptosuccinic acid (DMSA) has been shown to reduce lead levels in blood, brain and other tissues.[23] DMSA is one of the least toxic drugs and could be given through oral route, but the hydrophilic and lipophobic properties of DMSA do not allow it to cross the cell membrane. It was observed that monoesters of DMSA might be a more effective antidote for metal toxicity. It has been reported that monoisoamyl DMSA (MiADMSA) is more efficient than DMSA in mobilizing brain lead.[24] Some antioxidants such as *N*-acetylcysteine also function as chelators, and such dual behavior makes them strong candidates for their use against lead toxicity.

Before the availability of chelating drugs, as many as 45% of lead-poisoned children who presented with signs or symptoms of encephalopathy died, and more than one quarter of surviving patients experienced severe neurologic sequelae.[25] Cases of this nature are now extremely rare, and the concern with lead toxicity has shifted from symptomatic lead poisoning to sub-clinical effects.

In 1991, the Food and Drug Administration licensed succimer (dimercaptosuccinic acid) for the oral chelation of lead in children with blood lead levels at or above 45 μg dL^{-1} (2.17 μmol L^{-1}).[26] In the same year, the United States Centers for Disease Control and Prevention (CDC) reduced the action threshold from a blood lead concentration of 25 μg dL^{-1} (1.2−μmol L^{-1}) to

10 µg dL^{-1} (0.48 µmol L^{-1}). This recommendation was based on epidemiologic studies reporting cognitive impairments at blood lead levels below 25 µg dL^{-1} (1.2 µmol L^{-1}). Nevertheless, the CDC made no specific recommendations about chelation therapy of children with blood lead levels below 45 µg dL^{-1}. Although succimer reduced blood lead concentration in exposed children,[27] the effects of treatment on cognitive status were unknown. Thus, in 1994, a multi-center, randomized, placebo-controlled clinical trial of succimer was initiated. The Treatment of Lead-Exposed Children (TLC) trial was designed to test the hypothesis that children who had moderately elevated blood lead concentrations and were given succimer would attain higher scores on standardized tests of neurodevelopment than children who were given placebo.[28]

Three years after randomization, no salutary effects of treatment with succimer were observed on a battery of neuropsychological tests administered to TLC subjects when they were, on average, five years of age. These included measures of IQ, attention, language, sensorimotor acuities, visuospatial skills, memory and behavioral problems.

Because there were no detectable effects of pharmacologic treatment on neurodevelopment, the two treatment arms were combined in an analysis of the relationship between the rate of decline in blood lead levels and improvements in cognition.[29] This ancillary analysis was inspired by an earlier observational study in New York of children who were 1–7 years of age and had blood lead levels between 25 and 55 µg dL^{-1} (1.2–2.6 µmol L^{-1}).[30] In the New York study, children who were given ethylenediaminetetraacetic acid and/or therapeutic iron when clinically indicated were followed for six months. Regardless of the therapeutic regimen, children whose blood lead level fell the most had the largest improvement in cognitive test scores. In TLC, the change in cognitive test scores between the baseline and three-year follow-up was also correlated with decline in blood lead concentration. However, a closer examination of the data revealed that this was attributable only to an association in the placebo group. No relationship was observed between falling blood lead levels and improved cognition in the group that was treated with active drug.

Although results of the first wave of follow-up for TLC were consistently negative for drug effects on cognition and behavior, they were not necessarily conclusive. Lead may affect higher-level neurocognitive processes that are inaccessible, difficult to assess or absent in the pre-school-aged child. In older children, a wider and more differentiated range of abilities can be examined, scores on psychometric measures are more precise and reliable, and early academic performance and social functioning outside the home environment can be evaluated. Therefore, the cohort was followed into the first years of elementary education to determine whether these later emerging neurodevelopmental functions were spared the effects of lead in treated children compared with placebo control subjects. Chelation therapy with succimer was shown to lower average blood lead levels for ~6 months but resulted in no benefit in cognitive, behavioral and neuromotor endpoints.[31]

These new follow-up data confirmed the previous finding that the TLC regimen of chelation therapy is not associated with neurodevelopmental benefits in children with blood lead levels between 20 and 44 µg dL^{-1} (0.96–2.17 µmol L^{-1}). These results emphasize the importance of taking environmental measures to prevent exposure to lead. Chelation therapy with succimer cannot be recommended for children with blood lead levels between 20 and 44 µg dL^{-1} (0.96–2.12 µmol L^{-1}).

Since calcium and lead interactions are well documented and competition for binding to Ca-binding proteins may underlie a mechanism for lead absorption, other investigators have been searching for an eventual role, if any, of supplemental Ca at reducing blood lead levels (BPb) in moderately poisoned children.[32] Children aged 1–6 years with BPbs 10–45 µg dL^{-1} were enrolled in a double-blinded, placebo-controlled trial of the effects of Ca supplementation on BPbs. Children received either a Ca-containing liquid or an indistinguishable placebo. Dosage was adjusted biweekly on the basis of responses to a dietary Ca intake questionnaire to reach 1800 mg in the Ca-supplemented group. Samples for BPbs and measures to assess safety were collected before and after three months of supplementation and after an additional three months of follow-up. Bivariate and multiple regression analyses were performed. A total of 67 of 88 enrolled children with a mean age of 3.6 years completed three months of supplementation. There were no statistically significant differences between groups on hematological and biochemical measures, including serum and urinary Ca, at any time points. The average compliance rate was estimated to be 80% for each group during the three-month supplementation period. At enrollment, the average daily Ca intake in this group of inner-city children was greater than the recommended daily intake for age. Although BPbs declined during a three-month period in both groups, Ca supplementation aimed at providing 1800 mg of Ca per day had no effect on the change in BPbs. Ca supplementation should not be routinely prescribed for mild to moderately Pb-poisoned children who are dietary Ca sufficient.

12.3 Mercury

Mercury is a naturally occurring constituent of the Earth's crust. In its elemental (metallic) form, mercury is the only metal that exists in a liquid state at room temperature. Mercury readily volatilizes at standard temperature (O °C) and pressure (1 atm), and its presence in open containers can result in biologically significant air concentrations in unventilated or poorly ventilated spaces. Elemental mercury vapors are virtually odorless and very toxic. In recent years, elemental mercury has proven to be a potential source of toxicosis in children through either unintentional exposure or exposure resulting from inappropriate handling of liquid mercury obtained from school science laboratories, abandoned industrial facilities or warehouses. The shiny, silvery appearance of mercury in its liquid form makes it particularly enticing to children, and its insolubility in water and tendency to form beads when disturbed add to its mystique.

Exposure to metallic mercury can occur through either the inhalation, oral or dermal routes, with the particular route most dependent upon the specific type of mercury. In the case of metallic (liquid, elemental) mercury, only the inhalation route has proven to be biologically relevant in most instances. When taken orally, less than 0.01% is typically absorbed through the gastrointestinal (GI) tract. Skin contact normally results in even less absorption in most instances. In sharp contrast, however, up to 80% of inhaled mercury vapor can be expected to be absorbed through the lungs into the blood.

Another common exposure to mercury is to organic, alkyl mercurials. The typical sources of such exposures are through ingestion of contaminated seafood (methylmercury) and through multi-dose vials of vaccine, in which ethylmercury is used as a preservative (in thimerosal). Unlike inorganic forms of mercury, organic mercurials are readily absorbed through the digestive tract ($\sim$95%). For organomercurials, blood is a good indicator of exposure, and urine is a poor indicator, due to differences in the pharmacokinetics of these compounds. Hair is also a suitable indicator of a history of organic mercury exposure, since incorporation into the hair follicle of both methylmercury and ethylmercury is a known route of elimination of these organomercurials from the body.

Mercury is among the most toxic heavy elements,[33–35] and its compounds have been linked with a number of human health problems, including neurological problems,[27] myocardial infarction,[36] and a possible involvement in the development of some kinds of autism.[37] The effects of mercury on the body vary with the magnitude and duration of exposure, and with the age and overall health status of the exposed individual. Exposure to significant levels of metallic mercury can result in neurologic, respiratory, renal, reproductive, immunologic, dermatologic and a variety of other effects. However, neurologic effects are the most prominent feature of excessive exposure to mercury vapors, as well as organic mercury compounds, in most cases.[38,39]

Once in the blood, the half-life of metallic mercury is relatively short ($\sim$3 days for a single exposure), as it quickly partitions to other body compartments. The overall half-life of metallic mercury in the body averages approximately two months, depending on the duration and magnitude of exposure. Virtually, all of the absorbed metallic mercury is excreted in the urine. A 24-h urine specimen is used to provide the appropriate index of exposure.

In the case of methylmercury and ethylmercury, the biologic half-life is also about two months, but the biologic indicators are different than inorganic mercury exposure. In these cases, whole blood is the primary indicator of exposure, and only that portion of mercury that is oxidized to the cationic form is eliminated in the urine. Hair is a reliable indicator of prior or ongoing exposures to methylmercury and ethylmercury, since a small portion of it is incorporated into the hair follicle. Recent data from the National Health and Nutrition Examination Survey (NHANES) conducted by the CDC reported that 95% of sampled U.S. women between the ages of 16 and 49 have blood mercury levels of 7.1 μg L^{-1} or less and urine mercury concentrations of 5 μg L^{-1} or less (CDC, 2003). A urine mercury concentration of 20 μg L^{-1}

(again, prior to chelation) is widely considered to be without accompanying adverse health effects.[26]

Chelation therapy for mercury poisoning consists of the introduction of a charged molecule (typically containing one or more sulfhydryl groups) into the body for the purpose of elimination of the formed complex from the body in the urine.

Chelation therapy has historically been used in attempts to reduce the body burden of mercury and other toxic metals in highly symptomatic patients with elevated biological markers.[40,41]

Some physicians have also looked to mercury as a possible cause of undiagnosed health problems and applied chelation therapy as a treatment for those problems. As a result, the use of chelation has expanded in recent years to include the treatment of mildly symptomatic or asymptomatic patients with no documented history of mercury exposure, and it is becoming increasingly, and unfortunately, common for practitioners to make a diagnosis of mercury intoxication and begin treatment without carrying out an adequate clinical workup.

A number of chelating agents are currently either in practical use or under investigation for treating mercury poisoning. The available chelators differ in their efficacy for various forms of mercury, route of administration, side effects and route of excretion. Depending on the specific type of mercury and the health status of the patient, different chelators may be considered (Table 1). However, at present, no guidelines are available for physicians that specify the conditions under which chelation is medically indicated or contraindicated, thereby contributing to a growing confusion over the appropriate use of chelating agents.

DMSA and DMPS are vicinal dithiols related to the now outmoded British anti-Lewisite (BAL) 2,3-dimercaptopropanol, which was developed as an antidote for the arsenical war gas Lewisite (chlorovinylarsinedichloride).[42] BAL suffers from the disadvantages of low water solubility and a noxious smell, and so, it was modified to produce DMPS, which is water soluble and nearly odorless. DMSA was introduced later and has similar properties (*i.e.*, excellent water solubility and no odor) with lower toxicity. The clinical use of DMSA differs from DMPS (and BAL) in that it is usually administered by mouth, whereas the others are generally given intravenously.

Trivalent arsenicals such as Lewisite kill by interacting with essential thiols, and in particular, the dithiol lipoic acid in enzymes such as pyruvate dehydrogenase.[34] DMPS and DMSA are ideally suited for chelation therapy of As^{III} due to the stability of the five-membered rings formed upon chelation of arsenic (with a typical S–As–S bond angle of about 93°) (Scheme 1),[43] and it is widely believed that they bind a broad range of heavy metals with structures analogous to that shown for arsenic.

DMSA occurs in *rac*- and *meso*-diastereomers (*vide supra*), which apparently differ somewhat in their interaction with mercury.[44] Only the *meso* form is available commercially and used clinically. The chemistry of interaction of mercuric ions with DMSA and DMPS has been investigated by combining mercury L$_{III}$-edge X-ray absorption spectroscopy and density functional theory calculations.[45] Two possible diastereoisomers of Hg$_2$(DMSA)$_2$ occur in

Table 1 *Chelating agents used for mercury toxicity*

Chelating agent	Commercial name	Some adverse effects	Essential metals chelated	Route of excretion
2,3-dimercapto-propanol	British anti-lewisite (BAL); Dimercaprol	Urticaria; nausea, vomiting; abdominal pain; headache; convulsions; lacrimation; conjunctivitis		urine ($\sim 50\%$); bile and feces ($\sim 50\%$)
EDTA	EDTA (Versene) calcium disodium versenate edetate calcium disodium	fever; myalgia; arthralgia; tremors; headache; hypotension; nausea; vomiting; anemia; zinc deficiency	Cu, Fe, Zn, Mg, Ca	Primarily urine
D-penicillamine; N-acetyl-D,L-penicillamine (NAP)	NAP (Cupramine; Depen)	Fever; urticaria; nausea; vomiting; diarrhea; anemia	Cu, Fe, Zn	Primarily urine
meso-2,3-dimercaptosuccinic acid (DMSA) and sodium 2,3-dimercapto-1-propanesulfonate (DMPS)	DMSA (succimer; Chemet; Captomer) DMPS (Dimaval)	Nausea; vomiting; diarrhea; appetite loss; headache; chills; fever Skin rashes; nausea; weakness; vertigo (complete symptomatology not reported)	Cu, Zn Cu, Cr, Zn	Primarily urine Primarily urine

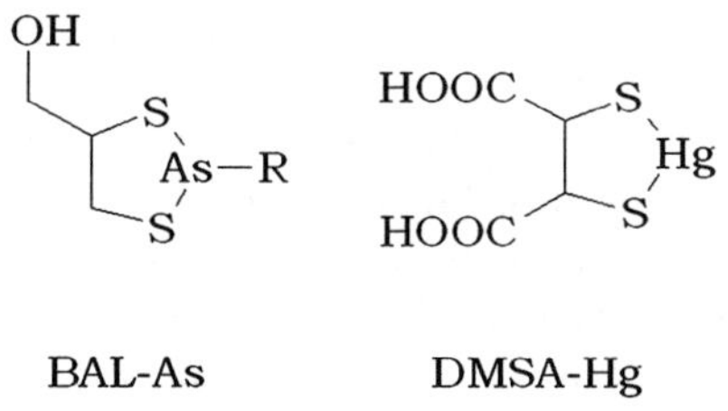

OH

As—R

BAL-As

HOOC

Hg

HOOC

DMSA-Hg

Scheme 1

solution and these are shown in Figure 1, where different Hg-Hg distances are apparent between the two.

DMPS forms a similar species with DMPS:Hg ratios of 1:1. The number of possible isomers (four conformers) is greater with DMPS than with DMSA because the carbon atom bearing the $-CH_2SO_3^-$ moiety is chiral, and commercially available DMPS is a racemic mixture of L- and D-forms. Computed structures for two possible constitutional isomers are shown in Figure 2.[37] The other forms are similar with slightly differing Hg$\cdots$Hg distances, as with

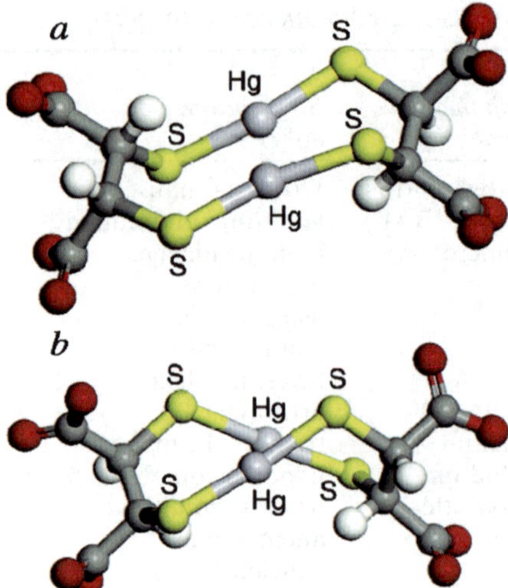

Figure 1 *Calculated structures of the two diastereomers of the smallest possible DMSA:Hg^{2+} complex. The carbon atoms are depicted as dark gray; oxygen atoms are depicted as red; hydrogen atoms are depicted as white; mercury atoms are depicted as light gray; and sulfur atoms are depicted as yellow* (Reproduced with permission from ref. 45.)

DMSA. At higher DMPS:Hg ratios (*e.g.*, 4:1), four-coordinate species are observed to form (Figure 2).[37] Density functional calculations indicate that true double chelate complexes, in which both thiolates of each of two DMPS molecules are bound to the metal, are stable.

DMSA does not form a four-coordinate species as observed for DMPS, most likely because the molecule would bear too large a negative charge to be stable at neutral pH. Since an effective drug for chelation therapy should interact in a stoichiometric manner with its target metal, this criterion is not fulfilled by the clinically used mercury chelation therapeutic drugs, DMSA and DMPS (Chemet and Dimaval).[37] Although DMSA and DMPS are effective in the clinical treatment of mercury poisoning, these compounds are likely sub-optimal for clinical chelation therapy by a considerable margin, and considerable improvements are therefore expected. Quantum chemical methods can provide the criteria for the design of a drug that will not only bind mercuric ions with considerable tenacity, but also with a high degree of specificity for mercuric ions above other cations-in effect a custom chelator.

The preferred coordination of mercuric ions is a linear, two-coordinate geometry. A custom chelator optimized for such linear coordination should present the metal with two thiol groups whose sulfurs are separated by twice the optimal Hg–S distance (*i.e.*, 2 × 2.345 Å). By using methyl mercury mercaptide ($CH_3SHgSCH_3$), the optimum binding geometry of Hg^{2+} to alkanethiolate donors was defined, and the energy of the molecule could be calculated as a function of Hg–S bond length, and of C–S–Hg and S–Hg–S bond angles (Figure 3).[37] A sharp minimum with an optimum bond length of 2.30 Å and an

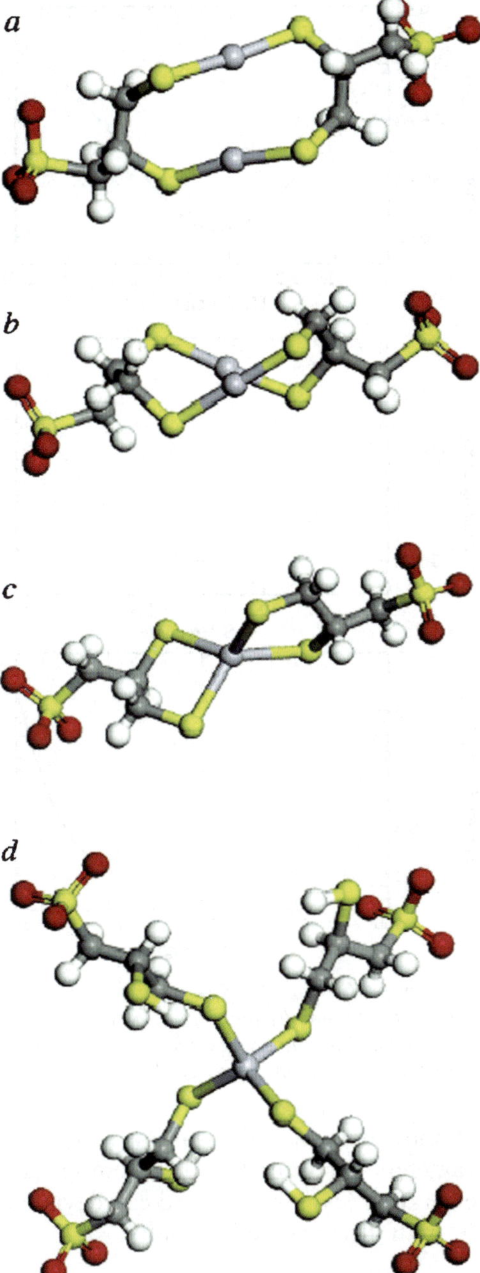

Figure 2 *Calculated structures of possible Hg^{2+}–DMPS complexes. (a) and (b) show two of the four possible isomers for linear Hg coordination, (c) shows the true double chelate and (d) shows the 4:1 complex. The color scheme for the atoms is the same as that used in Figure 1*
(Reproduced with permission from ref. 45.)

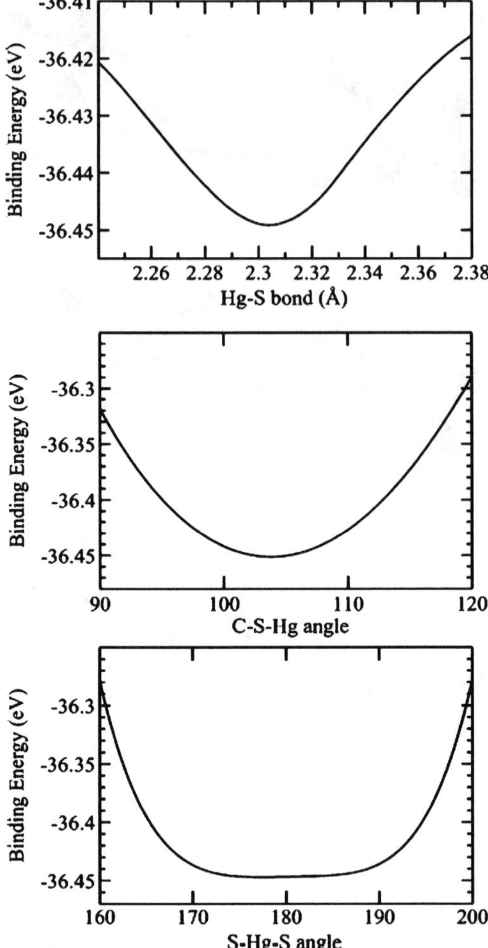

Figure 3 *Calculated energy profile for Hg–S bond length, C–S–Hg bond angle, and S–Hg–S bond angle for bis-methylmercaptan-mercury(II). The binding energy of the ordinates is defined as the energy required to separate all atoms to infinity* (Reproduced with permission from ref. 45.)

optimum C–S–Hg bond angle of 103° was evident, indicating that precise control of such parameters is required in a custom chelator. In contrast, the S–Hg–S bond angle is rather less sensitive and can deviate by nearly 10° about the optimum of 180° with little energetic penalty.

12.4 Iron

The importance of iron for survival, replication and differentiation of animals, plants and almost all microorganisms is well established,[46] and iron deficiency is a general problem in biology. However, iron in excess is toxic, particularly in man, where hemochromatosis is one of the most frequent genetic disorders. Excess iron accumulation within tissues, cells and even organelles, can result in toxicity and is associated with pathological disorders. In addition to the classic iron loading found in genetic hemochromatosis and in secondary iron overload disorders, such as the thalassemias, with their associated dysfunctions, there are many other diseases associated with excess iron. For example, iron accumulation in the brain has been associated with PD, AD, Huntington's chorea, HIV encephalopathy, basal ganglia disease, and Hallervorden Spatz syndrome, while in Friedrich ataxia, excessive mitochondrial iron accumulation occurs particularly in brain and cardiac tissue.

Brain cells, including neurons, astrocytes and microglia, show a decreased ability to respond to oxidative stress, particularly with respect to their levels of glutathione and glutathione peroxidase, such that alteration in their iron status may predispose them to iron-induced oxidative stress. It is clear that with aging there is a significant increase in iron stores in the brain which, if localized in susceptible regions, could contribute to the pathogenesis of various neurodegenerative diseases. Astrocytes provide protection and trophic support to neurons, but like neurons are susceptible to oxidative stress. Decreased function of astrocytes resulting from oxidative stress could contribute to neurodegeneration.

Extensive research effort has been applied to the development of iron chelators but, in spite of the many identified candidate agents, only few have been shown to satisfy the critical requirements for optimal performance, *i.e.*,

(i) high selectivity for Fe(III) in preference to Fe(II) in order to avoid redox cycling of the resulting complex and to minimize the generation of ROS;
(ii) absence of effects on the activity of iron-dependent enzymes such as, lipoxygenases, hydroxylases and ribonucleotide reductase;
(iii) high affinity for Fe(III) and efficient scavenging ability for non-specifically bound iron; and
(iv) kinetic stability of the complex such as to ensure efficient transport and excretion.

Desferrioxamine-B (DFO) is widely used for the treatment of transfusional iron overload since it efficiently scavenges iron from hepatocytes and extra-cellular pools.[47] Many possible alternative compounds have been investigated but most of them, as also DFO itself, suffer from poor absorption from the intestine or from excessively rapid metabolism or even from adverse toxic effects.

Intestinal absorption has been linked to membrane permeability through four parameters:[48]

(i) molecular weight (MW);
(ii) the water/*n*-octanol partition coefficient (log P);
(iii) the number of hydrogen bond donors (OH and NH groups); and
(iv) the number of hydrogen bond acceptors (N and O atoms).

Poor absorption follows from MW > 500, log $P > 5$, occurrence of more than five donors and 10 acceptors in the molecule.

The toxicity profile of DFO and the drug's efficacy are well known and several DFO-based regimens have been investigated. As an example, 24 h continuous infusion of DFO can reverse the cardiac complications in patients affected by iron overload.[49] The results of all tested regimens using intermittent high-dose DFO are intriguing,[50] and the duration of exposure to non-transferrin-bound iron may be relevant in causing cardiac damage. As a consequence, new iron chelators have been extensively investigated.

40SD02 (starch DFO) is synthesized by attaching DFO to a modified starch polymer. This drug is also sometimes referred to as starch Desferal. With Phase 1b clinical trials only recently completed, the drug is still early in its development. The Phase 1b trial, which examined the safety and effectiveness of the drug in patients, was held at two locations: the Children's Hospital and Research Center in Oakland, CA and the Weill Medical College of Cornell University in New York. Twelve thalassemia patients with transfusion-related iron overload received 1-h infusions of 40SD02 at three different doses. The patients were followed for 21 days and seemed to tolerate the drug well. At the highest dose tested, the amount of iron excreted following a 1-h infusion of 40SD02 was equivalent to that achieved from 3 to 5 days of conventional treatment with Desferal.

The presence of the starch polymer allows 40SD02 to persist in the body longer than traditional Desferal. Because the body retains the 40SD02 longer, it is expected that using this drug would reduce the frequency and duration of chelation therapy, compared to treatment using Desferal.

The orally active iron chelator deferiprone (1,2 dimethyl-3-hydroxypyrid-4-1, also known as L1, CP20, Ferriprox, or Kelfer) (Formula 9) has emerged from a long, extensive search for new therapies for iron overload. Deferiprone is a synthetic compound first designed in Professor R.C. Hider's laboratories at the University of Essex.[51] Iron excretion levels in the urine, in response to deferiprone in patients with heavy iron overload and with myelodysplasia and thalassemia major, were found to be similar to those obtained with therapeutic doses of deferoxamine. In those studies, iron excretion was found to be related to the dose of deferiprone within the range of 25–100 mg kg^{-1} body weight per day and to the iron load of the patient.

O
OH

N   CH$_3$
|
CH$_3$ **Deferiprone**

L1 displays a mean elimination half-life of 90–160 min, a molar ratio of binding of 3:1 and it is rapidly inactivated by glucuronidation, with consequent rapid decrease in the affinity for iron. The use of L1 is controversial: it is licensed in the EU but it is not approved in USA.

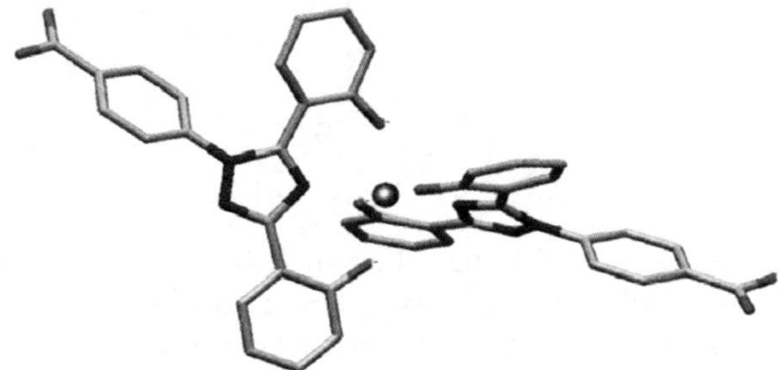

Figure 4 *Formation of a stable complex of iron and two molecules of the tridentate chelator ICL670*
(Reproduced with permission from ref. 53.)

Deferasirox (ICL670, Exjade, Novartis) is a member of a new class of tridentate iron chelators, the *N*-substituted *bis*-hydroxyphenyl-triazoles.[52] It is orally bioavailable and its terminal elimination half-life is between 8 and 16 h, allowing for once-daily administration. Metabolism and elimination of deferasirox and the iron chelate (Fe-[deferasirox]2) is primarily by glucuronidation followed by hepatobiliary excretion into the feces. No significant drug–drug interactions have been identified to date. Pre-clinical studies demonstrated the ability of deferasirox to enter and remove iron from cells. It forms a stable 1:2 complex with Fe(III) (Figure 4).[53]

ICL670 is a very promising iron chelator that has been studied in phase I, II and III and, most probably, it will be the favored drug if it proves non-toxic in long-term trials.

Other novel chelators in various stages of preclinical investigation include:

- HBED (*N,N'*-di-(2-hydroxybenzyl)ethylenediamine-*N,N'*-diacetic acid-HCl·H2O) (Formula 10), a hexadentate phenolic aminocarboxylate iron chelator that, when delivered subcutaneously, appears twice as efficient as DFO in iron excretion.
- PCIH (Pyridoxal isonicotinoyl hydrazone), orally active in rats, shows greater hepatic iron excretion but diminished cardiac iron mobilization when compared to DFO.
- GT56-252 (2-(2,4-dihydroxyphenyl)-4,5-dihydro-4-methyl-4-thiazocarboxylic acid) (Formula 11), an orally active iron chelator.

In can be concluded that, although the search for new oral chelation therapy or new combinations of chelators has not reached the ultimate goals so far, there are new drugs on the horizon that are very promising and could possibly further widen the number of patients that can be safely and suitably treated.

References

1. P. Bull *et al.*, *Nature Genet.*, 1993, **5**, 327.
2. G. Brewer, *Exp. Biol. Med.*, 2000, **223**, 39.
3. G. Brewer, *Wilson's Disease: A Clinician's Guide to Recognition, Diagnosis, and Management*, Boston, Kluwer Academic Publishers, 2001.
4. D. Cox and E. Roberts, *Wilson's disease. GeneClinics*, University of Washington, Seattle. On World Wide Web URL: http://www.geneclinics.org/profiles/wilson/details.html.
5. G. Brewer *et al.*, *J. Lab. Clin. Med.*, 1998, **132**, 264.
6. G. Brewer *et al.*, *Arch. Neurol.*, 1987, **44**, 490.
7. G. Brewer G, *Wilson's Disease for the Patient and Family: A Patients Guide to Wilson's Disease and Frequently asked Questions about Copper*, Philadelphia, Xlibris, 2001.
8. J. Mercer *et al.*, *Nat. Genet.*, 1993, **3**, 20.
9. R. Llanos and J. Mercer, *DNA Cell Biol.*, 2002, **21**, 259.
10. L. Mandinov *et al.*, *Proc. Natl. Acad. Sci. USA*, 2003, **100**, 6700.
11. F. Tarantini *et al.*, *J. Biol. Chem.*, 2001, **276**, 5147.
12. A. Mandinova, *J. Cell Sci.*, 2003, **116**, 2687.
13. M. Landriscina *et al.*, *J. Biol. Chem.*, 2001, **276**, 25549.
14. T.M. LaVallee *et al.*, *J. Biol. Chem.*, 1998, **273**, 22217.
15. W. Volker *et al.*, *Atherosclerosis (Berlin)*, 1997, **130**, 29.
16. G.J. Brewer *et al.*, *Clin. Cancer Res.*, 2000, **6**, 1.
17. Q. Pan *et al.*, *Cancer Res.*, 2002, **62**, 4854.
18. M.K. Schellenberger, *Neurotoxicology*, 1984, **5**, 177.
19. N. Ercal *et al.*, *Free Radic. Biol. Med.*, 1996, **21**, 157.
20. H. Gurer *et al.*, *Toxicology*, 1998, **128**, 181.
21. B. Halliwell, *Lancet*, 1994, **344**, 721.
22. H. Gurer and N. Ercal, *Free Radic. Biol. Med.*, 2000, **29**, 927.
23. M.M. Jones *et al.*, *Toxicology*, 1994, **89**, 91.
24. E.M. Walker Jr. *et al.*, *Toxicology*, 1992, **76**, 79.
25. J.R. Christian *et al.*, *Am. J. Public Health*, 1964, **54**, 1241.
26. S.L. Nightingale, *JAMA*, 1991, **265**, 1802.
27. J.H. Graziano *et al.*, *J. Pediatr.*, 1988, **113**, 751.

28. Treatment of lead-exposed children (TLC) trial group, *Pediatr. Res.*, 2000, **48**, 593.
29. X. Liu *et al.*, *Pediatrics*, 2002, **110**, 787.
30. H.A. Ruff *et al.*, *JAMA*, 1993, **269**, 1641.
31. K.N. Dietrich *et al.*, *Pediatrics*, 2004, **114**, 19.
32. M.E. Markowitz *et al.*, *Pediatrics*, 2004, **113**, e34.
33. T.W. Clarkson, *J. Trace Elem. Exp. Med.*, 1998, **11**, 303.
34. T.W. Clarkson, *Environ. Health Perspect.*, 2002, **110**(Suppl. 1), 11.
35. B. Weiss and T.W. Clarkson, *Environ. Health Perspect.*, 2002, **110** (Suppl. 5), 851.
36. M.D. Guallar *et al.*, *N. Engl. J. Med.*, 2002, **347**, 1747.
37. S. Bernard *et al.*, *Mol. Psych.*, 2002, **7**, S42.
38. ATSDR, *Case studies in environmental medicine: mercury toxicity*, U.S. Department of Health and Human Services, Agency for Toxic Substances and Disease Registry, Atlanta, GA, 1992.
39. ATSDR, *Toxicological profile for mercury (update)*, U.S. Department of Health and Human Services, Agency for Toxic Substances and Disease Registry, Atlanta, GA, 1999.
40. C.R. Baum, *Curr. Opin. Pediatr.*, 1999, **11**, 265.
41. H. Frumpkin *et al.*, *Environ. Health Perspect.*, 2001, **109**, 167.
42. M.G. Ord and L.A. Stocken, *Trends Biochem. Sci.*, 2000, **25**, 253.
43. A. von Döllen and H. Strasdeit, *Eur. J. Inorg. Chem.*, 1998, 61.
44. X. Fang *et al.*, *Chem. Res. Toxicol.*, 1996, **9**, 284.
45. G.N. George *et al.*, *Chem. Res. Toxicol.*, 2004, **17**, 999.
46. R.R. Crichton, *Inorganic Biochemistry of Iron Metabolism From Molecular Mechanisms to Clinical Consequences*, Wiley, Chichester, New York, 2001.
47. C. Herschko *et al.*, *Br. J. Haematol.*, 1998, **101**, 399.
48. C.A. Lipinski *et al.*, *Adv. Drug Delivery Rev.*, 1997, **23**, 3.
49. B.A. Davis and J.B. Porter, *Blood*, 2000, **95**, 1229.
50. M.J. Cunningham and D.G. Nathan, *Curr. Opin. Hematol.*, 2005, **12**, 129.
51. R.C. Hider *et al.*, Pharmaceutically active hydroxypyridones, GB patent 21–46989, 1984.
52. H.-P. Nick *et al.*, *Curr. Med. Chem.*, 2003, **10**, 1065.
53. M.D. Cappellini, *Best Pract. & Res. Clinical Haematol.*, 2005, **18**, 289.

Metal Complexes in the Brain Imaging and Diagnosis

Non-invasive study of a human body has always been a challenge to medical doctors, scientists and, later, designers of commercial devices. Discovery of X-rays at the end of nineteenth century has brought about a powerful and, as it seemed to be at the time, ultimate instrument for such study. Soon, however, it became clear that X-ray radiography may hardly be called non-invasive due to destructive effects on the tissues caused by ionizing radiation. Today's X-ray techniques, although much more safe and sophisticated than before, still employ the same kind of radiation and constitute the same kind of health risks as years ago.

Nuclear magnetic resonance employs low-intensity radiofrequency electromagnetic waves to study substances placed in a strong magnetic field. Neither radiation employed, nor the strong magnetic field have been proven to be harmful to living organisms in any way so far. Remarkable results have been achieved using this most non-invasive technique during the last decade, yet more horizons of its use still remain unexplored.

The main advantage of the magnetic resonance imaging (MRI) is that it is harmless to the patient. Of course, it would be of little use if MRI did not offer sufficient diagnostic power. In many cases, MRI is the only way to do unambiguous diagnosis, especially in detection of cerebral abnormalities, multiple sclerosis (MS) and lesions in sites often obscured by bone artifact on computed tomography (CT). MRI has inherently superior contrast scale compared to that of CT. Because of that, despite of relatively high cost of equipment and maintenance, it finds its way to more and more hospitals, clinics and research facilities in developed countries.

Plain MRI also has a few disadvantages when compared to contrast-enhanced CT. In head examination, tumors cannot be reliably distinguished from the surroundings, while in abdominal images it may be difficult to identify the loops of small bowel making the diagnosis of lesions uncertain. These shortcomings have led to much effort to develop ways to improve the MRI image quality.

Both CT and MRI are methods based on image generation by the computers. Therefore, choice of computational method becomes important to obtain quality images. These various methods constitute a whole separate area and will not be considered here.

The signal intensity in all X-ray-based methods reflects the electron density. Since the early years of X-ray radiography, a number of substances containing heavy elements with large number of electrons were employed to achieve greater contrast of images. In 1930s, a Nobel Prize was awarded for the development of contrast-enhanced X-ray angiography. The method involved injection of strontium salts into the neck artery and immediately taking X-ray pictures of the head. In such a way, the blood vessels of the brain become clearly visible on the image.

Since the signal intensity in MRI depends not only on the amount of water in a given site, but also on the magnetic relaxation times, T_1 and T_2, there is more opportunity to create a good picture. The overall quality of image is strongly dependent on hardware design, especially on transmitting and receiving coil design, and the pulse sequence employed to take a particular picture. For example, blood can appear black, gray or white – depending on the pulse sequence, velocity of flow and orientation of flow to the imaging plane.

It is hard to influence the proton density in the tissue. Because of that, the strongest effect on appearance of different tissues in the MR image can be achieved by the changes of the magnetic relaxation times T_1 and T_2 of protons in the tissue-contained water. Therefore, the application of the contrast substances in MRI may change tissue characteristics, as compared to application of the radiation-absorbing substances in CT.

Spin-lattice relaxation time T_1 and spin–spin relaxation time T_2 may be shortened considerably in presence of paramagnetic species. While shortening of T_1 leads to the increase of signal intensity, shortening of T_2 produces broader lines with decreased intensity. The net result is a non-linear relationship between signal intensity and the concentration of the contrast agent.[1] At low concentrations of contrast agent, a concentration increase provides an increase in signal intensity by affecting T_1 until the optimal concentration is reached. Then further increase in concentration reduces the signal because of the effect of contrast agent on T_2. Therefore, in clinical practice it is possible to achieve less than optimal contrast effect and even produce a negative effect. This dictates the use of agents that have a relatively greater effect on T_1 than on T_2, and additionally the use of pulse sequences that emphasize the changes in T_1.

Since free radicals generally cause serious damages to the living tissues and the paramagnetic effect of oxygen, although demonstrable, seems to be too weak to be practically applicable, the desired contrast enhancement can be suitably achieved by using paramagnetic metal ions. The paramagnetic effect depends on the number of unpaired electrons, as shown in Table 1.

There are many transition and lanthanide metals with unpaired spins, but for the metal to be effective as a relaxation agent the electron spin-relaxation time must match the Larmor frequency of the protons. This condition is met well for the Fe^{3+}, Mn^{2+} and Gd^{3+} ions.

The enhancement of the relaxation rates in the presence of paramagnetic species mainly results from the so-called "inner-sphere" contribution,

Table 1 *Some properties of selected metal ions*

Atomic number	Ion	Electronic configuration	Magnetic moment (Bohr magneton)
24	Cr^{3+}	$3d^3$	3.8
25	Mn^{2+}	$3d^5$	5.9
26	Fe^{3+}	$3d^5$	5.9
29	Cu^{2+}	$3d^9$	1.7–2.2
63	Eu^{3+}	$4f^8$	(6.9)
64	Gd^{3+}	$4f^7$	7.9
66	Dy^{3+}	$4f^9$	(5.9)

given by:[2]

$$\frac{1}{T_1^{\mathrm{IS}}} = \frac{qP_{\mathrm{m}}}{T_{1\mathrm{M}} + \tau_{\mathrm{M}}}$$

where q is the number of water molecules bound to the paramagnetic ion ("inner-sphere water molecules"), τ_{M} is the residence lifetime of water in the coordination sphere, and $T_{1\mathrm{M}}$ is the longitudinal relaxation time of coordinated water protons. qP_{m} is the mole fraction of water molecules bound to metal centers.

The dominant mechanism of T_1 relaxation is a through space dipole–dipole interaction between the unpaired electrons of a paramagnetic ion, such as Gd^{3+}, and the water molecules within the ion's inner-coordination sphere. Such paramagnetic dipole–dipole relaxation rate can be described by the Solomon–Bloembergen equation:[2]

$$\frac{1}{T_1^{\mathrm{dip}}} = \frac{2}{15}\left(\frac{\mu_0}{4\pi}\right)^2 \frac{\gamma_{\mathrm{N}}^2 g_{\mathrm{e}}^2 \mu_{\mathrm{B}}^2 S(S+1)}{r^6}\left[\frac{3\tau_{\mathrm{c}}}{1+\omega_{\mathrm{I}}^2\tau_{\mathrm{c}}^2} + \frac{7\tau_{\mathrm{c}}}{1+\omega_S^2\tau_{\mathrm{c}}^2}\right]$$

where μ_0 is the permeability of vacuum, γ_{N} is the proton magnetogyric ratio, g_{e} is the electronic g factor, μ_{B} is the Bohr magneton, ω_{I} and ω_{s} are the Larmor precession frequencies for the nuclear and electron spins, and r is the metal electron spin-water proton distance. S is the total electron spin of the metal ion. The correlation time, τ_{c}, modulating proton relaxation, is composed of contributions from rotational correlation time, τ_{r}, longitudinal electronic relaxation time, τ_{s}, and chemical exchange lifetime, τ_{M}, according to the equation:[2]

$$\frac{1}{\tau_{\mathrm{c}}} = \frac{1}{\tau_{\mathrm{s}}} + \frac{1}{\tau_{\mathrm{M}}} + \frac{1}{\tau_{\mathrm{r}}}$$

The overall paramagnetic relaxation rate enhancement caused by a 1 mM concentration of a given paramagnetic compound is called its relaxivity.

A major shortcoming of clinical and experimental MR contrast agents is that they are limited to reporting anatomical details only. Although a significant progress has been made on the design and preparation of MR agents that are brighter, selectively targeting and which remain longer *in vivo*, a new generation of agents that are capable of reporting on the physiological status and metabolic activity of cells or organisms have been prepared. These "procontrast" agents have been designed to exploit three fundamental physical properties of paramagnetic complexes that function as the switch or trigger to make them detectable by MRI:

(i) q, the number of water molecules coordinated to the paramagnetic ion;
(ii) τ_m, the lifetime of a water molecule bound to the paramagnetic ion; and
(iii) τ_r, the rotational correlation time of the complex.

Variation of q, τ_m or τ_r can increase or decrease the observed signal intensity, and therefore manipulation of these parameters in the design of an agent produces what is called "activatable" MR agents.

The main problem with paramagnetic heavy metal ions in their native form is their toxicity. Thus, research has been focused on the development of stable paramagnetic ion complexes. Both the metal ion and the ligand usually exhibit substantial toxicity in the unbound state. Together, however, they may create a thermodynamically and kinetically stable compound, which is much less toxic. Complexation of the metal ion with organic ligand, while considerably decreasing toxicity, may also alter paramagnetic properties of the metal ion.

Chromium–EDTA complex was the first such agent tried, but problems with the synthesis and long-term stability prevented its clinical application.[3]

Gadolinium–DTPA complex (diethylenetriamino-pentaacetic acid, Formula 1), a renally excreted chelate, with a very high formation constant ($\log K = 23$), had sufficiently favorable properties to be approved by Food and Drug Administration of USA for use in cranial disease diagnostics in mid-1988.

DTPA
(Diethylenetriamino-pentaacetic acid)

Another, relatively new type of paramagnetic contrast agents are the so-called superparamagnetic iron oxide (SPIO) based colloids. They consist of non-stoichiometric microcrystalline magnetite cores, which are coated with dextranes or siloxanes. Use of these colloids as tissue-specific contrast agents is now a well-established area of pharmaceutical development.

13.1 Gadolinium Compounds

The prominent feature of gadolinium(III) is the high number of unpaired electrons – seven. The Gd^{3+} ion retains a number of unpaired spins when bound to the organic ligand. The free Gd^{3+} ion is extremely toxic. Many of its complexes are thermodynamically and kinetically very stable and thus exhibit relatively low toxicity. As mentioned before, Gd-DTPA complex has been approved to clinical use and is now marketed in USA under the name "Magnevist".

The relationship between the thermodynamic stability of the complex and the acute toxicity *in vivo* is not a simple matter. The stability of a series of complexes of several ligands with Gd^{3+} has been investigated *vs.* acute toxicity on mice. Some of them are shown hereafter. The rate of transmetallation of Gd^{3+} by Cu^{2+} and selectivity of different ligands toward Gd^{3+} as compared to Zn^{2+}, Cu^{2+} and Ca^{2+} have also been determined. The main competitor to Gd^{3+} was found to be Zn^{2+}, and the most important thermodynamic criterion of toxicity is the selectivity of the ligand for Gd^{3+} over other endogenous metal ions. Zinc transmetallation was found to be the most likely mode of both acute and sub-chronic toxicity in experiments on rats. Complexes that undergo *in vivo* the transmetallation reactions much slower than the renal excretion rate have significantly improved toxicities than would be predicted by thermodynamics. Slower clearance from the body is likely to significantly increase the toxicity of any Gd^{3+} complex. Other ligands have been suggested and often used in gadolinium studies, such as DTPA-BM (diethylenetriamino-pentaacetic acid-*bis*(methylamide), Formula 2), DTPA-BP (*N*,*N'*-*bis*(2-pyridylmethyl) diethylenetriamine-*N*,*N'*,*N''*-triacetic acid, Formula 3), DOTA (1,4,7,10-tetra-azacyclododecane-*N*,*N'*,*N''*,*N'''*-tetraacetic acid, Formula 4), and CDTA (trans-1,2-diaminocyclohexane-*N*,*N'*,*N''*,*N'''*-tetraacetic acid, Formula 5).

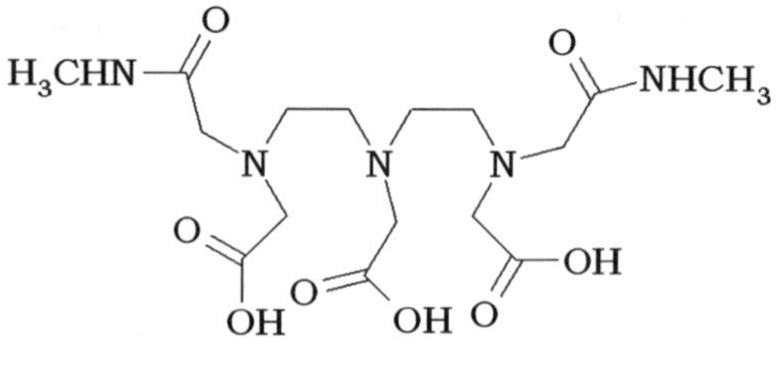

DTPA-BM
(Diethylenetriamino-pentaacetic acid bis(methylamide))

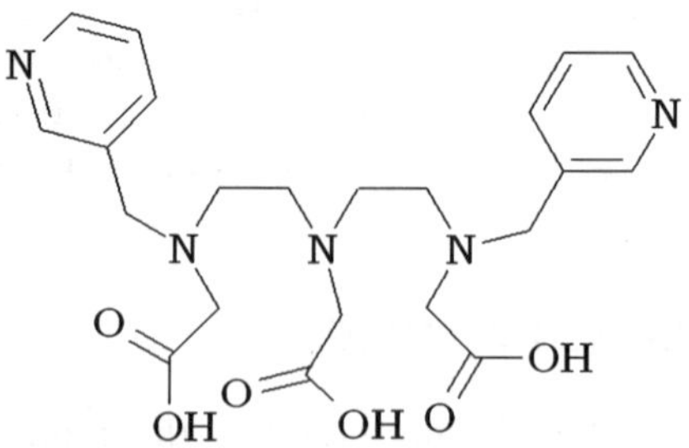

DTPA-BP
(N,N'-bis(2-pyridylmethyl) diethylenetriamine-N,N',N''-triacetic acid)

DOTA
(1,4,7,10-tetraazacyclododecane-N,N',N'',N'''-tetraacetic acid)

CDTA
(trans-1,2-diaminocyclohexane-N,N',N'',N'''-tetraacetic acid)

Gd^{3+} has been shown to inhibit Ca^{2+} binding to mammalian cardiac sarcoplasmic reticulum. The mechanism of toxicity could involve hemodynamic disruption.

Potential improvement could be achieved by covalently coupled ligand to protein to generate tissue-specific contrast agents.[4]

Understandably, search for new potential ligands for Gd^{3+} complexation is still a hot area of investigation. The complexes are evaluated against Gd–DTPA, the only approved compound for human use so far. Criteria include the thermodynamic stability, rates of excretion, toxicity, lipophilicity, biodistribution and the change in MR signal intensity. There are complexes that are slightly better than Gd–DTPA in some particular tests, while others are a little worse. However, neither of them got as much attention as Gd–DTPA itself. This complex is far from being the best choice, but it is relatively well known, widely used in many MRI facilities on a daily basis. Accumulated experience allows in many cases offset the shortcomings of Gd–DTPA, so the agent continues to play indispensable role in modern MR imaging.

13.2 Monocrystalline Iron Oxide Nanocompounds

Compounds called monocrystalline iron oxide nanocompounds (MION) constitute a relatively new but rapidly evolving area in the field of MRI contrast agents. As compared to the single approved Gd-containing complex, there is a variety of MION (also called SPIO – superparamagnetic iron oxide) reagents available on the market. Flashy names Feridex I.V, Endorem, Gastromark, Lumirem, Sinerem and more patents pending tell us that the last word in the area is yet to be said.

These compounds consist of non-stoichiometric microcrystalline magnetite cores, which are coated with dextranes (in ferumoxides) or siloxanes (in ferumoxils).[5] SPIO agents are much more effective in MR relaxation than their paramagnetic counterparts. Since they are non-stoichiometric, there was and still is much interest in studying these compounds with all the vast array of modern physical–chemical methods: single crystal X-ray diffraction, powder X-ray diffraction, Moessbauer spectroscopy, transmission electron microscopy, dynamic light scattering, atomic adsorption spectroscopy, spectrophotometry, electron microscopy, superconducting quantum interference devices, *etc.*

The compositions and physiochemical properties of non-stoichiometric magnetites are continuously variable between those of Fe_3O_4 and Fe_2O_3. Conceptually, these cation-deficient, inverse-spinel phases are formed by partial oxidation of Fe^{2+} in stoichiometric magnetite. The Fe^{2+} content is typically 8–15 mol%. The lattice parameters of these colloids also fall between those of Fe_3O_4 and Fe_2O_3.

The particles are usually of varying sizes from several to several hundred nanometers. They are superparamagnetic materials irregular in shape, and highly light absorbing.

SPIO compounds are the promising contrast agents since their properties may be fine-tuned for the specific application. They are non-toxic and rapidly cleared from the organism. Experiments have been successful in receptor-specific SPIO delivery.[6]

As an example of the use of SPIO compounds for imaging of neurological disorders, the burden of autoimmune encephalomyelitis (EAE), a commonly used animal model that in several respects mimics human MS, was studied in Lewis rats immunized by guinea pig myelin using different MRI techniques (macrophage tracking based on labeling cells *in vivo* by ultrasmall particles of iron oxide (USPIO), BBB breakdown and magnetization transfer imaging (MTI)), as well as immunohistological staining.[7] The resulting imaging data was compared with behavioral readouts. Animals were studied during the acute phase and the first re-lapse. Activated monocytes were detected during both episodes in the brain stem or cortex. These areas coincided in part with areas of BBB breakdown. Significant changes of the magnetization transfer ratios (MTRs) of up to 35% were observed in areas of USPIO accumulation. This suggests that infiltrating monocytes are the major source of demyelination in EAE, but monocyte infiltration and breakdown of the BBB are temporally or spatially independent inflammatory processes.

USPIO have been tested for brain and carotid atherosclerotic plaque imaging. As macrophages and the activation of local specific phagocytic microglia cells are involved in many pathologic tissue alterations of the CNS, USPIO labeling of these cells provides applications such as imaging of stroke, MS, atherosclerotic disease, spinal cord injury or even brain tumor characterization.

The physicochemical surface properties of USPIO allow their efficient internalization into macrophages and other phagocytic cells after intravenous injection, and make them appropriate for use as a contrast agent for macrophage MRI. USPIO are iron oxide nanoparticles composed of Fe_2O_3 and Fe_3O_4 stabilized by different coating agents. Ferumoxtran-10 are ultrasmall dextran-coated nanoparticles (15–30 nm) with a long blood-residence time (human blood half-life between 24 and 36 h), which allows the product to access macrophages present in deep and pathologic tissues (lymph node, brain, kidney, osteoarticular tissues, *etc.*).[8]

The route of transport of USPIO to the macrophages in tissues is not yet well defined. Several mechanisms have been suggested:[9]

(i) USPIOs are endocytosed by activated blood monocytes, which migrate into the pathologic tissues; for example, the long blood half-life of ferumoxtran-10 allows sufficient time for the blood monocytes to internalize the nanoparticles *via* endocytosis and for the progressive migration of these cells;

(ii) transcytosis of USPIOs through the endothelium and migration of the USPIOs into the tissue followed by progressive endocytosis of these USPIOs by *in situ* macrophages; and

(iii) transport of USPIOs into the pathologic tissue, in some cases *via* the inflammatory neovasculature irrigating the media and *adventitia* in atherosclerotic lesions.

Iron oxide nanoparticles cause a strong decrease in signal intensity (negative enhancement). USPIOs present the highest R_1 relaxivity, which results in an increase in signal intensity using a T_1-weighted sequence (positive enhancement). Thus, according to the sequence and the local concentration of USPIO, T_1 and T_2 enhancing effects can be observed independently. As high-field MRI becomes commonly used for animal imaging, the relaxivity properties of USPIO are particularly useful for T_2 imaging, since the R_1 relaxivity decreases at high field.

13.3 Delivery of MRI Contrast Agents

From the description of the BBB given in Chapter 2 of this book, it is clear that the brain endothelium will restrict the penetration of many agents that have proved useful in imaging/diagnosis in peripheral tissues. The following Figure 1 summarizes the different routes by which molecules can move across the BBB.[10]

Gases such as oxygen and carbon dioxide, water and small lipophilic agents (<600 Da) can generally penetrate the lipid membranes of the endothelial cells by passive diffusion. The hydrophilic paracellular (tight junctional) pathway across brain endothelium is relatively impermeable to solutes, so that even small ions experience significant restriction, giving rise to a measured transendothelial electrical resistance (TEER) of 1–6000 Ω cm^2, compared to 2–20 Ω cm^2 in

ROUTES ACROSS THE BRAIN ENDOTHELIUM

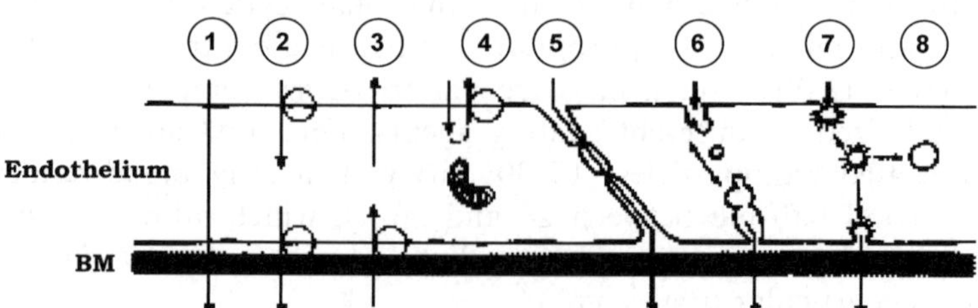

Figure 1   *Routes across the brain endothelium. (1) Lipophilic route through cell membranes, (2) carrier-mediated uptake, including "facilitated diffusion" aiding entry of glucose and large neutral amino acids (L system), (3) carrier-mediated efflux, including Na-dependent amino acid carrier (A system), (4) drug efflux transporters operating in the luminal membrane, including P-glycoprotein, (5) paracellular flux through the hydrophilic tight junctional pathway – negligible under normal conditions, but increased in some pathologies, (6) and (7) endocytosis and transcytosis via uncoated and coated vesicular pathways, expressed at low levels in brain endothelium, (8) some agents taken up by vesicular pathways end up in lysosomes and are not transcytosed*
(Reprinted with permission from ref. 20.)

capillaries of the systemic (peripheral) vasculature. Thus, solutes of the size of mannitol and sucrose that rapidly equilibrate with the interstitial space of most tissues by diffusing out of capillaries *via* the (leaky) tight junctions, show extremely low permeability through the endothelium of the brain. This means that traditional X-ray contrast agents used for the image intensification including organic iodides are effectively confined to the vascular compartment in normal brain. However, in several pathologies where imaging is required, including MS and brain tumors, BBB permeability may be increased, either transiently or chronically. The mechanisms underlying the increased permeability are not fully understood, but may involve opening of tight junctions by inflammatory mediators including locally produced cytokines and decline in production of barrier-inducing factors (with increase in production of permeability factors) in proliferating glial tumors. In either case, the permeability increase can be used for imaging sites of BBB breakdown, since contrast-enhancing agents (*e.g.*, Gd–DTPA used in MRI scanning) escape from the leaky vessels but remain confined in normal vessels.[11] A similar logic underlies the use of fluorescein to detect sites of BBB breakdown in the blood–retinal barrier by direct intraocular imaging.

There is a great deal of interest in the deliberate opening of the BBB for delivery of therapeutic agents (*e.g.*, to treat brain tumors). The best-established method, osmotic opening, involves a hypertonic bolus injection of arabinose or mannitol into the carotid artery, resulting in endothelial cell shrinkage and tight junction opening.[12] The barrier permeability stays elevated for a few hours, allowing time for treatment with cytotoxic drugs. Pharmacological barrier opening using analogues of bradykinin and nitric oxide, is less advanced, but has in some cases proved to give greater specificity, *e.g.*, increasing barrier permeability in tumors but not in normal brain. These techniques also hold promise for introduction of imaging agents, for accurate mapping of the extent of brain tumors, and to monitor the extent of spread of cytotoxic agents co-infused with a contrast-enhancing tracer. However, such barrier opening strategies still meet an upper size limitation: studies with larger particulate materials such as monocrystalline iron oxide nanoparticles, introduced intravascularly with barrier opening in an attempt to develop strategies for delivery of viral particles, showed that the endothelial basement membrane may act as a filter against particulate penetration.[13]

As the paracellular route is so tight in normal brain capillary endothelium, several carrier systems are present to ensure that the brain receives essential nutrients and substrates. Carriers have been identified for glucose, large neutral amino acids, nucleosides, carboxylic acids, choline and several other groups of compounds including peptides. However, the carriers have relatively restricted specificities, and while they can provide usefully specific transfer routes for targeting drugs to the brain, they have limited value for delivering diagnostic agents. The most obvious applications in this context are those for PET studies, using a bolus intravascular injection containing a positron-emitting tracer that can be used to follow particular aspects of brain function.

Several PET tracers have been used to determine BBB and cerebral function,[14] and in addition to following cerebral blood flow and volume, oxygen consumption and BBB permeability, it has been possible to monitor glucose and amino acid transport at the BBB and binding and utilization within the brain. Thus [^{18}F]L-DOPA, the dopamine precursor transported across the BBB *via* the L-system amino acid carrier, has been used to image dopaminergic pathways in Parkinson's patients,[15] and *O*-3-[^{11}C]methyl-D-glucose and 2-[^{18}F]fluoro-2-deoxy glucose have been used to map areas of increased glucose transport/phosphorylation as indicators of regional neural activity. PET scanning is thus a powerful tool for imaging brain function in normal physiology and in pathology.

Imaging the vascular system can be of great value for understanding a diverse set of pathophysiologies. Historically, interest has centered on obtaining a precise image of the vasculature tree of relatively large vessels, since occlusions and malformations on this level are known to cause the life-threatening sequels of end-organ ischemia and hemorrhage. This area, known generally as angiography and pioneered using X-ray contrast techniques, can now be accomplished with a wide variety of imaging methods, including MRI and ultrasound. More recently, the importance of capillary-level imaging of both microvascular perfusion and blood volume fraction has been recognized as an independent and important measurement of the organ-level response to a pathological state. Accurate measurement of the brain perfusion and blood volume is challenging because of the unique hemodynamic environment of the brain itself. While it possesses extremely high blood flow (20–100 ml/100 g min^{-1}), its parenchymal blood volume fraction is remarkably low (1–5%). Because most contrast media are confined to the intravascular space of the brain by the BBB, absolute signal changes are often small. In general, the techniques used to image this small but active vascular compartment can be divided into three categories:

(i) bolus injection of exogenous contrast agents that remain restricted to the vascular compartment so that rapid imaging of the first pass kinetics reflect cerebral perfusion;

(ii) injection of exogenous contrast agents that freely diffuse across the BBB and istribute in proportion to delivery rates or perfusion; and

(iii) methods that detect changes in endogenous blood components that reflect perfusion.

In the 1980s, several groups realized that such intrinsic measurements as perfusion (defined in units of ml blood g^{-1} min^{-1}) and blood volume fraction (ml g^{-1}) using intravascularly confined contrast agents were possible based on new ultrafast tomographic imaging techniques. The basic idea behind these advances was the application of dynamic imaging methods to monitor the vascular passage of a bolus of contrast agent and then to apply classical tracer kinetic techniques to extract hemodynamic parameters, especially blood flow and blood volume fraction, as well as their ratio, the mean transit time (MTT).

All dynamic contrast methods rely on the use of a tracer substance, which is introduced to the blood in a controlled manner upstream of the organ being studied. The tracer is then carried with the blood to the organ, and concentration-time profiles are measured, with the assumption that the signal change seen in each voxel is proportional to the amount of tracer present.

The purest expression of tracer kinetic methods applied to tomographic imaging is found in the application of highly radio-opaque contrast agents to high-speed CT. As the bolus passes through the brain, signal intensity increases proportional to the amount of contrast agent present within the voxel. More recently, newer technologies, both in scanning and in contrast media, have allowed ever better blood flow and volume imaging, both in animals and in human ischemic stroke.[16] Coupled with the ability to detect hemorrhage and obtain CT angiograms, these techniques may prove useful for the rapid assessment of cerebrovascular disease in acute strokes.

One significant early finding was that simple paramagnetic molecular chelates, such as Gd–DTPA, were intravascularly confined in the brain due to the existence of the BBB and that the relaxivity changes are linearly proportional to the amount of agent in the voxel. A good discussion of the physical properties of these agents can be found in an excellent review article by Lauffer.[17]

The second category of the imaging techniques involves the use of contrast agents that are lipophilic and freely diffusible across the BBB. These agents typically reach plateau concentrations in brain within 10–15 min post-injection. For single-photon emission computed tomography (SPECT), commonly used agents include ^{123}I-*N*-isopropyliodoamphetamine (^{123}I-IMP, Spectamine), ^{99m}Tc-hexamethyl propylene amine oxime (^{99m}Tc-HMPAO, Ceretec), and ^{99m}Tc-ethyl cysteinate dimer (^{99m}Tc-EDC, Neurlite). The advantage of these

SPECT methods is that they provide "snapshots" of perfusion at the time of injection. This may be useful in cases where cerebral perfusion needs to be measured during some events *e.g.*, epileptic seizure, the tracer is injected to catch the event and SPECT imaging can be conducted later after the patient has stabilized.

With PET, the cerebral perfusion can be quantitatively measured with a variety of freely diffusible tracers, including intravenous $H_2^{15}O$, ^{11}C and ^{15}O-butanol, and ^{18}F-fluomethane. The main assumption of these techniques is that the tracer is rapidly and completely extracted by brain tissue. Since this is only true for radiolabeled butanol, measurements with the other tracers tend to underestimate very high blood-flow levels. Cerebral blood volume can also be quantified with PET. Inhalation of $C^{15}O$ is performed in a single breath. Owing to the high specific activity of the label, the trace amount of gas administered is non-toxic. After a brief period of mixing to allow the tracer to bind to hemoglobin and distribute evenly within the blood pool, PET images are collected and normalized with activity in arterial blood samples to calculate cerebral blood volume maps.

For MRI, steady state or equilibrium approaches to measure blood volume have many advantages over bolus techniques. Particularly, they do not require the high-speed scanning, which is still not available on many MRI scanners, and offer the advantage of the longer imaging times with concomitant increases in signal-to-noise ratio and resolution.

Currently, no steady-state agents are approved for human use. However, in animals, both iron-oxide agents and lanthanide chelates attached to large macromolecules have shown sufficient relaxivity and biodurability. Recent cerebrovascular applications in animal studies have shown their utility, both for longitudinal dynamic studies of ischemia,[18] and the functional activation.[19]

An important and growing field involves the labeling of endogenous substances in the body, particularly water, as a means of imaging blood flow. Specifically, recent advances in MRI allows one to "label" and "detect" water protons in flowing blood by altering their spins. Most MRI experiments are to some extent flow-sensitive. That is, measurements taken in the presence of perfusion yield different signal intensities than do measurements in the same system devoid of blood flow. Normally, these effects are small, undesirable and ignored as "flow-artefacts". However, it is possible to take advantage of the flow-sensitivity of MRI to create quantitative images of perfusion.

All methods will be improved by moving to the higher field strengths (3.0 and 4.0 T) now becoming common in research settings. Together with the methods for imaging the blood volume previously described, images of cerebral blood flow obtained with ASL methods can be used to non-invasively assess cerebro-vascular function and disease (Figure 2).[20]

A principal barrier to the development of new contrast agents is the ability to transport the agents across cellular membranes. The majority of MRI contrast agents are restricted to the extracellular domains, and there are few examples of membrane permeable MR contrast agents reported in the literature.

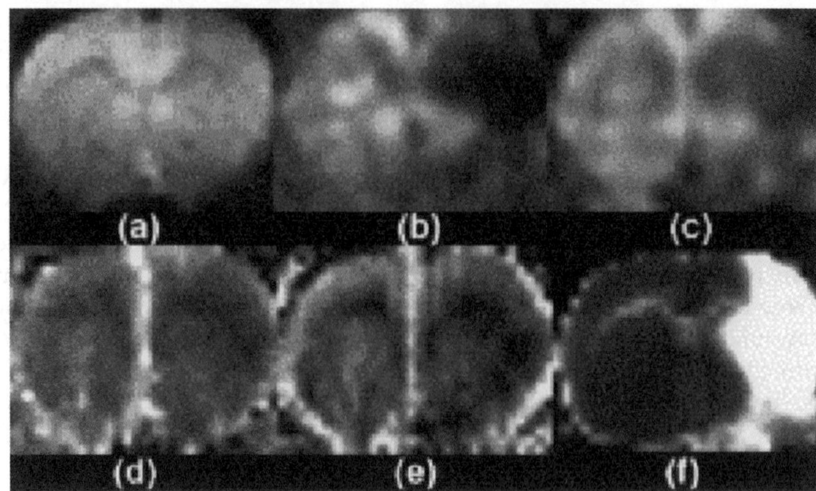

Figure 2 *Perfusion-sensitive MR images of rat brain after focal cerebral ischemia induced by occlusion of the middle cerebral artery. (a) Pre-contrast gradient echo image, (b) regional blood flow image obtained via arterial spin labeling method, (c) regional blood volume image obtained via dynamic susceptibility contrast MRI, (d) steady-state total blood volume image obtained via gradient echo scans after infusion of the blood pool agent MPEG-PL-DyDTPA, (e) steady-state microvascular blood volume image obtained via spin echo scans after infusion of the blood pool agent MPEG-PL-DyDTPA, (f) histological image of the infarct after staining with 2% 2-3-5-triphenyl tetrazolium chloride* (Reprinted with permission from ref. 20.)

Activatable MRI contrast agents were first developed with the aim of correlating developmental biological events with gene expression during imaging experiments.[21] A contrast agent was created having two distinct relaxation states, weak and strong, depending upon water exclusion or inclusion in the inner coordination sphere (Figure 3).[22]

The agent (4,7,10-tri[acetic acid]-1-[2-β-galactopyranosylethoxy]-1,4,7, 10-tetraazacyclododecane) gadolinium or "Egad" was designed to be activated by the enzyme β-galactosidase. The enzyme substrate (sugar) occupies all nine coordination sites, inhibiting water access to the paramagnetic ion. The contrast agent is irreversibly turned "on" when β-galactosidase cleaves the sugar and water becomes accessible to the ion. These agents have been successfully used *in vivo* to monitor gene expression in *Xenopus laevis*.[23]

Because of its role in signal transduction, intracellular messenger contrast agents have been created that are activated by micromolar concentrations of Ca^{2+}.[24] The Ca^{2+} sensitive contrast agent incorporates a calcium-binding domain BAPTA (1,2-*bis*-[*o*-aminophenoxy]-ethane-N,N,N,N-tetraacetic acid) linking two DO3A (1,4,7-*tris*(carboxymethyl)-1,4,7-10-tetraazacyclododecane) Gd^{3+} contrast agents. The mechanism of this agent involves the coordination of Ca^{2+} to BAPTA with one of the carboxylic acid arms of DO3A, which allows water access to the Gd^{3+} ion (Figure 4).[22] The use of this class of

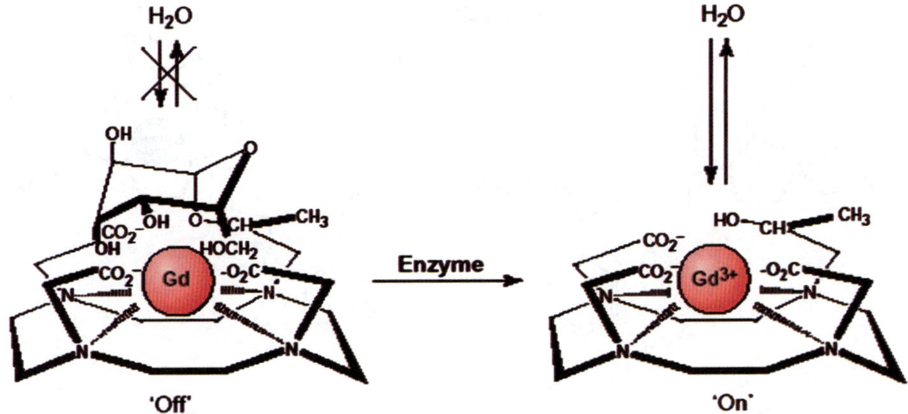

Figure 3 *Enzyme activated MR contrast agent. Schematic of the transition of methyl substituted Egad (EgadMe) from a weak to a strong relaxivity state. This diagram represents the site-specific placement of the galactopyranosyl ring on the tetraazamacro-cycle (side view). Upon cleavage of the sugar residue by β-galactosidase, an inner sphere coordination site of the Gd^{3+} ion becomes more accessible to water*
(Reproduced with permission from ref. 22.)

agents allows the mapping of the brain functions and signal transduction using MRI. A similar scheme has been used for detection of zinc ions, by using a ligand derived from diethylenetriaminepentaacetic acid (DTPA) with N,N,N,N-tetrakis (2-pyridylmethyl)ethylenediamine (TPEN) as a zinc-specific chelator.[25]

MR agents that exploit the difference in pH between extracellular tumors (pH 6.8–6.9) and healthy extracellular tissue (pH 7.4) have also been devised.[26] They consist of a DO3A-derived chelator with a sulfonamide nitrogen, which at a low pH is protonated and it is unable to chelate the paramagnetic ion. As a result, water access to the ion is restored creating a detectable signal. At a high pH the deprotonated amine binds Gd^{3+}, thereby preventing water access to the ion and resulting in a low MR signal.

Deoxyhemoglobin is by itself a paramagnetic MRI contrast agent: the MRI signal depends on the oxygenated state of the blood, and the blood oxygen level-dependent (BOLD) signal can be used for non-invasive mapping of human brain function.[27] Exploiting the BOLD method, activatable contrast agents sensitive to the partial pressure of O_2 (pO_2) have been synthesized,[28] where the oxidation state of a europium ion (Eu^{3+}) is varied to trigger the signal on and off by the environmental pO_2. Eu^{3+} is reduced to Eu^{2+} (isoelectronic with Gd^{3+}), and thereby enhances the observed MR signal upon reduction. The oxidation state of the metal is directly related to the pO_2 and allows for quantitative determination. An advantage of using Eu^{3+} is that its larger size allows for faster water exchange and therefore increases τ_M. The τ_M enhances the already established signal given by the redox-switch of the contrast agent.

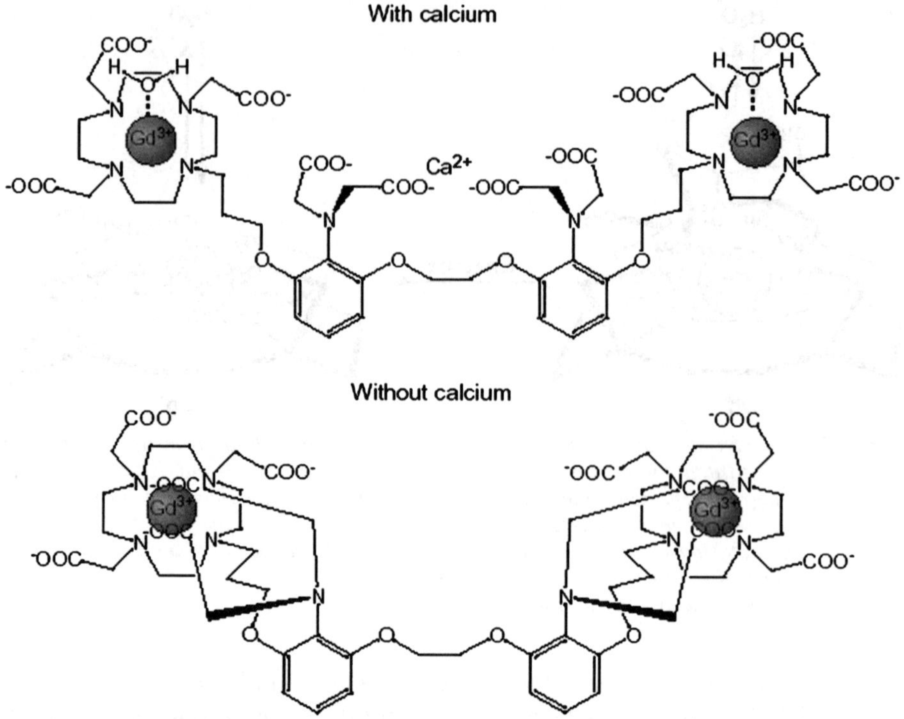

Figure 4   *Calcium activated MR contrast agent. Proposed conformational change of a*
Ca²⁺ activated MR contrast agent. The addition of calcium induces a confor-
mational shift allowing water access to Gd³⁺, and a change in observed
relaxivity
(Reproduced with permission from ref. 22.)

Relaxivity can be enhanced by an increase in rotational correlation time τ_R.
This property can be altered by the viscosity of the environment, or by covalent
and non-covalent interaction with a large species such as a protein. Typically,
τ_R agents are designed to bind to, and target specific proteins. A bioactivated
contrast agent has been thus developed that produces an increase in τ_R after
enzymatic cleavage of a peptide.[29] The contrast agent consisted of a human
serum albumin (HSA) binding inhibitor group, a HSA binding group and a
chelated Gd^{3+} ion. When carboxylpeptidase B (part of the thrombin-activata-
ble fibrinolysis inhibitor family) irreversibly cleaves the lysine masking groups,
the contrast agent facilitates the binding to HSA, thereby increasing the
relaxivity through τ_R. In this case, the ability to detect enzymatic cleavage is
not due to the modulation of q, as described above (EGad), but is dictated by
the agent binding to HSA. This dramatically shortens T_1 by increasing τ_r.

The recent development of MR contrast agents that can be used to detect
biological processes, such as β-galactosidase activity, has provided the means to
obtain direct physiological information in the form of a three-dimensional
image.[30,31] A major limitation of these bioactivated agents is their strictly

extracellular nature. The main goal in synthesizing new contrast agents that permeate cell membranes is to deliver these agents inside the cell. Accomplishing this goal will allow the detection of biologically important intracellular molecules with MRI, such as enzyme reporters and secondary messengers.

Recently, a polyarginine-based MRI contrast agent has been reported that is able to permeate cell membranes.[32] Standard peptide synthesis techniques were used to prepare a 16mer polyarginine chain and conjugate it to DOTA(*tris-t*-Bu ester). The conditions used to cleave the peptide from the resin simultaneously deprotected the *t*-butyl esters on the ligand (see scheme in Figure 5). The lanthanide ion was added and chelated by the ligand using the appropriate lanthanide hydroxide. Control compounds of Gd(III)-DO3A (**4**) and Eu(III)-DO3A (**5**) were prepared by chelating the appropriate lanthanide hydroxide to DO3A after converting the *t*-butyl esters of DO3A(*tris-t*-Bu ester) into carboxylic acids using TFA. Final products were purified on a Sephadex G-25 column or by HPLC using an Aquacil C-18 column and characterized by mass spectrometry and elemental analysis.

This first report has been then expanded through the synthesis and thorough testing of a series of membrane-permeable polyarginine oligomers of varying lengths covalently attached to a chelate framework that coordinates a lanthanide ion to the chelate 1,4,7,10-tetraazacyclo-dodecane-N,N',N'',N'''-tetraacetic acid (DOTA).[33] The selection of polyarginine (8, 12 and 16 monomer units) is due to published work that demonstrates the ability of these compounds to cross cell membranes.[34] The mechanism of this delivery is not completely understood; however, it is not a result of adsorptive- or receptor-mediated endocytosis.

The effects of varying polyarginine oligomer length on T_1 enhancement upon uptake by NIH/3T3 cells were evaluated and are shown in Figure 6.[33]

MRI studies demonstrated that cells treated with polyarginine-based contrast agents were showing increased contrast in an MR image. Importantly, the used concentration was well below the level toxic to the cells. Polyarginine-containing contrast agents are cell-type specific because they label some cell types more heavily than others. Additionally, the time necessary to label cells enough to be visualized by MRI is short (≤ 0.5 h). This is a critical experiment that demonstrates the ability of these agents to be employed in the study of biological phenomena using MRI.

As stated, the principle barrier to the development of new MR agents is the inherent lack of *in vivo* sensitivity. To obtain significant contrast over long periods of time, the observed relaxivity of MR agents must be increased. A common method of relaxivity enhancement is to increase the rotational correlation time (τ_r), which is accomplished by increasing the molecular weight of the agent by conjugation to proteins, polymers or the preparation of micellar structures.

Self-assembling peptide amphiphiles (PAs), developed as scaffolds for regenerative medicine,[35] can be utilized upon coupling to a modified MR agent. These PAs display biocompatibility and bioactivity, and self-assembly into cylindrical nanofibers and gels. Self-assembling PAs can be covalently linked to DOTA and form the supramolecular structures of peptide-amphiphile contrast

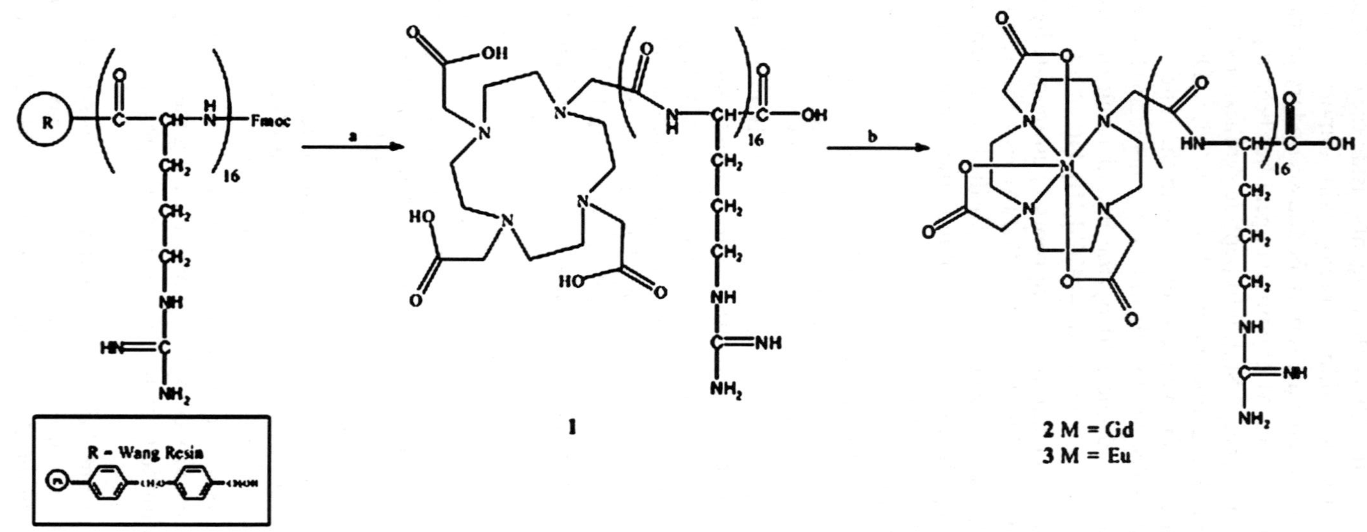

Figure 5 *Synthesis of polyarginine-containing lanthanide chelates. (a) (1) Piperidine, DMF (2) Fmoc-R(Pbf)-OH, HATU, DMF, DIPEA; (b) (1) piperidine, DMF (2) DOTA(tris-t-Bu ester), HATU, DMF, DIPEA (3) 95% TFA, 2.5% H_2O, 2.5% TIS; (b) Eu(OH)3 or Gd(OH)3 in water at 80 °C for 12 h*
(Reproduced with permission from ref. 33.)

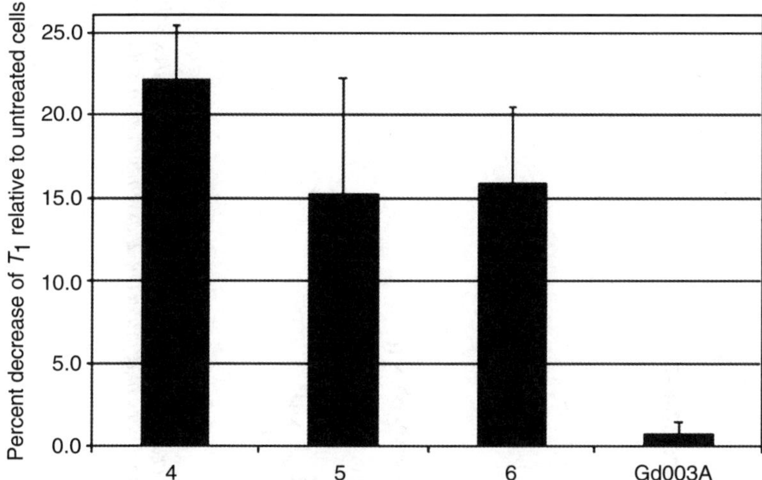

Figure 6 *Results of T_1 study of NIH/3T3 cells. Cells were incubated with 0.3 mM DOTA-(Arginine)$_8$-Gd(III) (4), DOTA-(Arginine)$_{12}$-Gd(III) (5), DOTA-(Arginine)$_{16}$-Gd(III) (6), and referred to Gd(III) DO3A (1,4,7-tris-Carboxymethyl-1,4,7,10-Tetraazacyclododecane). T_1 values were measured at 60 MHz and 37 °C. The graph depicts the percent change in T_1 from untreated NIH/3T3 cells. Error bars represent one standard deviation* (Reproduced with permission from ref. 33.)

agent (PACA) conjugates as either self-assembled nanofibers or spherical micelles. Moreover, the peptide sequence can be modified to enable tracking of the molecule by MRI (Figure 7).[36]

Ideally, one would like to extend the benefits of MRI to the PA gel scaffolds for three-dimensional non-invasive visualization. These PA monomers self-assemble into nanofibers with diameters on the order of six to eight nanometers and form self-supporting bioactive gels. The *in vivo* use of these gels would benefit greatly from the ability to detect and track their fate, migration and degradation by MRI.

Targeted radiotherapy using different vector molecules like monoclonal antibodies, peptides and others has made remarkable progress in recent years. The breakthrough begun with the first FDA-approved therapeutic radiolabeled monoclonal antibody, Zevalin, an anti-CD20 antibody labeled with ^{90}Y.[37]

Due to its overexpression in tumors, such as neuroblastomas, renal cell carcinomas, ovarian carcinomas and endometrial carcinomas, the L1 cell adhesion protein represents a target for tumor diagnosis and therapy with anti-L1-CAM antibody chCE7. Divalent fragments of this internalizing antibody labeled with $^{67/64}$Cu and ^{177}Lu were evaluated to establish a chCE7 antibody fragment for radioimmunotherapy and positron emission tomography imaging, which combines high-yield production with improved clearance and biodistribution properties.[38] chCE7F(ab′)$_2$ fragments were therefore produced in high amounts (0.2 g L^{-1}) in HEK-293 cells, substituted with the

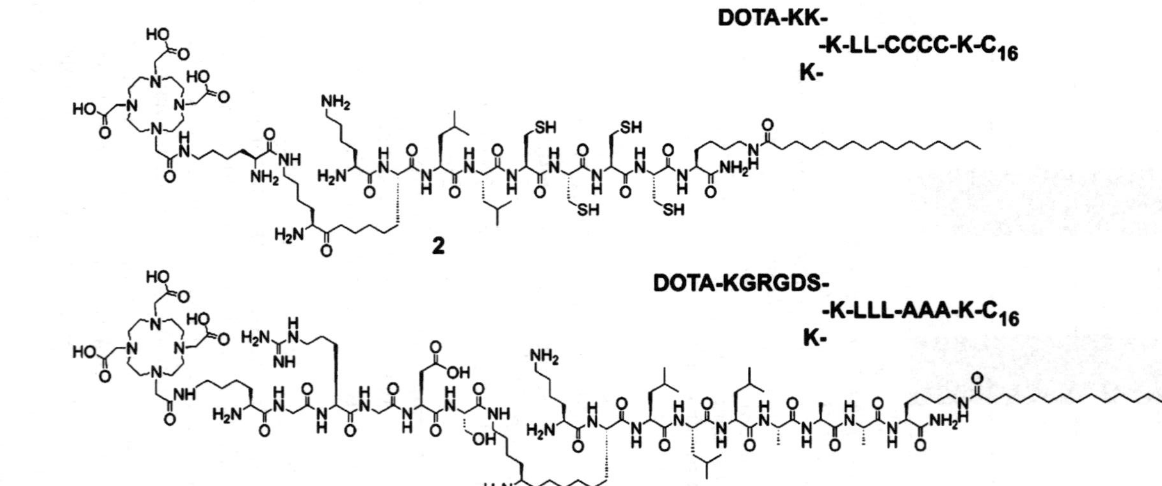

Figure 7 *Monomer structures of a chelate conjugated to* **1**, *an RGD bioactive epitope and* **2**, *a cross-linkable PA scaffold. Both structures self-assemble into nanostructures upon raising the pH above 7.0; molecule* **1** *assembles into fibers and molecule* **2** *assembles into spherical micelles*
(Reproduced with permission from ref. 36.)

peptide-linked tetraazamacrocycle 3-(*p*-nitrobenzyl)-1,4,7,10-tetraazacyclododecane-1,4,7,10-tetraacetate-triglycyl-L-*p*-isothiocyanato-phenylalanine, and labeled with ^{67}Cu and ^{177}Lu. The ^{177}Lu- and ^{67}Cu-labeled immunoconjugates reached maximal tumor accumulation at 24 h after injection. Positron emission tomography imaging allowed clear visualization of xenografts and peritoneal metastases and a detailed assessment of whole-body tracer distribution (Figure 8).

Nevertheless, peptides have several advantages over the antibodies (faster clearance, rapid tissue penetration, no antigenicity, readily synthesized and GMP produced, *etc.*) and the most commonly used receptor-targeting agents are a variety of somatostatin analogues. The molecular basis for the use of radiolabeled somatostatin analogues in peptide receptor-mediated radionuclide therapy is provided by the overexpression of the five somatostatin receptors (sstr1–sstr5) on a variety of human tumors, especially neuroendocrine tumors and their metastases. It is now more than a decade since the first radiolabeled analogue of somatostatin, [^{111}In-(DTPA)]-octreotide (OctreoScan), was approved for scintigraphy of patients with neuroendocrine tumors,[39] and it is still one of the best imaging agents. Because a β-particle emitter, such as ^{90}Y, in most cases seems more suitable for tumor therapy (peptide receptor-mediated radionuclide therapy) than the Auger electron emitter ^{111}In derivatives like DOTA-[Tyr^3]-octreotide (DOTA-TOC) have been developed, enabling stable labeling with ^{90}Y, ^{111}In or ^{177}Lu.[40] Numerous pre-clinical and clinical studies with [^{111}In]- and [^{90}Y]-labeled DOTA-TOC have shown the effective targeting and therapy using these conjugates. In addition, replacement of the alcohol group at the COOH terminus of the octapeptide by a carboxylic acid group led to increased sstr2 affinity when peptides are labeled with Y^{III}- and Cu^{II}-based radiometals and [^{177}Lu]-[DOTA^0,Tyr^3,Thr^8]-octreotide ([^{177}Lu-DOTA]-TATE) showed higher tumor uptake than [^{111}In-DTPA]-octreotide in six patients with somatostatin receptor–positive tumors.

13.4 MRI Staining of the Hippocampal System

In principle, functional MRI studies of animal brain may be based on hemodynamic techniques commonly employed for humans, that is, the monitoring of task-related MRI signal changes due to perfusion-induced alterations of the focal deoxyhemoglobin level. Extending this concept, however, more recent studies exploited the neuroaxonal uptake of paramagnetic Mn^{2+} ions, which shorten the T_1 relaxation properties of affected water protons and thereby enhance the signal of T_1-weighted MRI sequences.[41] The approach relies upon the fact that depending on brain function Mn^{2+} ions are taken up by neurons through voltage-gated Ca^{2+} channels and are axonally transported to respective projection fields. Accordingly, intracerebral injections of MnCl$_2$ bear the potential for unraveling neuronal connectivities *in vivo*.[42] Moreover, Mn^{2+} accumulation in the projection fields most likely reflects the electrophysiologic activity at the level of the injection site.[43] In contrast to deoxyhemoglobin-sensitive functional MRI sequences such as echo-planar imaging, which are

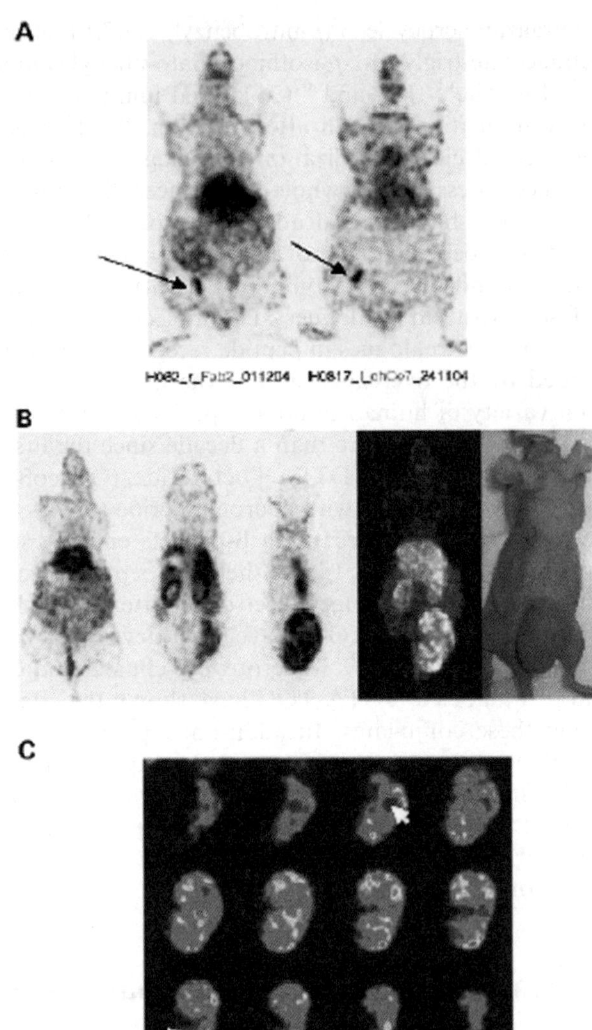

Figure 8   *PET imaging of tumor-bearing mice. Images were taken 21 h after injection of ^{64}Cu-DOTA-triglycine-chCE7F(ab')$_2$. (**A**) Comparative PET imaging of a mouse with a small peritoneal SKOV metastasis (arrow) injected with 20 MBq (100 µg) ^{64}Cu-chCE7 and seven days later with 10 MBq (130 µg) ^{64}Cu-chCE7F(ab')$_2$. Both whole body images show the same coronal plane through the tumor (slice thickness, 0.5 mm). (**B**) PET imaging of a mouse with a SK-N-BE2c tumor (right) injected with 10 MBq (130 µg) ^{64}Cu-chCE7F(ab')$_2$. The series of coronal whole body images show representative planes through the liver, kidneys and tumor. The image to the right is the projection image. (**C**) Series of coronal slices with a more detailed view of tracer distribution within the tumor in (**B**) (arrows depicting putatively necrotic sub-regions)*
(Reproduced with permission from ref. 38.)

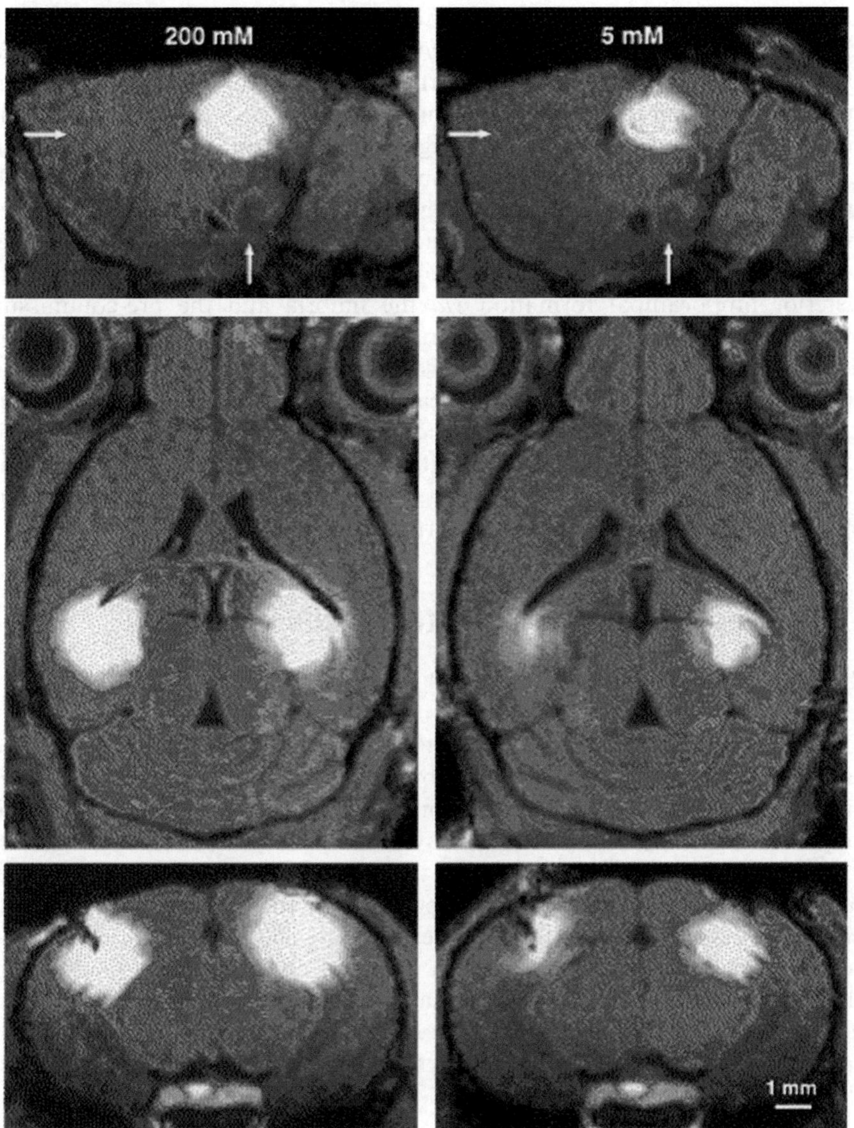

Figure 9 *T_1-weighted MR images of C57BL/6J mice. Images were taken 2 h after intra-hippocampal injection of (left) 200 mM $MnCl_2$ and (right) 5 mM $MnCl_2$. The parasagittal sections (2.5 mm lateral to the midline) and indicated (arrows) horizontal and coronal sections (top to bottom) reveal a pronounced signal enhancement at and around the injection site*
(Reproduced with permission from ref. 44.)

prone to magnetic susceptibility artifacts within the head, Mn^{2+}-enhanced MRI avoids such problems. Suitable T_1-weighted MRI sequences yield robust acquisitions at exquisite three-dimensional spatial resolution several hours after contrast administration.

As an example, morphology and function of the hippocampal system of C57BL/6J mice have been studied *in vivo* by using T_1-weighted 3D MRI after bilateral injection of $MnCl_2$ (0.25 μL, 5 or 200 mM) into the posterior hippocampal formation.[44] The neuronal uptake of the T_1-shortening Mn^{2+} ions resulted in a pronounced MRI signal enhancement within the CA3 sub-field and dentate gyrus with milder increases in CA1 and sub-iculum. This finding is in line with differences in the excitability of hippocampal neurons previously reported using electrophysiologic recordings. The subsequent ax-onal transport of Mn^{2+} highlighted the principal extrinsic projections from the posterior hippocampal formation *via* the fimbria and the pre-commissural fornix to the dorsal part of the lateral septal nucleus (Figure 9). A strong MRI signal enhancement was also observed in the ventral hippocampal com-missure. A time-course analysis revealed unsaturated conditions of Mn^{2+} accumulation at about 2 h after injection and optimal contrast-to-noise ratios at about 6 h after injection. The results using Mn^{2+}-enhanced 3D MRI may open new ways for studying the role of the hippocampal system in specific aspects of learning and memory in normal and mutant mice.

13.5   Contrast Agents in Stem Cells Therapy

The clinical application of stem cell therapy to remedy brain damage is dependent on the potential of grafts to promote sustained recovery after transplantation. At present, only the use of post-mortem tissue allows deter-mining how neural stem cells contribute to behavioral recovery. Although migratory properties and the integration of neural stem cells in damaged brains are considered to be beneficial to behavioral recovery, no indication has been provided how such features relate to the behavioral recovery. Several recent studies have demonstrated that incorporation of the contrast agents into the cells destined for cellular therapy before transplantation can afford their identification *in vivo* by MRI.[45,46] A bimodal contrast agent Gadolinium–RhodamIne Dextran (GRID) (Formula 6), detectable both *in vivo* by MRI and subsequently at post-mortem by fluorescent microscopy, has been suggested as a novel tool to track migrating neural stem cells.[47]

GRID

Pre-labeling neural stem cells with GRID has been shown to allow the following up *in vivo* of the transhemispheric migration of transplanted neural stem cells contralateral to a stroke lesion by serial MRI.[48]

13.6   Thallium Autometallography

Several methods can be used for imaging neuronal activity in the mammalian brain, such as

- the 2-deoxyglucose (2-DG) method,
- optical recording of intrinsic signals and
- functional MRI.

However, mapping neuronal activity with a spatial resolution of the single axons or synapses is impossible with the currently available methods. Because of the intrinsic complexity of the neocortex, where on average 40,000 neurons, 3.2 km of axons and 8×10^8 synapses are packed in 1 mm^3,[49] there is an obvious need for imaging techniques with cellular and sub-cellular resolution.

Visualizing neuronal activity with cellular resolution in large populations of neurons throughout the brain has been possible only by mapping the expression patterns of immediate early genes (IEGs).[50]

A novel method for mapping neuronal activity that is able to visualize activation patterns with light and electron microscopic resolution has been suggested,[51] based on the tight coupling of neuronal activity with K^+-uptake, mediated by the Na,K-ATPase. The rate of K^+-uptake increases with increasing activity, such that K^+ isotopes or K^+ analogues such as rubidium, thallium or cesium can be used as tracers for mapping neuronal activity.

Tl^+, a well-established K^+-probe, has been therefore suggested together with a modified histochemical method for the detection of heavy metals, the autometallographic method, to specifically detect thallium. This method, thallium autometallography (Tl-AMG), has been reported for analyzing the tonotopic columnar organization of the auditory cortex of Mongolian gerbils.

13.7   Targeting of Contrast Agents to AD Amyloid Plaques

Molecular imaging by magnetic resonance requires a molecular probe containing a contrast agent that is capable of selective tissue targeting in labeling a specific molecular entity within the tissue of interest, which can be detected by MRI. In the case of AD, the main pathological hallmark is the extracellular accumulation of amyloid-β (Aβ) peptide into plaques. These plaques are essential for the definitive diagnosis of AD, which is confirmed usually only post-mortem. Methods for direct imaging of individual β-amyloid plaques in humans would therefore provide a definitive pre-mortem diagnosis of this disease or a method of assessing disease progression. MRI has a spatial resolution of 30–50 μm, which at least theoretically, has the capacity to resolve individual plaques (the neuritic plaque size in an AD patient varies from 2 to 200 μm).

Radioiodinated human Aβ 40 has been used as a molecular probe, which binds to β-amyloid plaques both *in vitro* and *in vivo*.[52] The *in vivo* binding of plaques was demonstrated with radioiodinated, polyamine-modified, human Aβ40 following intravenous injection into a transgenic mouse model of AD. Furthermore, by covalently attaching Gd–DTPA to polyamine-modified Aβ, it has been possible to enhance selectively the individual plaques by MRI performed on the *ex vivo* AD mouse brain at 7 T with a spatial resolution approximating plaque size (62.5 μm^3).[53]

The ability to quantify the permeability of peptides and proteins at the BBB,[54] has allowed the evaluation of different protein modifications that might be used to enhance this permeability.[55,56] In particular, covalent modification with the naturally occurring polyamines, such as putrescine, spermidine or spermine, has resulted in dramatic increases in the BBB permeability of a number of proteins.[57] Indeed, polyamine modification of human Aβ40 as described above not only resulted in a significant increase in the BBB permeability but also resulted in enhanced binding to amyloid plaques in AD brain sections.

Polyamine–proteins have been further modified by targeting carboxyl groups of aspartic and glutamic acid residues utilizing water-soluble carbodiimide. This two-step reaction sequence involves the condensation between carboxyl groups of proteins with a nucleophile, such as the polyamine putrescine. In the initial reaction, the carbodiimide adds to ionized carboxyl groups to form an *O*-acylisourea intermediate. Subsequent reaction of the intermediate with the amine yields the corresponding amide. Because this reaction depends on the ionization of the individual carboxyl groups, the extent of modification can be controlled by maintaining a desired pH to limit the ionization of the carboxyl groups and, hence, preserve the bioactivity of the protein. One of the difficulties with this reaction is that the reactive intermediate also undergoes hydrolysis slowly, which may in turn react with other nucleophiles to form different carboxylated derivatives. In particular, reaction with an amino group from a second protein or the same protein may lead to a cross-link between the two proteins. This is less problematic with high-molecular mass proteins; however, it can be more problematic with low-molecular mass peptides or synthetic peptides lacking post-translational modifications, particularly those with unprotected *N*- and *C*-termini. This is particularly evident for a peptide such as Aβ40, which readily forms aggregates and fibrils leading to increased insolubility.

One way to avoid the problems associated with carbodiimide-mediated modification is to create unique protein isoforms by directly synthesizing the amine-modified carboxyl groups of glutamic and aspartic acid to create a glutamyl-4-aminobutane or asparagyl-4-aminobutane (Figure 10), which is then incorporated into the synthesis of the protein.[58]

By substitution of glutamic acid at positions 3, 11 and 22 with glutamyl-4-aminobutane and substitution of aspartic acid at positions 7 and 23 with

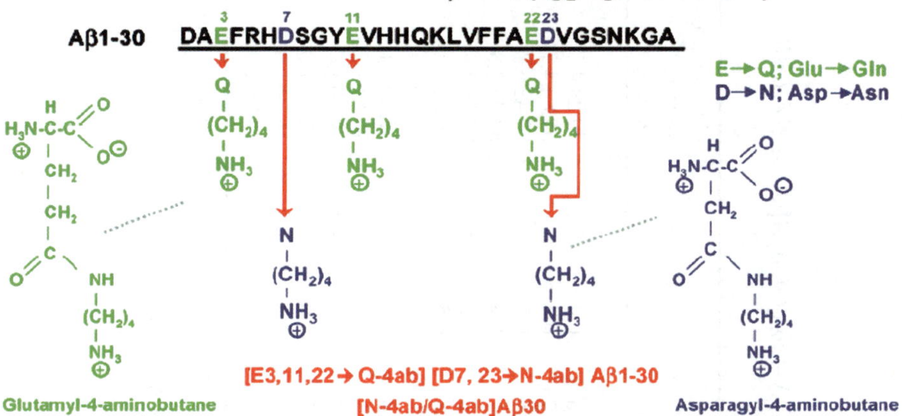

Figure 10 *Molecular structure of diamine-substituted Aβ1-30. Amino acid sequence of Aβ peptides and chemical structure of modified glutamine and asparagine residues*
(Reproduced with permission from ref. 58.)

asparagyl-4-aminobutane, the same derivative can be synthesized without the inherent problems of peptide cross-linking and decreased solubility. Of course, the number and position of this diamine substitution within the peptide will determine its BBB permeability and its ability to target the amyloid plaques. The diamine- and Gd–DTPA-substituted Aβ1-30 have been characterized by mass spectrometry, protein electrophoresis, the BBB permeability, the ability to bind plaques in AD tissue sections and also the ability to target amyloid plaques in the AD mouse after intravenous injection. The focus is on Aβ1-30, because the cell surface-binding domain (Aβ31–34) is excluded and the neuro-toxic domain (Aβ25–35) is truncated.[59] It is, in fact, particularly important to develop a derivative of Aβ that is non-toxic as this contrast agent might have clinical application for the definitive pre-mortem diagnosis of AD in human patients.

BBB permeability data can be obtained in terms of the product of the permeability coefficient and the surface area (PS, expressed in nanoliters per gram per second) and of the residual brain region blood volume (V_p, expressed in microliters per gram), as exemplified in Table 2.[58]

The diamine- and gadolinium-substituted derivative of Aβ yields enhanced binding *in vitro* to AD amyloid plaques and increased *in vivo* permeability at the BBB because of the unique Asp/Glu substitutions. In addition, specific *in vivo* targeting to AD amyloid plaques is demonstrated throughout the brain of an

Table 2 PS and V_p values of Aβ30, [N-4ab/Q-4ab]Aβ30, GdAβ30 and Gd[N-4ab/Q-4ab]Aβ30 at the BBB in the normal adult mouse (B6SJL)[a]

Brain region	Aβ30 (n = 11)	[N-4ab/Q-4ab]Aβ30 (n = 5)	RI	p	GdAβ30 (n = 9)	RI	p	Gd[N-4ab/Q-4ab]Aβ30 (n = 6)	RI	p
PS										
Cortex	117.5±4.7	139.6±9.4	1.2	ns[b]	49.8±5.8	0.4	e	79.2±4.1	0.7	e
Caudate-putamen	101.8±4.1	121.2±5.9	1.2	c	39.7±3.7	0.4	e	65.8±4.7	0.7	e
Hippocampus	113.8±5.2	139.6±10.3	1.2	ns[b]	50.7±6.5	0.4	e	89.4±6.2	0.8	ns[b]
Thalamus	136.0±6.1	147.0±9.2	1.1	ns[b]	61.0±8.7	0.4	e	84.4±3.8	0.6	e
Brain stem	134.4±6.1	149.0±9.2	1.1	ns[b]	62.0±8.0	0.5	e	92.2±5.9	0.7	d
Cerebellum	131.9±5.0	174.6±14.1	1.3	d	60.2±6.8	0.5	e	102.8±7.4	0.8	ns[b]
V_p										
Cortex	24.0±1.4	20.4±1.2	0.9	ns[b]	23.3±2.3	1.1	ns[b]	22.6±1.1	0.9	ns[b]
Caudate-putamen	12.3±0.7	11.2±0.8	0.9	ns[b]	14.6±4.2	1.2	ns[b]	14.5±1.2	1.2	ns[b]
Hippocampus	25.6±1.6	24.2±2.1	1.0	ns[b]	26.7±2.2	1.0	ns[b]	29.1±1.4	1.1	ns[b]
Thalamus	25.8±1.4	19.3±0.9	0.8	ns[b]	26.8±2.3	1.0	ns[b]	22.6±1.0	0.9	ns[b]
Brain stem	32.7±2.2	23.8±1.8	0.7	ns[b]	37.8±3.7	1.2	ns[b]	26.7±1.8	0.8	ns[b]
Cerebellum	31.0±2.2	31.7±2.3	1.0	ns[b]	36.5±3.4	1.2	ns[b]	34.9±1.0	1.1	ns[b]

[a] PS is the product of the permeability coefficient and the surface area ($\times 10^{-6}$ mL g^{-1} sec^{-1}, $\bar{x} \pm$ SEM) determined with[125I] Aβ30 derivatives over the course of 15 min and corrected for V_p. V_p is the residual brain region blood volume (microliters per gram, $\bar{x} \pm$ SEM) determined with [131I]Aβ30 derivatives given 15–30 sec prior to the end of the experiment. RI is the relative increase vs. Aβ30P. P is the ANOVA with Bonferroni multiple comparisons vs. Aβ30.
[b] Not significant ($p > 0.05$).
[c] $p < 0.05$.
[d] $p < 0.01$.
[e] $p < 0.001$.

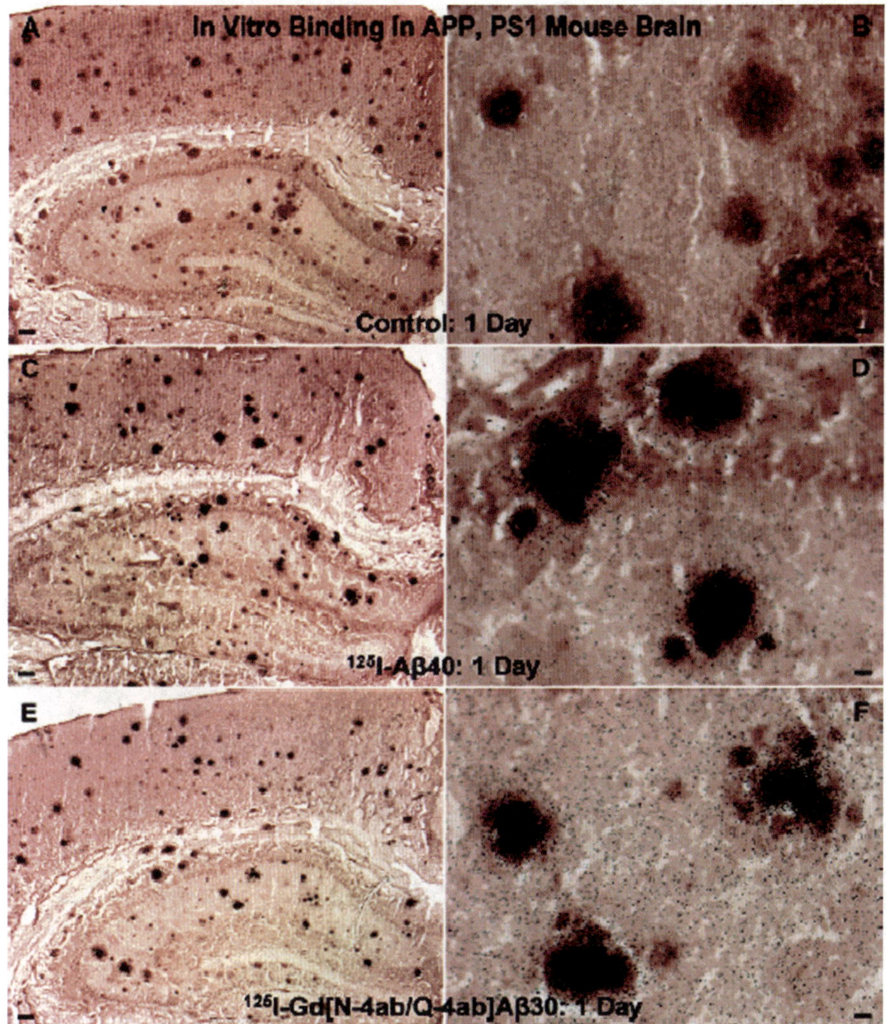

Figure 11 *Labeling of amyloid plaques in APP, PS1 transgenic mouse brain in vitro. **A** and **B** Unfixed APP, PS1 mouse brain section incubated with buffer only and processed for anti-Aβ IH and emulsion autoradiography. **C** and **D** section incubated with 100 pM [^{125}I]Aβ40 (exposed for one day). **E** and **F** section incubated with 100 pM [^{125}I]Gd[N-4ab/Q-4ab]Aβ30 (exposed for one day). For panels A, C, and E, the scale bar is 100 μm. For panels B, D and F, the scale bar is 10 μm*
(Reproduced with permission from ref. 58.)

APP, PS1 transgenic mouse after intravenous injection (Figures 11 and 12).[58] Because of the MRI contrast enhancement provided by gadolinium, this derivative should enable the *in vivo* MRI of individual amyloid plaques in the brains of AD animals or patients to allow for early diagnosis and also

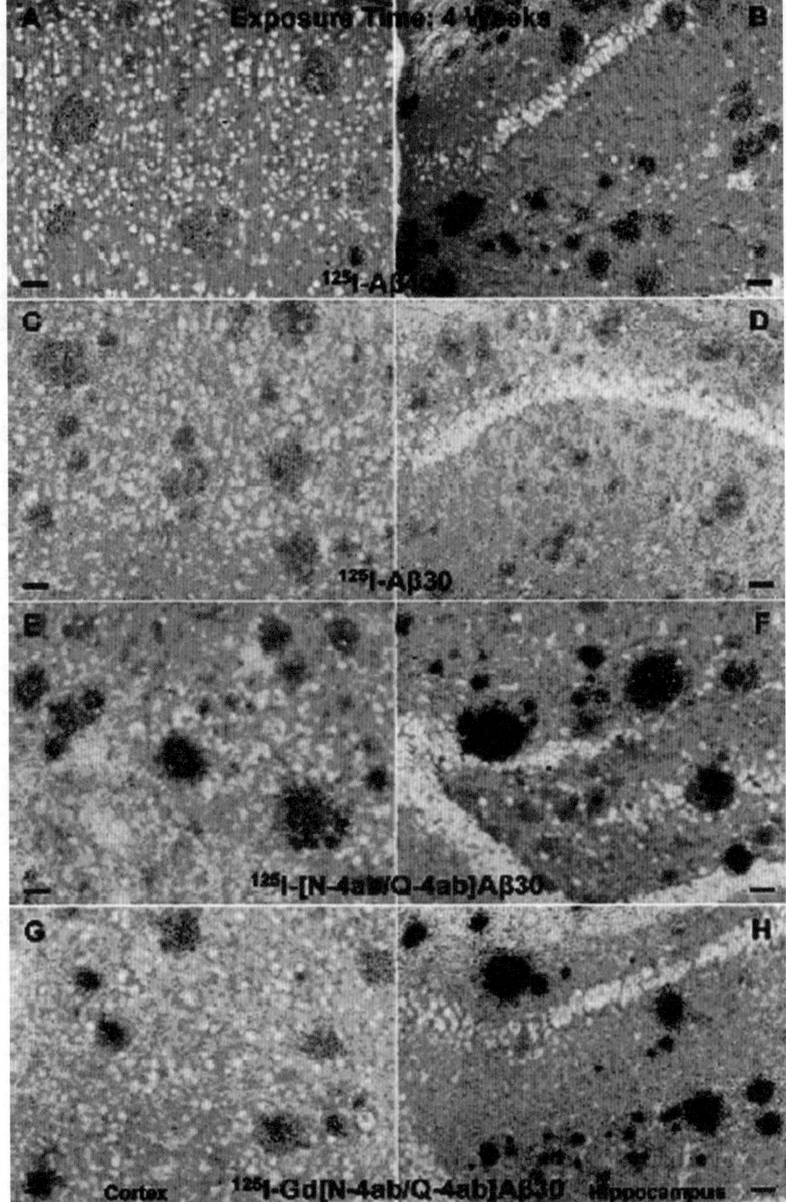

Figure 12 *Labeling of amyloid plaques in APP, PS1 transgenic mouse brain in vivo after exposure for four weeks. Fixed, frozen brain sections from a 21-month-old APP, PS1 mouse after intravenous injection with 750 μg of [^{125}I]Aβ40 (**A** and **B**), [^{125}I]Aβ30 (**C** and **D**), [^{125}I][N-4ab/Q-4ab]Aβ30 (**E** and **F**), or [^{125}I]Gd[N-4ab/Q-4ab]Aβ30 (**G** and **H**) and after being processed for anti-Aβ IH and emulsion autoradiography. For panels **A**, **C**, **E** and **G** (cortex), the scale bar is 50 μm. For panels **B**, **D**, **F** and **H** (CA1 sub-field of the hippocampus), the scale bar is 50 μm*
(Reproduced with permission from ref. 58.)

Table 3   *Agents for imaging of amyloid plaques in vivo*

Agent	Testing		
	In vitro	*Ex vivo*	*In vivo*
Congo red-based			
Congo red	Yes[a]	Yes	Not useful
Chrysamine G	Yes	Not useful	Not done
TcMAMA	Yes	Yes	Not useful
BSB	Yes	Yes	Not useful
Thioflavin-based			
Thioflavin T or S	Yes	Yes	Not useful
BTA	Yes	Yes	Not useful
Peptide-based			
10H3	Yes	Yes	Not useful
Amyloid β	Yes	Yes	In rats
Other			
FDDNP	Yes	Yes	Yes
FENE	Yes	Yes	Not useful

[a] 'Yes' indicates an agent successful in binding to amyloid plaques.

provide a direct measure of the efficacy of anti-amyloid therapies currently being developed.

Currently, most efforts at *in vivo* neuroimaging of amyloid plaques have focused on developing radioactive ligands that can be detected by PET or SPECT. These are described below, and summarized in Table 3.

There are several criteria for establishing an effective amyloid probe for use with PET or SPECT:

- the molecule must not be toxic, have fluorescent properties or be able to be tagged by a radioactive tracer;
- to maximize the signal-to-noise ratio, plasma clearance of the probe should be slow enough to allow binding to amyloid, yet fast enough to eliminate non-specific binding; and
- the molecular probe must be able to cross the BBB.

The majority cases the effort in developing cerebral amyloid probes has been focused on PET, using radiolabeled derivatives of Congo red (Formula 7) and thioflavin T (Formula 8). Both are dyes that bind specifically to amyloid in brain slices but do not cross the BBB. Derivatives of CR and thioflavin T are modified to enhance BBB crossing and amyloid labeling. Other efforts have concentrated on SPECT using modified proteins such as antibodies to amyloid, serum amyloid protein (SAP), as well as fragments of the Aβ peptide itself. However, there are exceptions to this generalization, such as SAP, used in both SPECT and PET, and rhenium complexes, nonproteinaceous derivatives of CR used in SPECT.

Congo red

Thioflavin T

^{18}FENE (Formula 9) and ^{18}FDDNP (Formula 10) have been also suggested since these compounds are highly lipophilic fluorescent molecules that bind to Aβ1–40 fibrils *in vitro*. Two kinetically distinguishable binding sites on Aβ1–40 were observed for each compound, with K_d 0.16–0.12 nM for the high-affinity binding site, and 71.2–1.86 nM for the low-affinity binding site. ^{18}FDDNP is thought superior to ^{18}FENE, as binding to weak-affinity binding sites will obscure signal from high-affinity sites. Both compounds specifically label plaques in AD brain slices, with non-specific binding in control brain slices. ^{18}FDDNP has been investigated *in vivo* in humans; although its pharmacokinetics is not available, it readily crosses the BBB due to its high lipophilicity.

FENE

FDDNP

References

1. V.M. Runge *et al.*, *Radiology*, 1984, **153**, 171.
2. I. Bertini *et al.*, *Solution NMR of Paramagnetic molecules*, Elsevier, Amsterdam, 2001.
3. V.M. Runge *et al.*, *AJR*, 1983, **141**, 1209.
4. M. Spanoghe *et al.*, *Magn. Reson. Imaging*, 1992, **10**, 913.

5. T. Shen *et al.*, *Magn. Reson. Med.*, 1993, **29**, 599.
6. A. Bader *et al.*, *Magn. Reson. Imaging*, 1995, **13**, 991.
7. M. Rausch *et al.*, *Magn. Reson. Med.*, 2003, **50**, 309.
8. W. Leenders, *Drugs*, 2003, **6**, 987.
9. J. Bulte and R. Brooks, in *Magnetic Nanoparticles as: Contrast Agent for MR Imaging Scientific and Clinical Applications of Magnetic Carriers*, Häfeli *et al.* (eds), New York, NY, Plenum Press, 1997, 527–543.
10. N.J. Abbott and I.A. Romero, *Mol. Med. Today*, 1996, **2**, 106.
11. D. Barnes, in *Physiology and Pharmacology of the Blood Brain Barrier*, M.W.B. Bradbury (ed), Springer, Berlin, 1992, 301–312.
12. R.A. Kroll and E.A. Neuwelt, *Neurosurgery*, 1998, **42**, 1083.
13. E.A. Neuwelt *et al.*, *Neurology*, 1998, **50**, 5021.
14. D.J. Brooks, in *Physiology and Pharmacology of the Blood–Brain Barrier*, M.W.B. Bradbury (ed), Springer, Berlin, 1992, 313–325.
15. L. Farde, *Trends Neurosci.*, 1996, **19**, 211.
16. G.J. Hunter *et al.*, *Am. J. Neuroradiol.*, 1998, **19**, 29.
17. W. Lauffer *et al.*, *J. Comput. Assist. Tomogr.*, 1985, **9**, 431.
18. L.M. Hamberg *et al.*, *Magn. Reson. Med.*, 1996, **35**, 168.
19. J.B. Mandeville *et al.*, *Magn. Reson. Med.*, 1998, **39**, 615.
20. N.J. Abbott *et al.*, *Advanced Drug Delivery Rev.*, 1999, **37**, 253.
21. R.E. Jacobs *et al.*, *Trends Cell Biol.*, 1999, **9**, 73.
22. T.J. Meade *et al.*, *Curr. Opin. Neurobiol.*, 2003, **13**, 597.
23. A.Y. Louie *et al.*, *Nat. Biotechnol.*, 2000, **18**, 321.
24. W.-h. Li *et al.*, *J. Am. Chem. Soc.*, 1999, **121**, 1413.
25. K. Hanaoka *et al.*, *J. Chem. Soc. Perkin*, 2001, **1**, 1840.
26. M.P. Lowe *et al.*, *J. Am. Chem. Soc.*, 2001, **123**, 7601.
27. S. Ogawa *et al.*, *Proc. Natl. Acad. Sci. USA*, 1992, **89**, 5951.
28. L. Burai *et al.*, *Chem. Commun.*, 2002, **20**, 2366.
29. A.L. Nivorozhkin *et al.*, *Angew. Chem. Int. Ed. Engl.*, 2001, **40**, 2903.
30. A.Y. Louie *et al.*, *Nat. Biotechnol.*, 2000, **18**, 321.
31. R.A. Moats *et al.*, *Angew. Chem. Int. Ed. Engl.*, 1997, **36**, 726.
32. M.J. Allen and T.J. Meade, *J. Biol. Inorg. Chem.*, 2003, **8**, 746.
33. M.J. Allen *et al.*, *Chem. Biol.*, 2004, **11**, 301.
34. S. Futaki *et al.*, *J. Biol. Chem.*, 2001, **276**, 5836.
35. J.D. Hartgerink *et al.*, *Science*, 2001, **294**, 1684.
36. S.R. Bull *et al.*, *Nano Lett.*, 2005, **5**, 1.
37. T.E. Witzig *et al.*, *J. Clin. Oncol.*, 2002, **20**, 2453.
38. J. Grunberg *et al.*, *Clin. Cancer Res.*, 2005, **11**, 5112.
39. E.P. Krenning *et al.*, *Eur. J. Nucl. Med.*, 1993, **20**, 716.
40. M. de Jong *et al.*, *Eur. J. Nucl. Med.*, 1997, **24**, 368.
41. Y.J. Lin and A.P. Koretsky, *Magn. Reson. Med.*, 1997, **38**, 378.
42. K.S. Saleem *et al.*, *Neuron*, 2002, **34**, 685.
43. A. Van der Linden *et al.*, *Neuroscience*, 2002, **112**, 467.
44. T. Watanabe *et al.*, *NeuroImage*, 2004, **22**, 860.
45. M. Hoehn *et al.*, *Proc. Natl. Acad. Sci. USA*, 2002, **99**, 16267.
46. Z.G. Zhang *et al.*, *Ann. Neurol.*, 2003, **53**, 259.

47. M. Modo *et al.*, *NeuroImage*, 2002, **17**, 803.
48. M. Modo *et al.*, *NeuroImage*, 2004, **21**, 311.
49. M. Abeles, in *Corticonics Neural Circuits of the Cerebral Cortex*, Cambridge University Press, Cambridge, CA, 1991, 1–64.
50. A. Chauduri and S. Zangenehpour, in *Immediate Early Genes and Inducible Transcription Factors in Mapping of the Central Nervous System Function and Dysfunction Handbook of Chemical Neuroanatomy*, L. Kaczmarek and H.J. Robertson, (eds), vol. 19, Elsevier, Amsterdam, 2002, 103–145.
51. J. Goldschmidt *et al.*, *NeuroImage*, 2004, **23**, 638.
52. T.M. Wengenack *et al.*, *Nat. Biotechnol.*, 2000, **18**, 868.
53. J.F. Poduslo *et al.*, *Neurobiol. Dis.*, 2002, **11**, 315.
54. J.F. Poduslo *et al.*, *Proc. Natl. Acad. Sci. USA*, 1994, **91**, 5705.
55. J.F. Poduslo and G.L. Curran, *Proc. Natl. Acad. Sci. USA*, 1992, **89**, 2218.
56. J.F. Poduslo and G.L. Curran, *J. Neurochem.*, 1996, **66**, 1599.
57. J.F. Poduslo *et al.*, *Ann. Neurol.*, 2000, **48**, 943.
58. J.F. Poduslo *et al.*, *Biochemistry*, 2004, **43**, 6064.
59. K. Yamada and T. Nabeshima, *Pharmacol. Ther.*, 2000, **88**, 93.

Subject Index